DISCRETE MATHEMATICS
THROUGH APPLICATIONS

Discrete Mathematics Through Applications

SECOND EDITION

Nancy Crisler

Patience Fisher

Gary Froelich

W. H. Freeman and Company

New York

ACQUISITIONS EDITOR: Patrick Farace
PUBLISHER: Michelle Russel Julet
MARKETING MANAGER: Kimberly Manzi
PROJECT EDITOR: Christopher Miragliotta
COVER AND TEXT DESIGNER: Victoria Tomaselli
COVER ILLUSTRATION: Jon Conrad
ILLUSTRATION COORDINATOR: Bill Page
PHOTO RESEARCHERS: Jennifer MacMillan, Elyse Rieder
PRODUCTION COORDINATOR: Julia DeRosa
COMPOSITION: The PRD Group, Inc.
MANUFACTURING: RR Donnelley & Sons Company

COVER PHOTOS:

GRAPHICS CALCULATOR: Courtesy of Hewlett Packard Company; INTERNATIONAL SPACE STATION: NASA; IMMIGRANTS ON BOAT: Reproduced from the Collections of the Library of Congress; DOMINOES: Lambert/Archive Photos; CAKE: © 1997 PhotoDisc; PARACHUTERS: Promotion Voile/ Liaison.

Library of Congress Cataloging-in-Publication Data
Crisler, Nancy.
    Discrete mathematics through applications/Nancy Crisler,
Patience Fisher, and Gary Froelich.—2nd ed.
        p.   cm.
    Includes index.
    ISBN 0-7167-3652-7
    1. Mathematics.   2. Computer science—Mathematics.   I. Fisher,
Patience.   II. Froelich, Gary W.   III. Title.
    QA39.2.C75   1999
    510—dc21
                                                                99-32013
                                                                    CIP

Printed in the United States of America

Third printing 2002

W. H. Freeman and Company
41 Madison Avenue, New York, NY 10010
Houndmills, Basingstoke RG21 6XS, England

Contents

Preface

The first edition of *Discrete Mathematics Through Applications* appeared in 1994. The purpose of the second edition remains the same as that of its predecessor: to introduce students to discrete mathematics and its importance in today's world. Since writing the first edition, we have heard from many teachers who have used the book. The feedback has been overwhelmingly positive. Teachers have reported that students find the book interesting, readable, and challenging. Teachers have also commented that some students say they are enjoying a mathematics course for the first time.

As we worked on the first edition, our estimate of the appropriate audience was based primarily on our own experience. It seems that we were close to the mark. Most users of the book report that students in their discrete mathematics courses are college-bound students who plan majors in areas such as business, education, social science, and law but that students interested in math, science, or engineering careers also take the course. Indeed, discrete mathematics courses are often established to help prepare students for majors that require courses in discrete or finite mathematics.

No doubt most people who have worked to implement discrete mathematics courses have found it necessary to educate people on the nature and importance of the subject. Often such education starts by pointing out the different definitions of *discrete* and *discreet,* perhaps accompanied by a witticism about "discreet mathematics."

A good short description of discrete mathematics offers another kind of comparison—one that contrasts discrete and continuous. The real numbers, for example, are a continuous set because between any two of them there are infinitely many others. Thus, the study of functions in algebra, trigonometry, and calculus is not discrete mathematics because functions are defined on the real numbers. On the other hand, discrete topics like permutations and combinations, which treat only whole numbers, can be found in traditional courses but in relative isolation.

The growing emphasis on discrete mathematics can be attributed to the changing nature of human society. The United States of the early twentieth century secured its place in the world through science and engineering, and calculus was the most important mathematical tool. Today, however, mathematics is being applied to important problems in the social sciences and, of course, to problems that arise in the design of computer systems. Most of these applications involve discrete mathematics. Moreover, educators find many discrete topics accessible to a broad range of students because an understanding of important problems requires little specialized background—a stark contrast to the study of calculus, which traditionally has required several years of preparation.

Support for the implementation of discrete mathematics courses in schools in the United States has come from a number of organizations. In 1989, the National Council of Teachers of Mathematics (NCTM) gave considerable attention to discrete topics in its *Curriculum and Evaluation Standards for School Mathematics*. In the years that followed, state curriculum frameworks recommended the inclusion of discrete topics. Seminars designed to help teachers acquire the knowledge and skills needed to teach unfamiliar discrete topics were conducted by Rutgers University, Boston College, the Council of Presidential Awardees in Mathematics, and others.

The authors of this book were members of an NCTM task force that developed guidelines for the implementation of discrete topics. Indeed, the content of this book was inspired by the work of that task force. It was the generous offer of financial support for the development of a suitable text by the Consortium for Mathematics and Its Applications (COMAP), a nonprofit organization with a long-standing interest in the inclusion of discrete topics in mathematics curricula, that enabled us to write the first edition. We were delighted to be asked by COMAP and W. H. Freeman and Company to do a second edition.

As teachers, we have found a great deal of satisfaction in teaching discrete mathematics. As writers, we are pleased to hear users of our book reporting similar experiences. The applicability of the subject is so obvious that students rarely ask, What is this good for? Indeed, the mathematics often derives from real-world problems.

Features of this Book

It has been extremely satisfying to be able to write a book that reflects our own interest in discrete mathematics and the way in which we teach it and to hear from others that our approach works well for them. Among the features that have received much favorable response are:

- Treatment of a topic usually begins with an exploration activity. We believe that exploration of a problem is critical to understanding it and its solution.

- Lessons are written in an informal style that is designed to be read easily by the student. Users of the book report that students who are absent usually have little trouble reading material they missed.

- Exercises often contain new ideas. Students learn by doing. In most cases, students are expected to do all the exercises.

- The themes of mathematical modeling, appropriate use of technology, and decision making are consistently emphasized.

- The strands of algorithmic thinking, recursive thinking, and mathematical induction are woven throughout the book.

Supplements

In addition, there are several supplements to the text that teachers have found helpful. They are:

- An *Instructor's Manual*. It includes our suggestions for teaching individual lessons, comments on exercises, masters for creating transparencies and handouts, and a complete answer key.

- A cross-platform CD-ROM (Macintosh and Windows). It includes computer and calculator programs designed for use with the text.

- A *Test Bank*. New to this edition, the test bank was developed by Sue Ann McGraw of Lake Oswego High School and Robert Owen of Lewis and Clark College and contains three tests per chapter. It is available in printed form and on CD-ROM.

Information on these supplements is available from the W. H. Freeman and Company Web site, **http://www.whfreeman.com**. You can also contact your local Freeman representative or e-mail your request to **faculty-services@sasmp.com**.

In addition, there are several videos that we and other users of the text have found valuable. They can be found in each chapter's materials list in the *Instructor's Manual* and are available from either the Annenberg Corporation for Public Broadcasting, 1-800-LEARNER, **http://www.learner.org**, or the Consortium for Mathematics and Its Applications, 1-800-77COMAP, **http://www.comap.com**.

Changes in the Second Edition

Needless to say, in preparing this edition, we were careful not to change what worked well. No content has been omitted from the first edition. Among the changes are:

- Significant updated data throughout

- Additional suggestions for projects and activities

- Chapter extensions that discuss topics related to the content of each chapter and can serve as starting points for class activities and projects

- More and updated news articles, in response to favorable comments from users

- Sample calculator screens that call attention to problems for which calculators are particularly suited and blackline masters in the *Instructor's Manual* that give detailed calculator instructions

- Biographical captions to accompany photos of historical figures

- Listings of Internet resources in the *Instructor's Manual*

- New treatment of fractals in Chapter 8

Acknowledgments

We want to thank COMAP for financial and technical support of both the first and the second editions. The faith in our abilities shown by Dr. Solomon Garfunkel, COMAP's executive director, is particularly appreciated.

Our appreciation goes, too, to W. H. Freeman and Company for giving us the opportunity to do a second edition and to members of their editorial and production staff for their outstanding efforts and many long days that kept the book on schedule. Among them are Patrick Farace, Christopher

Miragliotta, Brian Stuss, Victoria Tomaselli, Bill Page, Julia DeRosa, Jennifer MacMillan, and Christopher Granville. In addition, we thank Carol Loomis for copyediting the manuscript for this new edition.

We also want to thank the many users of the first edition who gave constructive comments and shared their teaching experiences. We are delighted that the book helped them build a positive experience for their students, but we know that the book could not have been successful without their outstanding efforts and dedication to their students. To our colleagues from high schools who took their time to review chapters of the manuscript, we also extend our thanks: Charles Biehl, Carter School of Wilmington; Stacy Fitch, C. E. Jordan High School; Julie Hicks, Woodbridge Senior High School; and Donna B. Wilder, Stonewall Jackson High School.

Last, but certainly not least, we want to thank the students who used the book in their study of discrete mathematics and all those who will do so in the future. We cannot help mention again the students we specifically thanked in the preface to the first edition—Kari Bauer, Dan Froelich, and Tom Hehre—for writing computer programs that accompany the text. Kari, Dan, and Tom have since graduated from high school and college and are now pursuing their own careers.

We have been privileged to share our interest in discrete mathematics with others by writing this book and by participating in institutes, seminars, and workshops around the country. We know that in so doing, we have helped you incorporate discrete topics into your own courses and curricula. That knowledge has made the many weeks and months spent writing or on the road worthwhile, and we hope that, in turn, the new edition of this book proves useful to you in your efforts.

Foreword

Discrete mathematics and its applications is one of the most rapidly expanding areas in the mathematical sciences. The modeling and understanding of finite systems is central to the development of the economy, computer science, the natural and physical sciences, and mathematics itself. This rapid growth of discrete topics applications has made the definition and development of course work in discrete mathematics a more difficult task than the development of materials and courses of study in more established areas of the mathematical sciences. This second edition of *Discrete Mathematics Through Applications* has moved forward with the field. It contains added material describing aspects of discrete mathematics that were not so central to the field at the time of the first edition five years ago.

Some versions of discrete mathematics materials have attempted to address business applications, others applications for the computer sciences. The present text provides a sound introduction to social choice, to matrices and their uses, to graph theory and its applications, and to counting and finite probability. As such, it directly responds to the call made for discrete mathematics by the National Council of Teachers of Mathematics' *Curriculum and Evaluation Standards for School Mathematics* (1989). But, more than that, it provides a solid introduction to the processes of optimization, existence, and algorithm construction within those areas.

Central to students' development in mathematics is mastery of the processes of problem-solving, communication, reasoning, and representing. All these processes receive ample treatment in this text. However, teachers must expend special effort to see that students are actually required to develop their abilities in these areas. Discrete mathematics provides an excellent platform for allowing students to develop these skills, as the fundamental structure for analyzing situations in the positive integers. As such, students are free to analyze specific subcases and generalize their hypotheses to larger settings. Work with matrices, graphs, and combinatorics provides rich settings for these processes to develop and blossom. Exercises in the text are tied to real-world settings to help students connect their mathematics study with life outside of the classroom. Extension pieces at the end of chapters allow them to further explore these links in a variety of ways.

The initial work with social choice allows students to begin their study of discrete mathematics in a realm rich with applications, modeling real-world events. When used as the initial entry to the course, this helps break the mold of studying and learning mathematics only to do more of the same. Students suddenly see mathematics as a useful tool in human decision making. They also come to see that not all problems have immediate, if any, solutions. This realization sets the course not only for the text but also for the students' future careers as users of mathematics in their personal and professional lives.

John A. Dossey
Distinguished University Professor of Mathematics Emeritus
Illinois State University

Election Theory

Throughout your life you are faced with decisions. As a student, you must decide which courses to take and how to divide your time among school work, activities, social events, and, perhaps, a job. As an adult, you will be faced with many new decisions, including whether to vote for one candidate or another.

The decisions you make are important. In the case of Nielsen television ratings, the decisions of viewers across the country determine whether a show will survive to another season. Because of the consequences of their work, organizations like Nielsen Media Research have a formidable responsibility: to combine the preferences of all the individuals in their survey into a single result and to do so in a way that is fair to all television programs.

How are the wishes of many individuals combined to yield a single result? Do the methods for doing so always treat each person fairly? If not, is it possible to improve on these methods? This chapter examines a process that is fundamental to any democratic society: group decision making. You may be surprised to learn that the study of this process is considered a part of mathematics, but the boundaries of mathematics were extended considerably in the last half of the twentieth century. Election theory is one of the most recent inclusions.

ER Tops Nielsens

New York, November 20, 1998 (Associated Press)

Prime-time ratings as compiled by Nielsen Media Research for Nov. 9–15. Listings include the week's ranking, with season-to-date rankings in parentheses.

1. (1) ER, NBC
2. (2) Friends, NBC
3. (6) NFL Monday Night Football, ABC
4. (7) Touched By An Angel, CBS
5. (3) Frasier, NBC
6. (4) Jesse, NBC
7. (5) Veronica's Closet, NBC
8. (8) 60 Minutes, CBS

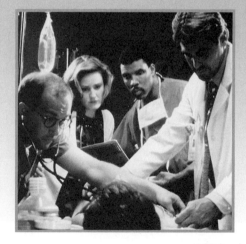

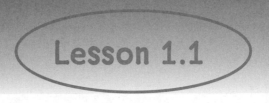

An Election Activity

Every democratic institution must have a process by which the preferences of individuals are combined to produce a group decision. For example, the preferences of individual voters must be combined in a fair way in order to fill political offices.

An excellent way to begin an exploration of group decision making is to give the process a try. Therefore, in this lesson you will combine the preferences of the individuals in your class into a single result by a method of your own invention. Before you begin, a word of reassurance and a preview of things to come: Many important problems in election theory (and other topics in discrete mathematics) can be understood and solved without a lot of background knowledge, and mathematicians know that there is no single right way to reach a group decision.

Explore This

On a piece of paper write the names of the following soft drinks, in the order given:

Coke

Dr. Pepper

Mountain Dew

Pepsi

Sprite

Rank the soft drinks. That is, beside the name of the soft drink you like best, write "1." Beside the name of your next favorite soft drink, write "2." Continue until you have ranked all five soft drinks.

As directed by your instructor, collect the ballots from all members of your class and share the results by, for example, writing all ballots on a chalkboard. Since everyone has written the soft drinks in the same order, you should be able to record quickly only the digits from each ballot.

Your task in this activity is to devise a method of combining the rankings of all the individuals in your class into a single class ranking . Your method should produce a first-, second-, third-, fourth-, and fifth-place soft drink for the entire class.

If you are working in a small group, your group should agree on a single method. After everyone finishes, each group or individual should present the ranking to the class and describe the method used to obtain it. Clear communication of the method used to obtain a result is important in mathematics, so everyone should strive for clarity when making the presentation.

As each group (or individual) makes its presentation, record the ranking in your notebook for use in this lesson's exercises.

Exercises

1. Did all the group rankings produced in your class have the same soft drink ranked first? If not, which soft drink was ranked first most often?

2. Repeat Exercise 1 for the soft drink ranked second.

3. Repeat Exercise 1 for the soft drink ranked third.

4. Repeat Exercise 1 for the soft drink ranked fourth.

5. Repeat Exercise 1 for the soft drink ranked fifth.

6. Write a description of the method you used to achieve a group ranking. Make it clear enough that someone else could use the method. You may want to break down the method into numbered steps.

7. Did anyone in your class use a method similar to yours? Explain why you think they are similar.

8. Did your method result in any ties? How could you modify your method to break a tie?

9. Mathematicians often find it convenient to represent a situation in a compact way. A good representation conveys all the essential information about a situation. A **preference schedule** is a way to represent the preferences of one or more individuals. The preference schedule shown below displays four choices, called A, B, C, and D. This preference schedule indicates that the individual whose preference it represents ranks B first, C second, D third, and A fourth.

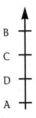

Since there are often several people who have the same preferences, mathematicians write the number of people or the percentage of people who expressed that preference under the schedule. The preferences in a group of 26 people are represented by the preference schedules shown in the following figure.

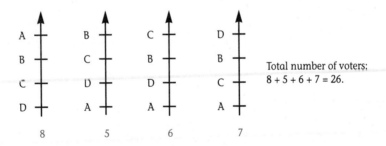

a. Apply the method you used to determine your class's soft drink ranking to this set of preferences. List the first-, second-, third-, and fourth-place rankings that your method produces. If your method cannot be applied to this set of preferences, then explain why it cannot and revise it so that it can be used here.

b. Do you think the ranking your method produces is fair? If you worked in a group, do all members of your group think the result is fair? In other words, do the first-, second-, third-, and fourth-place rankings seem reasonable, or are there reasons that one or more of the rankings seem unfair? Explain.

c. Would preference schedules be a useful way to represent the individual preferences for soft drinks among the members of your class? Explain.

10. When your class members voted, they ranked the soft drinks from first through fifth. However, voters in most U.S. elections do not get to rank the candidates. Do you think allowing voters to vote by ranking candidates would be a good practice? Explain.

Point of Interest

1996 Presidential Election Results

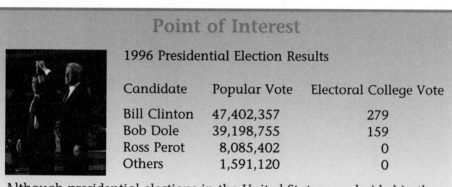

Candidate	Popular Vote	Electoral College Vote
Bill Clinton	47,402,357	279
Bob Dole	39,198,755	159
Ross Perot	8,085,402	0
Others	1,591,120	0

Although presidential elections in the United States are decided in the electoral college, it is rare for the winner of the popular vote to lose in the electoral college.

11. There are three choices in a situation that permits individuals to rank the choices. Call the choices A, B, and C. The following figure gives the six possible preferences that an individual could express.

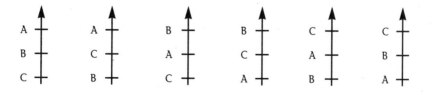

A fourth choice, D, enters the picture. If D is attached to the bottom of each of the previous schedules, there are six schedules with D at the bottom. Similarly, there are six schedules with D third, six with D second, and six with D first, or a total of 4(6) = 24 schedules. Thus, the total number of schedules with four choices is four times the total number of schedules with three choices.

a. There are 24 possible schedules with four choices. How many are there with five choices? With six?

b. Mathematicians use symbols to represent this relationship. The symbol S_n represents the number of schedules when there are n choices. You have seen that $S_n = nS_{n-1}$. Write an English translation of the mathematical sentence $S_n = nS_{n-1}$.

12. The mathematical sentence in Exercise 11b is a **recurrence relation**, a verbal or symbolic statement that describes how one number in a list can be derived from the previous number (or numbers). If, for example, the first number in a list is 7 and the recurrence relation states that to obtain any number in the list you must add 4 to the previous number, the second number in the list is $7 + 4$, or 11. This recurrence relation is stated symbolically as $T_n = 4 + T_{n-1}$. Another example of a recurrence relation is $T_n = n + T_{n-1}$. Complete the following table for the recurrence relation $T_n = n + T_{n-1}$.

n	T_n
1	3
2	$2 + 3 = 5$
3	
4	
5	

13. Complete the following table for the recurrence relation $A_n = 3 + 2A_{n-1}$.

n	A_n
1	4
2	$3 + 2(4) = 11$
3	
4	
5	

Group-Ranking Methods and Algorithms

If the soft drink data for your class are typical, you know that the problem of establishing a group ranking is not without controversy. The reason is that there is seldom complete agreement on the correct way to achieve a group ranking. This lesson examines several common methods of determining a group ranking from a set of individual preferences. As you examine these methods, consider whether any of them are similar to the one you devised in the previous lesson.

Consider the preferences of Exercise 9 of the previous lesson, which are shown again in Figure 1.1.

Many group-ranking situations, such as elections in which only one office is to be filled, require the selection of a single winner. In the set of preferences shown in Figure 1.1, choice A is ranked first on eight schedules,

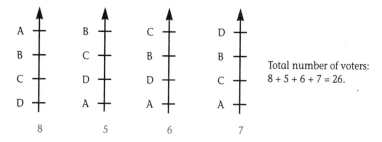

Total number of voters:
8 + 5 + 6 + 7 = 26.

Figure 1.1 Preferences of 26 voters.

more often than any other choice. If A wins on this basis, A is called the **plurality winner.** The plurality winner is based on first-place rankings only. The winner is the choice that receives the most votes. Note, however, that A is first only on about 30.8% of the schedules. Had A been first on over half the schedules, A would be a **majority winner.**

Point of Interest

The system used to determine the president of the United States does not require that the winner receive a majority of the popular vote. Among the presidents who have served at least one term without getting a majority of the popular vote are John F. Kennedy, Woodrow Wilson, and Abraham Lincoln.

Mathematician of Note

Jean-Charles de Borda (1733–1799), a French cavalry officer, naval captain, mathematician, and scientist, preferred a method that assigned points to the rankings of individuals because he was dissatisfied with the plurality method.

The Borda Method

Did anyone in your class determine the soft-drink ranking by assigning points to the first, second, third, and fourth choice of each individual's preference and obtaining a point total? If so, these groups used a type of **Borda count.**

The most common way of applying the Borda method to a

ranking of n choices is to assign n points to a first-place ranking, $n - 1$ to a second-place ranking, $n - 2$ to a third-place ranking, . . . , and 1 point to a last-place ranking. The group ranking is established by totaling each choice's points.

In the example of Figure 1.1, A is ranked first by 8 people and fourth by the remaining 18, so A's point total is $8(4) + 18(1) = 50$. Similar calculations give totals of 83, 69, and 58 for B, C, and D, respectively, as summarized below.

$$A: 8(4) + 5(1) + 6(1) + 7(1) = 50.$$
$$B: 8(3) + 5(4) + 6(3) + 7(3) = 83.$$
$$C: 8(2) + 5(3) + 6(4) + 7(2) = 69.$$
$$D: 8(1) + 5(2) + 6(2) + 7(4) = 58.$$

In this case, the plurality winner does not fare well under the Borda system.

Point of Interest

```
8*4+5*1+6*1+7*1
              50
8*3+5*4+6*3+7*2
```

Borda counts can be done quickly on a graphing calculator: after the first calculation is typed and entered, it is replayed and edited.

The Runoff Method

Many elections in the United States and other countries require a majority winner. If there is no majority winner, a runoff election between the top two candidates is held. Runoff elections are expensive because of the cost of holding another election and time-consuming because they require a second trip to the polls. However, if voters are allowed to rank the candidates, these inconveniences can be avoided.

To conduct a runoff, determine the number of firsts for each choice. In the example of Figure 1.1, A is first eight times, B is first five times, C is first six times, and D is first seven times.

Eliminate all choices except the two with the highest totals: Choices B and C are eliminated, and A and D are retained. Now consider each of the preference schedules on which the eliminated choices were ranked first. Choice B was first on the second schedule. Of the two remaining choices, A and D, D is ranked higher than A, so these 5 votes are transferred to D. Similarly, the 6 votes from the third schedule are transferred to D. The totals are now 8 for A and 7 + 5 + 6 = 18 for D, and so D is the runoff winner (see Figure 1.2).

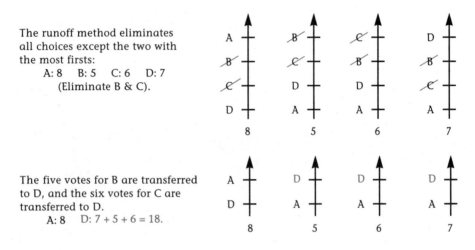

The runoff method eliminates all choices except the two with the most firsts:
A: 8   B: 5   C: 6   D: 7
(Eliminate B & C).

The five votes for B are transferred to D, and the six votes for C are transferred to D.
A: 8   D: 7 + 5 + 6 = 18.

Figure 1.2 The runoff method.

The Sequential Runoff Method

Some elections, such as the voting to determine the site of the Olympic Games, are conducted by a variation of the runoff method that eliminates only one choice at a time. Although the members of the Olympic committee do not find it inconvenient to vote several times in the course of a few days, it would be impractical to do so when millions of voters are involved. Fortunately, as with the runoff method, if voters rank the candidates, they need vote only once.

In the example of Figure 1.1, B is eliminated first because it is ranked first the fewest times. The 5 first-place votes for B are transferred to C. The point totals are now 8 for A, 5 + 6 = 11 for C, and 7 for D.

There are three choices remaining. Now D's total is the smallest, so D is eliminated next. The 7 votes are transferred to the remaining choice that is ranked highest by these 7. Thus, C is given an additional 7 votes and so defeats A by 18 to 8 (see Figure 1.3).

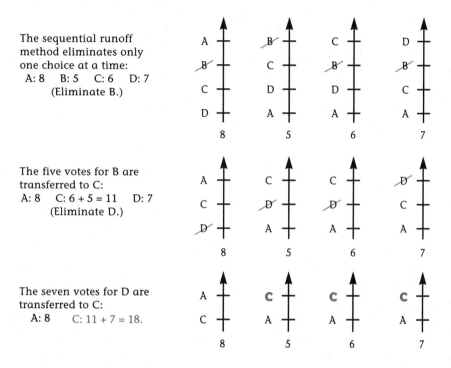

The sequential runoff method eliminates only one choice at a time:
A: 8   B: 5   C: 6   D: 7
    (Eliminate B.)

The five votes for B are transferred to C:
A: 8   C: 6 + 5 = 11   D: 7
    (Eliminate D.)

The seven votes for D are transferred to C:
    A: 8   C: 11 + 7 = 18.

Figure 1.3 The sequential runoff method.

Exercises

1. Which soft drink is the plurality winner in your class? Is it also a majority winner? Explain.

2. Which soft drink is the Borda winner in your class?

3. Which soft drink is the runoff winner in your class?

4. Which soft drink is the sequential runoff winner in your class?

2004 Olympics Awarded to Athens

Lausanne, Switzerland,
September 6, 1997

Athens took the lead in the first round of voting to decide the 2004 Olympic Games host on Friday and maintained that lead until the decisive fourth round. Athens won the right to stage the Games with 66 votes in the final round to 41 for Rome.

How the International Olympic Committee members voted:

Athens	32	38	52	66
Rome	23	28	35	41
Cape Town	16	22	20	
Stockholm	20	19		
Buenos Aires	16			

Cape Town won a tie-break against Buenos Aires 62–44 to advance to the second round.

5. For the example of Figure 1.1, determine the percentage of voters that ranked each choice first and last.

   a. Enter the results in a table like the following:

Choice	First	Last
A		
B		
C		
D		

   b. On the basis of these percentages only, which choice do you think would be the most objectionable to voters? The least objectionable? Explain your answers.

   c. Which choice do you think best deserves to be ranked first for the group? Explain your reasoning.

   d. Give at least one argument against your choice.

6. The 1998 race for governor of Minnesota had three strong candidates. The following are unofficial results from the general election.

Jesse Ventura	768,356
Norm Coleman	713,410
Hubert Humphrey III	581,497
Others	12,017

   a. What percentage of the vote did the winner receive? Is the winner a majority winner?

   b. What is the smallest percentage the plurality winner can receive in a race with exactly three candidates? Explain.

7. Determine the plurality, Borda, runoff, and sequential runoff winners for the following set of preferences.

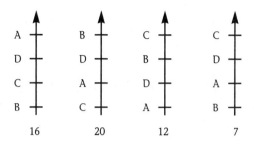

A	B	C	C
D	D	B	D
C	A	D	A
B	C	A	B
16	20	12	7

8. The Borda method determines a complete group ranking, but the other methods examined in this lesson produce only a first. Each of these methods may be extended, however, to produce a complete group ranking.

    a. Describe how the plurality method could be extended to determine a second, third, and so forth. Apply this to the example in Figure 1.1 and list the second, third, and fourth that your extension produces.

    b. Describe how the runoff method could be extended to determine a second, third, and so forth. Apply this to the example in Figure 1.1 and list the second, third, and fourth that your extension produces.

    c. Describe how the sequential runoff method could be extended to determine a second, third, and so forth. Apply this to the example in Figure 1.1 and list the second, third, and fourth that your extension produces.

9. Each year the Heisman Trophy recognizes one of the country's outstanding college football players. The year 1997 marked the first time a defensive player received the award. The results of the voting follow. Each voter selects a player to rank first, another to rank second, and another to rank third.

Charles Woodson.

	1st	2nd	3rd	Points
1. Charles Woodson, Michigan	433	209	98	1,815
2. Peyton Manning, Tennessee	281	263	174	1,543
3. Ryan Leaf, Washington State	70	205 ·	174	861
4. Randy Moss, Marshall	17	56	90	253
5. Ricky Williams, Texas	4	31	61	135
6. Curtis Enis, Penn State	3	18	20	65
7. Tim Dwight, Iowa	5	3	11	32
8. Cade McNown, UCLA	0	7	12	26
9. Tim Couch, Kentucky	0	5	12	22
10. Amos Zerouoe, West Virginia	3	1	10	21

a. How many points are awarded for a first-place vote? For a second-place vote? For a third-place vote?

b. Would the ranking produced by this system have differed if the plurality method had been used? Explain.

10. When runoff elections are used in the United States, voters do not rank the candidates and therefore must return to the polls to vote in the runoff. In some countries, such as Ireland, a method commonly called "instant runoff" is used. In an instant runoff, the voters rank the candidates and do not return to the polls. Examine the vote totals in the two runoffs shown below. Do the totals tell you anything about the merits of the instant runoff? Explain.

President of Ireland: 1997 Results

	General Election	Runoff
Mary Banotti	372,002	497,516
Mary McAleese	574,424	706,259
Derek Nally	59,529	
Adi Roche	88,423	
Dana Scallon	175,458	

U.S. House Texas District 9: 1996 Results

	General Election	Runoff
Nick Lampson	83,781	59,217
Steve Stockman	88,171	52,853
Geraldine Sam	17,886	

11. In the sequential runoff method, the number of choices on a given round is 1 less than the number of choices on the previous round. Let C_n represent the number of choices after n rounds and write this as a recurrence relation.

12. A procedure for solving a problem is called an **algorithm.** This section has presented various algorithms for determining a group ranking from individual preferences. Algorithms are often written in numbered steps in order to make them easy to apply. The following is an algorithmic description of the runoff method.

  1. For each choice, determine the number of preference schedules on which the choice was ranked first.

  2. Eliminate all choices except the two that were ranked first most often.

  3. For each preference schedule, transfer the vote total to the remaining choice that ranks highest on that schedule.

  4. Determine the vote total for the preference schedules on which each of the remaining choices is ranked first.

  5. The winner is the choice ranked first on the most schedules.

  a. Write an algorithmic description of the sequential runoff method.

  b. Write an algorithmic description of the Borda method.

13. The number of first-, second-, third-, and fourth-place votes for each choice in an election can be described in a table, or *matrix,* as shown below.

The preferences:

A	B	C	C
D	D	B	D
C	A	D	A
B	C	A	B
20	10	12	15

The matrix:

	A	B	C	D
1st	20	10	27	0
2nd	0	12	0	45
3rd	25	0	20	12
4th	12	35	10	0

The number of points that a choice receives for first, second, third, and fourth place can be written in a matrix, as shown below.

	1st	2nd	3rd	4th
Points	4	3	2	1

A new matrix that gives the Borda point totals for each choice can be computed by writing this matrix alongside the first, as shown below.

$$[4 \quad 3 \quad 2 \quad 1] \begin{bmatrix} 20 & 10 & 27 & 0 \\ 0 & 12 & 0 & 45 \\ 25 & 0 & 20 & 12 \\ 12 & 35 & 10 & 0 \end{bmatrix}$$

The new matrix is computed by multiplying each entry of the first matrix by the corresponding entry in the first column of the second matrix and finding the sum of these products:

$$4(20) + 3(0) + 2(25) + 1(12) = 142.$$

This number is the first entry in a new matrix that gives the Borda point totals for choices A, B, C, and D:

$$\begin{array}{cccc} & A & B & C & D \\ \text{Point totals:} & [142 & \underline{\quad} & \underline{\quad} & \underline{\quad}] \end{array}$$

a. Calculate the remaining entries of the new matrix.
b. If you have the matrix but not the preference schedules, by which methods is it possible to determine the winner? Explain.

Computer/Calculator Explorations

14. Enter the soft drink preferences of your class members into the election machine computer program that accompanies this book. Compare the results given by the computer to your answers to the first four exercises of this lesson. Resolve any discrepancies.

Projects

15. Write a short report on the history of any of the methods discussed in this section. Look into the lives of people who were influential in developing the method. Discuss factors that led them to propose the method.

16. Find at least two examples of group rankings that are currently used somewhere in the world but not discussed in this section. Describe how the group ranking is determined. Compare each new method with the

methods described in this section. Are any of these new methods the same as the ones described in the lesson? What are some advantages and disadvantages of each new method?

17. Select one or more countries that are not discussed in this section and report on the methods they use to conduct elections.

Nebraska Ranked Number One

USA TODAY CNN,
January 3, 1998

The final USA Today CNN football coaches' poll was released today. First-place votes are in parentheses, followed by record and total points based on 25 points for a first-place vote through one point for a 25th-place vote.

1. Nebraska (32)	13–0	1,520
2. Michigan (30)	12–0	1,516
3. Florida State	11–1	1,414
4. North Carolina	11–1	1,292
5. UCLA	10–2	1,239
6. Florida	10–2	1,209
7. Kansas State	11–1	1,192
8. Tennessee	11–2	1,122
9. Washington State	10–2	1,076
10. Georgia	10–2	1,007

More Group-Ranking Methods and Paradoxes

Different methods of determining a group ranking often give different results. This fact led the Marquis de Condorcet to propose that a choice that could obtain a majority over every other choice should win.

Again consider the set of preference schedules used as an example in the last lesson (see Figure 1.4).

Mathematician of Note

The Marquis de Condorcet (1743–1794) was a French mathematician, philosopher, and economist who shared an interest in election theory with his friend, Jean-Charles de Borda.

To examine the data for a Condorcet winner, compare each choice with every other choice. For example, begin by comparing A with B, then with C, and finally with D. In the figure A is ranked higher than B on 8 schedules and lower on 18. (An easy way to see this is to cover C and D on all the schedules.) Because A cannot obtain a majority against B, A cannot be the Condorcet winner. Therefore, there is no need to compare A to C or to D.

Now consider B. You have already seen that B wins against A, so begin by comparing B with C. You can see that B is ranked higher than C on 8 + 5 + 7 = 20 schedules and lower than C on 6.

Now compare B with D. You see B is ranked higher than D on 8 + 5 + 6 = 19 schedules and lower than D on 7. Therefore, B can obtain a majority over each of the other choices and so is the Condorcet winner.

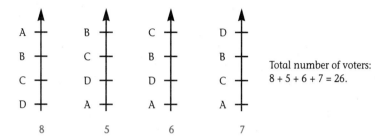

Total number of voters:
8 + 5 + 6 + 7 = 26.

Figure 1.4 Preferences of 26 voters.

Since B is the Condorcet winner, it is unnecessary to make comparisons between C and D. Although all comparisons do not have to be made, it can be helpful to organize them in a table:

	A	B	C	D
A		L	L	L
B	W		W	W
C	W	L		W
D	W	L	L	

To see how a choice does in one-on-one contests, read across the row associated with that choice. You see A, for example, loses in one-on-one contests with B, C, and D.

Although the Condorcet method may sound ideal, it sometimes fails to produce a winner. Consider the set of schedules shown in Figure 1.5.

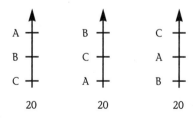

Figure 1.5 Preferences of 60 voters.

Notice that A is preferred to B on 40 of the 60 schedules but that A is preferred to C on only 20. Although C is preferred to A on 40 of the 60, C is preferred to B on only 20. Therefore there is no Condorcet winner.

You might expect that if A is preferred to B by a majority of voters and B is preferred to C by a majority of voters, then a majority of voters would prefer A to C. But the example shows that this need not be the case.

In other mathematics classes you have learned that many relationships are transitive. The relation "greater than" ($>$), for example, is transitive because if $a > b$ and $b > c$, then $a > c$.

You have just seen that group-ranking methods may violate the transitive property. Because this intransitivity seems contrary to intuition, it is known as a **paradox**. This particular paradox is sometimes referred to as the **Condorcet paradox**. There are other paradoxes that can occur with group-ranking methods, as you will see in this lesson's exercises.

Exercises

1. Determine the Condorcet winner in the soft drink ballot your class conducted in Lesson 1.1.

2. Propose a method for resolving situations in which there is no Condorcet winner.

3. In a system called **pairwise voting**, two choices are selected and a vote taken. The loser is eliminated, and a new vote is taken in which the winner is paired against a new choice. This process continues until all choices but one have been eliminated. An example of the use of pairwise voting occurs in legislative bodies in which bills are considered two at a time. The choices in the set of preferences shown in the following figure represent three bills being considered by a legislative body.

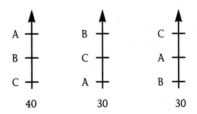

   a. Suppose you are responsible for deciding which two will appear on the agenda first. If you strongly prefer bill C, which two bills would you place on the agenda first? Why?
   b. Is it possible to order the voting so that some other choice will win? Explain.

4. A panel of sportswriters is selecting the best football team in a league, and the preferences are distributed as follows.

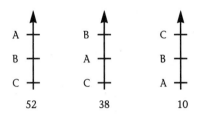

a. Determine the winning team using a 3-2-1 Borda count.
b. The 38 who ranked B first and A second decide to lie in order to improve the chances of their favorite and so rank C second. Determine the winner using a 3-2-1 Borda count.

5. When people decide to vote differently from the way they feel about the choices, they are said to be *voting insincerely*. People are often encouraged to vote insincerely because they have some idea of the election's result beforehand. Explain why advance knowledge is possible.

6. Many political elections in the United States are decided by the plurality method. Construct a set of preferences with three choices in which the plurality method would encourage insincere voting. Identify the group of voters that would be encouraged to vote insincerely and explain the effect of their doing so on the election.

7. Many people consider the plurality method unfair because it sometimes produces a winner that a majority of voters do not like.
a. What percentage of voters rank the plurality winner last in the preferences shown below?

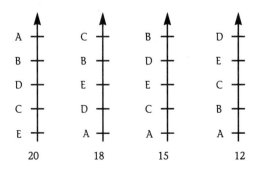

    b. Runoffs are sometimes used to avoid the selection of a controversial winner. Is the runoff winner an improvement over the plurality winner in this set of preferences? Explain.

    c. Do you consider the sequential runoff winner an improvement over the plurality and runoff winners? Explain.

8. a. Use a runoff to determine the winner in the set of preferences shown below.

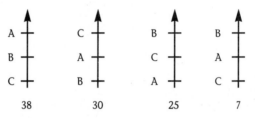

A	C	B	B
B	A	C	A
C	B	A	C
38	30	25	7

    b. In some situations, votes are made public. For example, people have the right to know how their elected officials vote on issues. Suppose these schedules represent such a situation. Because they expect to receive some favors from the winner and because they expect A to win, the seven voters associated with the last schedule decide to change their preferences from

B	to	A
A		B
C		C

and to "go with the winner." Conduct a new runoff and determine the winner.

    c. Explain why the results are a paradox.

9. a. Use a 4-3-2-1 Borda count to determine a group ranking for the following set of preferences.

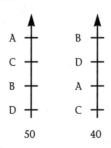

A	B
C	D
B	A
D	C
50	40

b. These preferences represent the ratings of four college athletic teams, and team C has been disqualified because of a recruiting violation. Write the schedules with team C removed and use a 3-2-1 Borda count to determine the group ranking.

c. Explain why these results are a paradox.

10. Write a brief summary of the five methods of achieving a group ranking (plurality, Borda, runoff, sequential runoff, and Condorcet) that have been discussed in this and the previous lesson. Include at least one example of why each method can lead to unfair results.

11. The Condorcet method requires that in theory each choice be compared with every other one, although in practice many of the comparisons do not have to be made in order to determine the winner. Consider what the number of comparisons would be if every comparison were made.

   Mathematicians sometimes find it helpful to represent the choices and comparisons visually. If there are only two choices, a single comparison is all that is necessary. In the diagram that follows, the choices are represented by points, or *vertices,* and the comparisons by line segments, or *edges.*

a. Add a third choice, C, to the diagram. Connect it to A and to B to represent the additional comparisons. How many new comparisons are there? What is the total number of comparisons that must be made?

b. Add a fourth choice, D, to the diagram. Connect it to each of A, B, and C. How many new comparisons are there, and what is the total number of comparisons?

c. Add a fifth choice to the diagram and repeat. Then add a sixth choice and repeat. Complete the following table.

Number of Choices	Number of New Comparisons	Total Number of Comparisons
1	0	0
2	1	1
3		
4		
5		
6		

12. Let C_n represent the total number of comparisons necessary when there are n choices. Write a recurrence relation that expresses the relationship between C_n and C_{n-1}.

Computer/Calculator Explorations

13. Use the preference schedule program that accompanies this book to find a set of preferences with at least four choices that demonstrate the same paradox found in Exercise 8 when the sequential runoff method is used.

14. Use the preference schedule program to enter several schedules with five choices. Use the program's features to alter your data in order to produce a set of preferences with several different winners. Can you find a set of preferences with five choices and five winners? If so, what is the minimum number of schedules with which this can be done? Explain.

Projects

15. Research and report on paradoxes in mathematics. Try to determine whether the paradoxes have been satisfactorily resolved.

16. Research and report on paradoxes outside mathematics. In what way have these paradoxes been resolved?

17. Select an issue of current interest in your community that involves more than two choices. Have each member of your class vote by writing a preference schedule. Compile the preferences and determine the winner by five different methods.

18. Investigate the contributions of Charles Dodgson (Lewis Carroll) to election theory. Was he responsible for any of the group-ranking procedures you have studied? What did he suggest doing when the Condorcet method fails to produce a winner?

19. Investigate the system your school uses to determine academic rankings of students. Is it similar to any of the group-ranking procedures you have studied? If so, could it suffer from any of the same problems? Propose another system and discuss why it might be better or worse than the one currently in use.

20. Investigate elections in your school (class officers, officers of organizations, homecoming royalty, and so forth). Report on the type of voting and the way winners are chosen. Recommend alternative methods and explain why you think the methods you recommend would be more fair.

Arrow's Conditions and Approval Voting

Paradoxes, unfair results, and insincere voting are some of the problems that have caused people to look for better ways to reach group decisions. In this lesson you will learn of some recent and important work that has been done to improve the group-ranking process. First, consider an example involving pairwise voting.

Ten representatives of the language clubs at Central High School are meeting to select a location for the clubs' annual joint dinner. The committee must choose among a Chinese, French, Italian, or Mexican restaurant (see Figure 1.6).

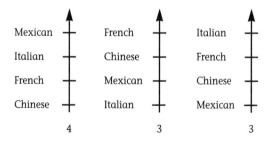

Figure 1.6 Preferences of 10 students.

Racquel suggests that because the last two dinners have been held at Mexican and Chinese restaurants, this year's dinner should be at either an Italian or a French restaurant. The group votes 7 to 3 in favor of the Italian restaurant.

Martin, who doesn't like Italian food, says that the community's newest Mexican restaurant has an outstanding reputation. He proposes that the group choose between Italian and Mexican. The other members agree and vote 7 to 3 to hold the dinner at the Mexican restaurant.

Sarah, whose parents own a Chinese restaurant, says that she can obtain a substantial discount for the event. The group votes between the Mexican and Chinese restaurants and selects the Chinese by a 6 to 4 margin.

Look carefully at the group members' preferences. Note that French food is preferred to Chinese by all, yet the voting has resulted in the selection of the Chinese restaurant!

Mathematician of Note

Kenneth Arrow (1921–   ) received a degree in mathematics before turning to economics. His application of mathematical methods to election theory brought him worldwide recognition.

In 1951, paradoxes such as this led Kenneth Arrow, a U.S. economist, to formulate a list of five conditions that he considered necessary for a fair group-ranking method. These fairness conditions today are known as **Arrow's conditions.**

One of Arrow's conditions says that if every member of a group prefers one choice to another, then the group ranking should do the same. According to this condition, the choice of the Chinese restaurant when all members rated French food more favorably is unfair. Thus, Arrow considers pairwise voting a flawed group-ranking method.

Arrow's Conditions

1. Nondictatorship: The preferences of a single individual should not become the group ranking without considering the preferences of the others.

2. Individual Sovereignty: Each individual should be allowed to order the choices in any way and to indicate ties.

3. Unanimity: If every individual prefers one choice to another, then the group ranking should do the same. (In other words, if

every voter ranks A higher than B, then the final ranking should place A higher than B.)

4. Freedom from Irrelevant Alternatives: The winning choice should still win if one of the other choices is removed. (The choice that is removed is known as an irrelevant alternative.)

5. Uniqueness of the Group Ranking: The method of producing the group ranking should give the same result whenever it is applied to a given set of preferences. The group ranking should also be transitive.

Arrow inspected the common methods of determining a group ranking for adherence to his five conditions. He also looked for new methods that would meet all five. After doing so, he arrived at a surprising conclusion.

In this lesson's exercises, you will examine a number of group-ranking methods for their adherence to Arrow's conditions. You will also learn Arrow's surprising result.

Exercises

1. Your teacher decides to order soft drinks for your class on the basis of the soft drink vote conducted in Lesson 1.1 but, in so doing, selects the preference schedule of a single student (the teacher's pet). Which of Arrow's conditions are violated by this method of determining a group ranking?

2. Instead of selecting the preference schedule of a favorite student, your teacher places all the individual preferences in a hat and draws one. If this method were repeated, would the same group ranking result? Which of Arrow's conditions does this method violate?

3. Do any of Arrow's conditions require that the voting mechanism include a secret ballot? Is a secret ballot desirable in all group-ranking situations? Explain.

4. Examine the paradox demonstrated in Exercise 9 of Lesson 1.3 on pages 22 and 23. Which of Arrow's conditions are violated?

5. Construct a set of preference schedules with three choices, A, B, and C, showing that the plurality method violates Arrow's fourth condition.

In other words, construct a set of preferences in which the outcome between A and B depends on whether C is on the ballot.

6. There are often situations in which insincere voting results. Do any of Arrow's conditions state that insincere voting should not be part of a fair group-ranking method? Explain.

7. Suppose that there are only two choices in a list of preferences and that the plurality method is used to decide the group ranking. Which of Arrow's conditions could be violated? Explain.

8. After failing to find a group-ranking method for three or more choices that always obeyed all his fairness conditions, Arrow began to suspect that such a method might not exist. He applied logical reasoning to the problem and proved that no method, known or unknown, can always obey all five conditions. In other words, any group-ranking method will violate at least one of Arrow's conditions in some situations.

   Arrow's proof demonstrates how mathematical reasoning can be applied to areas outside mathematics. This and other achievements earned Arrow the 1972 Nobel Prize in economics.

   Although Arrow's work means that a perfect group-ranking method will never be found, it does not mean that current methods cannot be improved. Recent studies have led several experts to recommend a system called **approval voting.**

   In approval voting, you may vote for as many choices as you like, but you do not rank them. You mark all those of which you approve. For example, if there are five choices, you may vote for as few as none or as many as five.

   a. Write a soft drink ballot like the one you used in Lesson 1.1. Place an "X" beside each of the soft drinks you find acceptable. At the direction of your instructor, collect the ballots from each member of your class. Count the number of votes for each soft drink and determine the winner.

   b. Determine a complete group ranking.

   c. Was the approval winner the same as the earlier plurality winner in your class?

   d. How does the group ranking in part b compare with the earlier Borda ranking?

9. Examine Exercise 4 of Lesson 1.3 on page 21. Would any members of the panel of sportswriters be encouraged to vote insincerely if approval voting were used? Explain.

10. What is the effect on a group ranking of casting approval votes for all choices? Of casting approval votes for none of the choices?

11. The voters whose preferences are represented below all feel strongly about their first choices but are not sure about their second and third choices. They all dislike their fourth and fifth choices. Since the voters are unsure about their second and third choices, they flip coins to decide whether to give approval votes to their second and third choices.

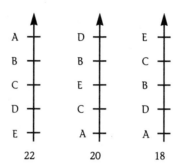

a. Assuming the voters' coins come up heads about half the time, how many approval votes would you expect each of the five choices to get? Explain your reasoning.

b. Do the results seem unfair to you in any way? Explain.

12. Approval voting offers a voter many choices. If there are three candidates for a single office, for example, the plurality system offers the voter four choices: vote for any one of the three candidates or for none of them. Approval voting permits the voter to vote for none, any one, any two, or all three.

To investigate the number of ways in which you can vote under approval voting, consider a situation with two choices, A and B. You can represent voting for none by writing { }, voting for A by writing {A}, voting for B by writing {B}, and voting for both by writing {A, B}.

a. List all the ways of voting under the approval system when there are three choices.

b. List all the ways of voting under the approval system when there are four choices.

c. Generalize the pattern by letting V_n represent the number of ways of voting under the approval system when there are n choices and writing a recurrence relation that describes the relationship between V_n and V_{n-1}.

13. Listing all the ways of voting under the approval system can be difficult if not approached systematically. The following algorithm describes one way of finding all the ways of voting for two choices. The results are shown applied to a ballot with five choices, A, B, C, D, and E.

	List 1	List 2
1. List all choices in order in List 1.	A B C D E	
2. Draw a line through the first choice in List 1 that doesn't already have a line drawn through it and write it as many times in List 2 as there are choices in List 1 without lines through them.	A̶ B C D E	A
		A
		A
		A
3. Beside each item you wrote in List 2 in step 2, write a choice in List 1 that does not have a line through it.		AB
		AC
		AD
		AE

4. Repeats steps 2 and 3 until each choice has a line through it. The items in the second list show all the ways of voting for two items.

Write an algorithm that describes how to find all the ways of voting for three choices. You may use the results of the previous algorithm to begin the new one.

14. Many patterns can be found in the various ways of voting when the approval system is used. The following table shows the number of ways of voting for exactly one item when there are several choices on the ballot. For example, in Exercise 12, you listed all the ways of voting when there are three choices on the ballot. Three of these, {A}, {B}, and {C}, are selections of one item.

Number of Choices on the Ballot	Number of Ways of Selecting Exactly One Item
1	1
2	2
3	3
4	—
5	—

Complete the table.

15. Let $V1_n$ represent the number of ways of selecting exactly one item when there are n choices on the ballot and write a recurrence relation that expresses the relationship between $V1_n$ and $V1_{n-1}$.

16. The following table shows the number of ways of voting for exactly two items when there are from one to five choices on the ballot. For example, your list in Exercise 12 shows that when there are three choices on the ballot, there are three ways of selecting exactly two items: {A, B}, {A, C}, and {B, C}.

Number of Choices on the Ballot	Number of Ways of Selecting Exactly One Item
2	1
3	3
4	—
5	—

Complete the table.

17. Let $V2_n$ represent the number of ways of selecting exactly two items when there are n choices on the ballot and write a recurrence relation that expresses the relationship between $V2_n$ and $V2_{n-1}$. Can you find more than one way to do this?

Computer/Calculator Explorations

18. Design a computer program that lists all possible ways of voting when approval voting is used. Use the letters A, B, C, . . . to represent the choices. The program should ask for the number of choices and then display all possible ways of voting for one choice, two choices, and so forth.

Projects

19. Investigate the number of ways of voting under the approval system for other recurrence relations (see Exercises 14 through 17). For example, in how many ways can you vote for three choices, four choices, and so forth?

20. Arrow's result is an example of an impossibility theorem. Investigate and report on other impossibility theorems.

21. Research and report on Arrow's theorem. The theorem is usually proved by an indirect method. What is an indirect method? How is it applied in Arrow's case?

Weighted Voting and Voting Power

ValueVision Merger in Trouble

Minneapolis, March 9, 1998 (Reuters)

Two shareholders of ValueVision International Inc said Monday they will vote against a proposed merger with Nation Media Corp and instead have offered their own merger proposal.

The shareholders, Michael Blake and Brian Danzis, who control about 200,000 shares, said their VV Acquisition Corp has offered $5.50 a share for home shopping outfit ValueVision.

National Media and ValueVision have agreed to merge in a stock swap in which ValueVision holders would get 1.19 shares of a new holding company for each share exchanged and National Media holders would get one share for each share swapped.

The first four lessons of this chapter examined situations in which all voters are considered equals. In some voting situations there are voters who have more votes than others; essentially, the vote of some voters carries more weight than the vote of others. This lesson examines such situations, beginning with a simple example.

A small high school has 110 students. Because of recent growth in the size of the community, the sophomore class is quite large. It has 50 members, and the junior and senior classes each have 30 members.

The school's student council is composed of a single representative from each class. Each of the three members is given a number of votes proportionate to the size of the class represented. Accordingly, the sophomore representative has five votes, and the junior and senior representatives each have three. The passage of any issue that is before the council requires a simple majority of six votes.

The student council's voting procedure is an example of **weighted voting**. Weighted voting occurs whenever some members of the voting body have more votes than others have.

Weighted voting is fairly common in the United States. For example, it is used in local government in some parts of the country and in corporate shareholder meetings. In recent years, several people have questioned whether weighted voting is fair. Among them is John Banzhaf III, a law professor at George Washington University who has initiated several legal actions against weighted voting procedures used in local government.

To understand Banzhaf's reasoning, consider the number of ways that voting on an issue could occur in the student council example.

Mathematician of Note

John Banzhaf III, a law professor who also holds an engineering degree, is a well-known consumer rights advocate.

It is possible that the issue is favored by none of the members, one of them, two of them, or all three. In which cases would the issue be passed? The following list gives all possible ways of voting for an issue and the associated number of votes.

{; 0} {So; 5} {Jr; 3} {Sr; 3} {So, Jr; 8} {So, Sr; 8} {Jr, Sr; 6} {So, Jr, Sr; 11}

For example, {Jr, Sr; 6} indicates that the junior and senior representatives could vote for an issue and that they have a total of six votes between them.

Each of these collections of voters is known as *a coalition*. Those with enough votes to pass an issue are known as **winning coalitions.** The winning coalitions in this example are those with six or more votes and are listed below along with their respective vote totals.

{So, Jr; 8} {So, Sr; 8} {Jr, Sr; 6} {So, Jr, Sr; 11}

The last winning coalition is different from the other three in one important respect: If any one of the members decides to vote differently, the coalition will still win. No one member is essential to the coalition. Banzhaf reasoned that the only time a voter has power is when the voter belongs to a coalition that needs the voter in order to pass an issue. The coalitions for which at least one member is essential are

{So, Jr; 8}   {So, Sr; 8}   {Jr, Sr; 6}

Notice that the sophomore representative is essential to two of the coalitions. This is also true of the junior and senior representatives. In other words, about the same number of times, each of the representatives can be expected to cast a key vote in passing an issue.

A paradox: Although the votes have been distributed to give greater power to the sophomores, the outcome is that all members have the same amount of power!

Since distributing the votes in a way that reflects the distribution of the population does not result in a fair distribution of power, mathematical procedures can be used to find a way to measure actual power when weighted voting is used.

A measure of the power of a member of a voting body is called a **power index**. In this lesson, a voter's power index is the number of winning coalitions to which the voter is essential. For example, in the student council situation, each representative is essential to two winning coalitions and thereby has a power index of 2, as do the junior and senior representatives.

A Power Index Algorithm

1. List all coalitions of voters that are winning coalitions.

2. Select any voter, and record a 0 for that voter's power index.

3. From the list in step 1, select a coalition of which the voter selected in step 2 is a member. Subtract the number of votes the voter has from the coalition's total. If the result is less than the number of votes required to pass an issue, add 1 to the voter's power index.

4. Repeat step 3 until you have checked all coalitions for which the voter chosen in step 2 is a member.

5. Repeat steps 2 through 4 until all voters have been checked.

Exercises

1. Consider a situation in which A, B, and C have 3, 2, and 1 votes, respectively, and in which 4 votes are required to pass an issue.
   a. List all possible coalitions and all winning coalitions.
   b. Determine the power index for each voter.
   c. If the number of votes required to pass an issue is increased from 4 to 5, determine the power index of each voter.

2. In a situation with three voters, A has 7 votes, B has 3, and C has 3.
   a. Determine the power index of each voter.
   b. A *dictator* is a member of a voting body who has all the power. A *dummy* is a member who has no power. Are there any dictators or dummies in this situation?

3. The student council example in this lesson depicted a situation with three voters that resulted in equal power for all three. In Exercises 1 and 2, power was distributed differently. Find a distribution of votes that results in a power distribution among three voters that is different from the ones you have already seen. How many new power distributions in situations with three voters can you find?

4. In this lesson's student council example, can the votes be distributed so that the members' power indices are proportionate to the class sizes? Explain.

5. In this lesson's student council example, suppose that the representatives of the junior and senior classes always differ on issues and never vote alike. Does this make any practical difference in the power of the three representatives? Explain.

6. (See Exercise 12 of Lesson 1.4 on page 29.) Let C_n represent the number of coalitions that can be formed in a group of n voters. Write a recurrence relation that describes the relationship between C_n and C_{n-1}.

7. One way to determine all winning coalitions in a weighted voting situation is to work from a list of all possible coalitions. Use A, B, C, and D to represent the individuals in a group of four voters and list all possible coalitions.

8. Weighted voting is commonly used to decide issues at meetings of corporate stockholders. Each member is given one vote for each share of stock held.
   a. A company has four stockholders: A, B, C, and D. They own 26%, 25%, 25%, and 24% of the stock, respectively, and more than 50%

of the vote is needed to pass an issue. Determine the power index of each stockholder. Use your results from Exercise 7 as an aid.

b. Another company has four stockholders. They own 47%, 41%, 7%, and 5% of the stock. Find the power index of each stockholder.

c. Compare the percentage of stock owned by the smallest stockholder in parts a and b. Do the same for the power index of the smallest stockholder in each case.

Nassau Districting Ruled Against Law

NEW YORK TIMES,
January 15, 1970

Albany—The Court of Appeals ruled today that the present "weighted voting" plan of the Nassau County Supervisors was unconstitutional but that a new plan was not necessary until after the 1970 federal census.

In a unanimous opinion, the state's highest court said the county's present charter provision is a clear violation of the one-man-one-vote principle, in that it specifically denied the town of Hempstead representation that reflected its population.

The town, the court pointed out, constituted 57.12 percent of the county's population, but because of the weighted voting plan its representatives on the board could cast only 49.6 percent of the board's vote.

9. A landmark court decision on voting power involved the Nassau County, New York, Board of Supervisors. In 1964, the board had six members. The number of votes given to each was 31, 31, 21, 28, 2, and 2.

a. Determine the power index for each member.

b. The board was composed of representatives of five municipalities with the populations shown in the following table.

Hempstead	728,625
North Hempstead	213,225
Oyster Bay	285,545
Glen Cove	22,752
Long Beach	25,654

The members with 31 votes both represented Hempstead. The others each represented the municipality listed in the same order as in the table. Compare the power indices of the municipalities with their populations.

10. A *minimal winning coalition* is one in which all the voters are essential.

a. Give an example of a weighted voting situation with a winning coalition for which at least one but not all of the voters is essential. Identify the minimal winning coalitions in this situation.

b. Would defining a voter's power index as the number of minimal winning coalitions be equivalent to the definition used in this lesson? Explain.

Computer Explorations

11. Use the weighted voting program that accompanies this book to experiment with different weighted voting systems when there are three voters. Change the number of votes given to each voter and the number of votes required to pass an issue. How many different power distributions are possible? Do the same for weighted voting systems with four voters.

Projects

12. The Security Council of the United Nations is composed of five permanent members and ten others who are elected to two-year terms. For a measure to pass, it must receive at least nine votes that include all five of the permanent members. Determine a power index for a permanent member and for a temporary member. (Assume that all members are present and voting.)

13. Research and report on other power indices. What, for example, is the Shapley-Shubik power index?

14. The president of the United States is chosen by the electoral college. What does this system do to the power of voters in different states in selecting the president? Research the matter and report on the relative power of voters in different states.

15. Research and report on court decisions about weighted voting.

EU Nations Tackle Tough Issues

Noordwijk, the Netherlands, April 15, 1997

At a meeting behind closed doors European Union member states started serious negotiations on the most divisive institutional issues which the Intergovernmental Conference on the revision of the Maastricht Treaty so far avoided. The IGC ministerial session showed how far apart member states are on some of these issues.

The weighting of votes in the EU Council is a delicate issue which will certainly be settled at the last minute. The main problem is that, at present, a decision taken by a qualified majority in the Council must be backed by member states representing 60% of the EU population. The fear is that, in an enlarged Union, it might be possible for decisions to be taken by a qualified majority which in fact does not represent the majority of the EU population. Bigger member states also fear that they might be outvoted (but smaller states, like Luxemburg, tell them that, in fact, there have never been "coalitions" of "small" countries against "large" member states). In Noordwijk, Italy suggested to give Germany, Britain, France and Italy two more votes in the Council (they have ten each now) and one more to Spain (which would have nine instead of eight), in order to appease these fears.

Proportional Representation

Democracies are founded on the principle that all people should have representation in government. Most democratic countries have minority populations who feel they should be represented by one of their own members, and courts have agreed.

Unfortunately, ensuring minority representation in a legislative body like the U.S. House of Representatives is not always easy. If, for example, a state has five representatives in the U.S. House and a minority is 40% of the state's population, it seems reasonable that the minority should hold $0.4 \times 5 = 2$ of the seats. However, depending on how the boundaries of the state's five congressional districts are drawn, the minority might hold no seats.

Historically, ensuring minority representation in the U.S. House has been accomplished when district boundaries are redrawn after each census: in states with a significant minority population, some districts are established in which the minority has over half the population. Unfortunately, this practice sometimes has produced districts with a shape so unusual that courts have declared the districts unconstitutional.

How, then, does a democracy provide fair representation in government? Many democracies use an election procedure called **proportional representation**. Although there are several proportional representation systems in use, they all have a common goal: to ensure that minorities and/or political parties have representation in government proportionate to their numbers in the general population.

One form of proportional representation is achieved through a practice called **cumulative voting**. In this system, several representatives are elected from a single district. If, for example, the district has three representatives, each voter has three votes. The voter can split the three votes in any way, and can even cast more than one vote for a single candidate. Cumulative voting is used in some local elections in the United States.

Another system, the **party-list system**, is used in many European countries. In this system, each party has a list of candidates on the ballot. Each voter votes for one of the parties. When the election is over, the party receives a number of seats proportionate to the vote it received. The seats are usually assigned to the names on the party's ballot by taking the names in order from the top of the ballot until the correct number is obtained.

The **mixed member system** has voters vote for a party and a candidate. A portion of the seats are assigned to candidates and another portion to parties. All individual winning candidates receive seats. The remaining seats are awarded to members of parties that do not have a number of individual seats proportionate to the vote they received. In 1994, New Zealand voters abandoned a plurality system like the one currently used in the United States in favor of the mixed member system, which is also used in several European countries.

In the **preference vote system**, voters rank the candidates. A threshold is established, and all candidates with a vote total over the threshold are elected. Remaining seats are distributed by conducting a form of instant runoff among the remaining candidates.

Voting System Violates Law, Court Rules

WASHINGTON POST,
Saturday, September 17,
1994

Richmond, Sept. 16—A federal appeals court agreed today that the election system used in a county on Maryland's Eastern Shore diluted the voting strength of black residents, but the court failed to endorse the proposed remedy.

Worcester County, in which Ocean City is located, had appealed a lower court order requiring it to use a "cumulative voting" system to elect county commissioners.

Under cumulative voting, all five of the county's commissioners would be elected at large and each voter would have five votes. A voter could cast all five votes for one candidate or split them in any way.

The system, used in only a few places across the nation as a way of increasing minority representation, had been proposed by the plaintiffs in a voting rights lawsuit against the county.

1. Write a summary of what you think are the important points of this chapter.

2. Consider the following set of preferences.

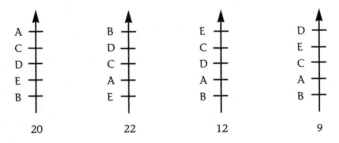

A	B	E	D
C	D	C	E
D	C	D	C
E	A	A	A
B	E	B	B
20	22	12	9

   a. Determine the winner using a 5-4-3-2-1 Borda count.
   b. Determine the plurality winner.
   c. Determine the runoff winner.
   d. Determine the sequential runoff winner.
   e. Determine the Condorcet winner.
   f. Suppose that this election is conducted by the approval method and all the voters decide to approve of the first two choices on their preference schedules. Determine the approval winner.

3. Complete the following table for the recurrence relation $B_n = 2B_{n-1} + n$.

n	B_n
1	3
2	$2(3) + 2 = 8$
3	
4	
5	

4. In this chapter, you have encountered many paradoxes involving group-ranking methods.
   a. One of the most amazing paradoxes occurs when a winning choice becomes a loser when its standing actually improves. In which group-ranking method(s) can this occur?
   b. Discuss at least one other paradox that occurs with group-ranking methods.

5. In the 1912 presidential election, polls showed that the preferences of voters were as follows.

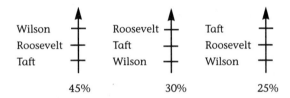

Wilson	Roosevelt	Taft
Roosevelt	Taft	Roosevelt
Taft	Wilson	Wilson
45%	30%	25%

   a. Who won the election? Was he a majority winner?
   b. How did the majority of voters feel about the winner?
   c. How might one of the groups of voters have changed the results of the election by voting insincerely?
   d. Discuss who might have won the election if a different method had been used.

6. Your class is ranking soft drinks and someone suggests that the names of the soft drinks be placed in a hat and the group ranking be determined by drawing them from the hat. Which of Arrow's conditions does this method violate?

7. State Arrow's theorem. In other words, what did Arrow prove?

8. Can the point system used to do a Borda count affect the ranking (for example, a 5-3-2-1 system instead of a 4-3-2-1 system)? Construct an example to support your answer.

9. The 1992 presidential election was unusual because of a strong third-party candidate. In that election Bill Clinton received 43% of the popular vote, George Bush 38%, and Ross Perot 19%.

   Steven Brams and Samuel Merrill III used polling results to estimate the percentage of those voting for one candidate who also approved of another.

Approximately 15% of Clinton voters approved of Bush and approximately 30% approved of Perot.

Approximately 20% of Bush voters approved of Clinton and approximately 20% approved of Perot.

Approximately 35% of Perot voters approved of Clinton and approximately 30% approved of Bush.

a. Estimate the percentage of approval votes each candidate would have received if approval voting had been used in the election.

b. Find the total of the three percentages you gave as answers in part a. Explain why the total is not 100%.

10. Choose an election method from those you have studied in this chapter that you think best to use to determine the winner for the following preferences. Explain why you think your choice of method is best.

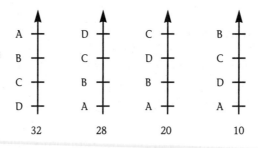

11. Consider a situation in which voters A, B, C, and D have 4, 3, 3, and 2 votes, respectively, and 7 votes are needed to pass an issue.

a. List all winning coalitions and their vote totals.

b. Find the power index for each voter.

c. Do the power indices reflect the distribution of votes? Explain.

d. Suppose the number of votes necessary to pass an issue is increased from 7 to 8. How does this change the power indices of the voters?

Bibliography

Amy, Douglas J. 1993. *Real Choices/New Voices: The Case for Proportional Representation Elections in the United States.* New York: Columbia University Press.

Brams, Steven J. 1995. "Approval Voting on Bills and Propositions." *The Good Society* 5(2): 37–39.

Brams, Steven J. 1976. *Paradoxes in Politics.* New York: Free Press.

Brams, Steven J., and Peter Fishburn. 1983. *Approval Voting.* Boston: Birkhauser.

Brams, Steven J., and Samuel Merrill III. 1994. "Would Ross Perot Have Won the 1992 Presidential Election under Approval Voting?" *PS: Politics and Political Science* 27(1): 39–44.

Bunch, Brian. 1982. *Mathematical Fallacies and Paradoxes.* Princeton, NJ: Van Nostrand-Reinhold.

COMAP. 1997. *For All Practical Purposes: Introduction to Contemporary Mathematics.* 4th ed. New York: W. H. Freeman.

Cole, K. C. 1998. *The Universe and the Teacup: The Mathematics of Truth and Beauty.* New York: Harcourt, Brace.

Davis, Morton. 1980. *Mathematically Speaking.* New York: Harcourt Brace Jovanovich.

Falletta, Nicholas. 1983. *The Paradoxican.* Garden City, NY: Doubleday.

Gardner, Martin. 1980. "From Counting Votes to Making Votes Count: The Mathematics of Elections." *Scientific American* 251(4): 16.

Hoffman, Paul. 1988. *Archimedes' Revenge.* New York: W. W. Norton.

Lucas, William F. 1992. *Fair Voting: Weighted Votes for Unequal Constituencies.* Lexington, MA: COMAP.

Lum, Lewis, and David C. Kurtz. *Voting Made Easy: A Mathematical Theory of Election Procedures.* Greensboro, NC: Guilford College.

Meyer, Rochelle Wilson, and Walter Meyer. 1990. *Play It Again Sam: Recurrence Equations in Mathematics and Computer Science.* Arlington, MA: COMAP.

Niemi, Richard G., and William H. Riker. 1976. "The Choice of Voting Systems." *Scientific American* 234(6): 21.

Pakhomov, Valery. 1993. "Democracy and Mathematics." *Quantum* January–February: 4–9.

Taylor, Alan D. 1995. *Mathematics and Politics: Strategy, Voting, Power and Proof.* New York: Springer-Verlag.

Fair Division

Whether the parties involved are individuals, organizations, communities, states, or nations, a joint endeavor raises questions of fairness. Massive undertakings like the international space station raise many fairness issues. For example, how will time aboard the station be divided among the nations that are paying for it? What is a fair way to divide the cost of building and maintaining the station among the nations involved?

Fair division questions arise in the simplest of situations. For example, you have no doubt experienced a feeling of unfairness when, as a child, someone else received a piece of cake or portion of ice cream you felt was better than yours. How can a portion of food be divided fairly among two or more children? Is the meaning of fairness when food is divided among children different from the meaning of fairness when an estate is divided among heirs or when seats in Congress are divided among states? Are the methods that are commonly used to divide food, estates, and legislatures necessarily the fairest methods? Discrete mathematics plays an important role in answering these questions.

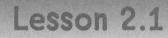
A Fair Division Activity

1st Module of New Space Station Set

Moscow,
November 24, 1998
(Associated Press)

Russian and U.S. ground controllers completed a series of maneuvers today to allow the first module of the international space station to dock with the next segment early next month.

The Zarya, or Sunrise, module that was built by a Russian company for NASA has been smoothly orbiting Earth since its launch Friday.

The space station involves 16 nations, with the United States and Russia playing the biggest roles. It will consist of more than 100 elements that will take 45 assembly flights to complete, and is scheduled to be finished by 2004.

The United States plans to pay $24 billion of the space station's $40 billion price tag. It will not be inhabitable until at least early 2000, following the launch of the Russian-made crew module that is set to blast off next summer.

There are many ciorcumstances in which the division of an object in a fair way is important to those involved. Three of the most common are the division of food among children, a house in an estate among heirs, and the seats in a governmental body among districts. Each has characteristics that make it different from the others.

In this lesson, you will consider an example of each of these three situations and propose a solution of your own. As is the case in election theory, fair division is an area of discrete mathematics in which important problems can be understood and solved without a lot of background knowledge.

Explore This

Below is a set of three fair division problems. You will find it convenient to discuss one or more of them with a few other people, as time permits. The following is a description of one way to divide the problems among small groups in your class. At the direction of your instructor, divide your class into groups of three people. Write the numbers 1, 2, or 3 on each of several slips of paper. Have someone from each group draw one of the slips from a bag or box. Each group should consider the fair division problem listed below that corresponds to the number drawn by the group. Allow about 15 minutes for each group to discuss the problem.

After all groups have finished their discussions, a spokesperson for each group should present the group's decision to the class. Each group that discussed Problem 1 should report first, and so forth.

In your notebook, record the method used by each group. You will need the record for this lesson's exercises.

1.   Martha and Ray want to divide the last piece of the cake that their mother baked yesterday. Propose a method of dividing the piece of cake that will seem fair to both Martha and Ray.

2.   Juan and Mary are the only heirs to their mother's estate. The only object of significant value is the house in which they were raised. Propose a method of resolving the issue of the disposition of the house that will seem fair to both Juan and Mary.

3.   The sophomore, junior, and senior classes of Central High School have 333, 288, and 279 members, respectively. The school's student council is composed of 20 members divided among the three classes. Determine a fair number of seats on the council for each class.

If your class is typical, the class members may not have reached consensus on the best way to solve each of the three problems. In this lesson's exercises you will consider some of the important fairness issues in each of the three.

Exercises

1.   a.   Did any groups resolve Problem 1 by relying on the mother's authority? In what way?
     b.   Does the resolution of such a problem by the mother or other authority figure always produce a solution that seems fair to both children? Explain.
     c.   Cite at least two examples of situations in which fair division problems are resolved by an authority.

2. a. Did any of the groups use a random event such as a coin flip to resolve Problem 1? In what way?
   b. Does the use of randomness in such a problem always produce a solution that seems fair to both children? Explain.
   c. Cite at least two examples of situations in which randomness is used to resolve an issue.

3. a. Did any of the groups use a means of measuring the piece of cake to resolve Problem 1? In what way?
   b. Does the use of measurement in such problems always produce a solution that seems fair to both children? Explain.
   c. Give an example of a situation in which measurement is likely to result in an agreeable solution to a fair division problem.

4. A common way of resolving Problem 1 is to have one of the children cut the cake into two pieces and to have the other choose first.
   a. If Martha cuts the cake. How will she feel about the two pieces?
   b. If Ray gets to choose one of the pieces that Martha cut, how will he feel about the two pieces?
   c. If you were one of the participants in this scheme, would you rather be the cutter or the chooser? Why?

5. Write a description of what you consider to be desirable results of a process that fairly divides a cake among any number of people.

6. Did any of the groups resolve Problem 2 by selling the house and dividing the cash? Why might the results of such a process be unsatisfactory to one or more of the heirs?

7. Did any of the groups use a method that considers the possibility that the heirs might not agree on the value of the house? In what way?

8. Suppose that Juan thinks the house is worth $100,000 and Mary feels it is worth $120,000.
   a. Who do you think should receive the house? Explain.
   b. How might the person who doesn't get the house be compensated?

9. Write a description of what you consider to be desirable results of a process that fairly divides a house among several heirs.

10. How might the possibility of lying about the value of the house affect the result of a division process?

11. Did all the groups that discussed Problem 3 divide the seats among the classes in the same way? If not, describe the differences.

12. If some of the groups that discussed Problem 3 obtained different results, which of the methods do you think is the fairest? If all groups produced the same result, do you agree that the result is fair? Why or why not?

13. Write a description of what you consider to be desirable results of a process that fairly divides the seats in a student council among a school's classes.

14. Summarize the similarities and differences in the meaning of fairness in this lesson's three problems. For each of the three problems, explain why you think it is or is not possible to achieve fairness.

Projects

15. Pick a situation in which individuals or groups have developed a procedure for settling fair division problems other than the division of food, an estate, or legislative seats. Report on the method used. Compare it with methods developed in this lesson and later in this chapter.

Estate Division

A fair division problem can be either *discrete* or *continuous.* The problem of dividing a house among heirs and that of dividing a student council among classes are examples of the discrete case. Discrete division occurs whenever the objects of the division cannot be meaningfully separated into pieces. Dividing a cake is an example of the continuous case because the cake can be divided into any number of pieces.

This lesson considers fair division of an estate among heirs. In your discussions in Lesson 2.1, you may have decided that fairness is difficult to define in some situations because different people place different values on the same object. However, it is sometimes possible to use such differences of opinion to the advantage of all those involved. The following estate division algorithm produces an appealing paradox: Each of the heirs receives a share that is larger than he or she thinks is fair.

The Three Stooges. Left to right: Moe, Curly, and Larry.

An Algorithm for Dividing an Estate

1.   Each heir submits a bid for each item in the estate. (Bids are not made on cash in the estate because it can be divided equally without controversy.)

2.   A fair share is determined for each heir by finding the sum of his or her bids and dividing this sum by the number of heirs.

3.   Each item in the estate is given to the heir who bid the highest on that item.

4.   Each heir is given an amount of cash from the estate that is equal to his or her fair share (from step 2) less the amount the heir bid on the objects he or she received. If this amount is negative, the heir pays that amount into the estate.

5.   The remaining cash in the estate is divided equally among the heirs.

Stooges' Heirs Win Ruling

DETROIT NEWS,
December 29, 1995

The heirs of Three Stooges Larry Fine, Joe "Curly Joe" DeRita and Moe Howard own the rights to the legendary comic trio, a federal bankruptcy judge has ruled. The ruling ends years of litigation involving the heirs, which stalled film and merchandising deals worth millions of dollars a year. The legal challenges began in July 1993 when Fine's four grandchildren and DeRita's widow claimed the heirs of Moe Howard, who owned the rights to the characters, owed them more than $5 million in profits from merchandising and marketing deals. A court last year ordered Howard's heirs to pay $4.3 million, of which $1.6 million was to go to Jean DeRita, and to hand over all rights to the characters to a company controlled by the heirs of all three. The Howard heirs, consisting of his daughter, Joan Maurer, and grandson Jeffrey Scott, subsequently declared bankruptcy.

The first four steps of the algorithm give each heir goods and cash whose total value equals what the heir feels is a fair share of the estate. The extra cash awarded in the fifth step is a bonus.

Whenever you encounter a new algorithm, a sample application of the algorithm is helpful. After considering the following example, you will have the opportunity to apply the algorithm to several different situations in this lesson's exercises.

Estate Division Example

Amanda, Brian, and Charlene are heirs to an estate that includes a house, a boat, a car, and $150,000 in cash.

Step 1

Each heir submits bids for the house, boat, and car. The bids are summarized in the following table, or matrix.

	House	Boat	Car
Amanda	$80,000	$5,000	$8,000
Brian	$70,000	$9,000	$11,000
Charlene	$76,000	$7,000	$13,000

For example, the entries in Amanda's row indicate the value to Amanda of each item in the estate.

Step 2

A fair share is determined for each heir.

Amanda: ($80,000 + $5,000 + $8,000 + $150,000) ÷ 3 = $81,000.
Brian: ($70,000 + $9,000 + $11,000 + $150,000) ÷ 3 = $80,000.
Charlene: ($76,000 + $7,000 + $13,000 + $150,000) ÷ 3 = $82,000.

Point of Interest

```
(80000+5000+8000
+150000)/3
            81000
(70000+9000+1100
0+150000)/3
```

The graphing calculator features used to do Borda counts can also be used to determine fair shares quickly. Replay the calculation for the first heir and edit it by moving the cursor and typing new digits when necessary. Delete and Insert features should be used when digits must be eliminated or introduced.

Step 3

The house is given to Amanda, the boat to Brian, and the car to Charlene.

Step 4

Cash equal to the difference between the fair share and the value of the awarded items is given to each heir.

Amanda: $81,000 − $80,000 = $1,000.

Brian: $80,000 − $9,000 = $71,000.

Charlene: $82,000 − $13,000 = $69,000.

Step 5

The cash given to the heirs totals $141,000, which leaves $150,000 − $141,000 = $9,000 cash in the estate. Each heir receives a bonus of $9,000 ÷ 3 = $3,000.

The results of this example can be summarized in a matrix:

	Amanda	Brian	Charlene
Total of bids and cash	$243,000	$240,000	$246,000
Fair share	$81,000	$80,000	$82,000
Items received	House	Boat	Car
Value of items received	$80,000	$9,000	$13,000
Initial cash received	$1,000	$71,000	$69,000
Share of remaining cash	$3,000	$3,000	$3,000

For each heir, totaling the last three rows of the matrix gives the value the heir attaches to the items and cash received. For example, Amanda

For Better or For Worse®   **by Lynn Johnston**

Lynn Johnston Productions, Inc./Distributed by United Feature Syndicate, Inc.

Magness Feud Settled

DENVER POST, January 8, 1998

A year-long Wild West struggle over the largest contested fortune in Colorado history came to a peaceful end Monday as parties feuding over the $1 billion estate of Tele-Communications Inc. founder Bob Magness announced a broad settlement that avoids a nasty court battle.

There were hugs and kisses all around as a notable cast of characters from Colorado business and society holstered their lawyers and banked their millions.

"This is what Pop would have wanted," said Kim and Gary Magness in a statement, referring to their father, Bob Magness. His death in 1996 triggered a battle for his fortune, and for control of TCI, the nation's largest cable TV company, based in the Denver Tech Center.

feels the value of her share of the estate is $80,000 + $1,000 + $3,000 = $84,000, which is more than the $81,000 that Amanda feels is a fair share.

The final settlements for each heir are:

Amanda: the house and $4,000

Brian: the boat and $74,000

Charlene: the car and $72,000

Exercises

1. The application of any fair division algorithm requires certain assumptions, or *axioms*. For example, the success of the estate division algorithm requires that each heir be capable of placing a value on each object in the estate. If any heir considers an object priceless or is otherwise incapable of placing a dollar value on an object, the algorithm fails. Give at least one other axiom that you think is necessary for the success of this algorithm.

2. Garfield and Marmaduke are heirs to an estate that contains only a summer cottage. Garfield bids $70,000, and Marmaduke bids $60,000.
   a. What does Garfield feel is a fair share? Marmaduke?
   b. What is the difference between Garfield's fair share and Garfield's bid for the cottage?
   c. Because the value Garfield assigned to the cottage is more than Garfield's fair share, Garfield must pay cash into the estate. How much cash must Garfield pay?
   d. Marmaduke is given an amount of cash from Garfield's payment equal to Marmaduke's fair share. How much does Marmaduke receive? If the remaining cash is divided equally, what will be the final value of Marmaduke's settlement? Of Garfield's?

e. Garfield must borrow money in order to pay into the estate, and the interest on this loan is $2,000. Do you think this should be considered when arriving at a settlement? If so, suggest how the settlement should be revised.

f. If the division between Garfield and Marmaduke were settled by another algorithm that is frequently used to divide estates, Marmaduke would be given half of Garfield's bid. Compare the final settlements for Garfield and Marmaduke by this method with the settlements in part d. Which result do you think is fairest? Explain.

3. Amy, Bart, and Carol are heirs to an estate that consists of a valuable painting, a motorcycle, a World Series ticket, and $5,000 in cash. They submit the bids shown in the following matrix.

	Painting	Motorcycle	Ticket
Amy	$2,000	$4,000	$500
Bart	$5,000	$2,000	$100
Carol	$3,000	$3,000	$300

a. Use the algorithm of this lesson to divide the estate among the heirs. For each heir, state the fair share, the items received, the amount of cash, and the final settlement. Summarize your results in a matrix.

A helpful hint: It is relatively easy to lose track of the estate's cash as payments are made into and out of the estate. Errors can be avoided by tracking the cash with a table designed for that purpose:

Cash in the estate	$5,000
Received from Amy	———
Received from Bart	———
Paid to Carol	———
Cash remaining	———

b. It is common for one or more heirs to pay into an estate. This lesson's algorithm fails if an heir who must pay into the estate cannot do so. Suggest a way the algorithm could be modified to account for situations in which one or more heirs cannot raise the cash necessary to complete the division.

4. Suppose that in the division of Exercise 3, Amy had received previous financial support from the estate in the form of a loan to pay college tuition. A will states that she is to receive only 20% of the estate, whereas Bart and Carol are to receive 40% each. Adapt the algorithm of this lesson to this situation and describe a fair division of the estate.

5. If two heirs submit an identical highest bid for an item, how would you resolve the tie?

6. Alan, Betty, and Carl are heirs to an estate and have submitted the bids shown in the following table.

	House	Boat	Car
Alan	$90,000	$4,000	$10,000
Betty	$95,000	$5,000	$8,000
Carl	$92,000	$4,000	$9,000

The awarding of items in the estate can be indicated in a matrix, as shown below.

	Alan	Betty	Carl
House	0	1	0
Boat	0	1	0
Car	1	0	0

The entries in Alan's column indicate the items that he received. For example, the 1 in Alan's column and the car's row indicates that Alan received the car. Each of the other two entries in Alan's column is a 0; this indicates that Alan received neither the house nor the boat.

A new matrix can be computed by writing the second matrix beside the first, as shown below.

$$\begin{bmatrix} \$90{,}000 & \$4{,}000 & \$10{,}000 \\ \$95{,}000 & \$5{,}000 & \$8{,}000 \\ \$92{,}000 & \$4{,}000 & \$9{,}000 \end{bmatrix} \begin{bmatrix} 0 & 1 & 0 \\ 0 & 1 & 0 \\ 1 & 0 & 0 \end{bmatrix}$$

The new matrix is computed by multiplying each entry in the first row of the first matrix by the corresponding entry in the first column of the second matrix and finding the sum of these products:

$$\$90{,}000(0) + \$4{,}000(0) + \$10{,}000(1) = \$10{,}000.$$

Because the result, $10,000, was obtained from the first row of the first matrix and the first column of the second matrix, it is written in the first row and the first column of the new matrix.

The entry for the first row and the second column of the new matrix is found by performing a similar calculation with the first row of the first matrix and the second column of the second matrix:

$$\$90{,}000(1) + \$4{,}000(1) + \$10{,}000(0) = \$94{,}000.$$

a. Calculate the remaining entries of the new matrix.
b. The $10,000 in the first row and the first column of the new matrix can be interpreted as the value to Alan of the items received by Alan. Write an interpretation of the number in the first row and the second column of the new matrix.
c. Write an interpretation of the number in the second row and the second column of the new matrix.

7. Could the estate division algorithm of this lesson encourage insincerity by any of the heirs? Explain.

8. In 1998, the U.S. Supreme Court settled a dispute between New York and New Jersey over control of Ellis Island by dividing the island between the two states. Is the problem of how to divide an island among two or more parties a continuous or discrete problem? Explain.

9. Two friends have decided to share an apartment in order to obtain a nicer apartment than either could afford individually. They choose a two-bedroom apartment that rents for $900 monthly, including utilities. One bedroom is larger and sunnier than the other. Propose a procedure for deciding which of the friends gets the nicer bedroom.

High Court Awards New Jersey Most of Ellis Island

NEW YORK TIMES,
May 27, 1998

New Jersey won the battle of Ellis Island today.

Five years after New Jersey sued New York in the Supreme Court to establish sovereignty over the largely artificial mix-ture of rock and land-fill in New York Harbor, the Justices swept aside New York's appeal to history and sentiment and ruled by a vote of 6 to 3 that the entire landfill part, nearly 90 percent of the island, is New Jersey's territory.

Computer/Calculator Exploration

10. It can be instructive to examine the results of an estate division when one or more of the bids is changed. However, it is tedious to redo all the calculations several times over. Fortunately, this lesson's estate division algorithm can be implemented on a computer spreadsheet, which simplifies changing values and inspecting the results. Use a computer spreadsheet to perform this lesson's estate division algorithm. A sample output and the formulas that generated it are shown below. In this case, the results are those of Exercise 3. Once your spreadsheet is complete, use it to answer the questions that follow.

	A	B	C	D	E	F
1		Estate Division Spreadsheet				
2						
3		Amy	Bart	Carol		
4	Painting	2000.00	5000.00	3000.00		
5	Motorcycle	4000.00	2000.00	3000.00		
6	Ticket	500.00	100.00	300.00		
7						
8		Amy	Bart	Carol		Cash:
9	Bid total	11500.00	12100.00	11300.00		5000.00
10	Share	0.333333	0.333333	0.333333		666.67
11	Fair share	3833.33	4033.33	3766.67		966.67
12	Object value	4500.00	5000.00	0.00		−3766.67
13	Cash received	−666.67	−966.67	3766.67		
14	Extra cash	955.56	955.56	955.56		2866.67
15						
16	Final total	4788.89	4988.89	4722.23		
17						

	A	B	C	D	E	F
1			Estate Division Spreadsheet			
2						
3		Amy	Bart	Carol		
4	Painting	2000	5000	3000		
5	Motorcycle	4000	2000	3000		
6	Ticket	500	100	300		
7						
8		Amy	Bart	Carol		Cash:
9	Bid total	=SUM(B4:B6)+$F$9	=SUM(C4:C6)+$F$9	=SUM(D4:D6)+$F$9		5000
10	Share	=1/3	=1/3	=1/3		=−B13
11	Fair share	=B9*B10	=C9*C10	=D9*D10		=−C13
12	Object value	=B5+B6	=C4	=0		=−D13
13	Cash received	=B11−B12	=C11−C12	=D11−D12		
14	Extra cash	=$F$14*B10	=$F$14*C10	=$F$14*D10		=SUM(F9:F12)
15						
16	Final total	=SUM(B12:B14)	=SUM(C12:C14)	=SUM(D12:D14)		

a. What would happen if the amount of cash in the estate were 0? Change the amount in cell F9 to 0 and see. Describe the result.

b. What would happen if Bart lied about the value he placed on the motorcycle and said he felt it was worth $5,000? Change the amount in cell C5 to 5,000 and see. (Change the cash back to 5,000 before doing this one.) Describe the result.

c. What would happen if Bart really did feel that the motorcycle was worth $5,000 but accepted a $2,000 bribe from Amy to bid $2,000? How would this collusion between Bart and Amy change the value of the final settlements for Bart and Amy?

d. Explain how to change the formulas in the spreadsheet to account for the situation in Exercise 4.

Projects

11. Matrix calculations like the multiplication shown in Exercise 6 are useful in programming computers to do tedious calculations. Research and report on the use of matrix applications in computer science.

12. Research division procedures that are used at auctions. What are Dutch and English auctions? Why do some auctions award the contract to the second-highest bidder? When are closed and open bids used?

Apportionment Algorithms

The problem of fairly dividing an estate involves discrete objects, but also involves cash. When a fair division problem is strictly discrete, the problem can be impossible to solve in a way that treats all parties fairly.

The fair allocation of discrete objects occurs in a variety of situations. For example, your school's administrators must decide a fair way to allocate teaching positions to the school's various departments and equipment such as computers to classrooms. One of the most politically charged fair distribution problems in the United States involves the apportionment of seats in the U.S. House of Representatives among the states. (The House is reapportioned every ten years after a new census is taken.)

Unlike estate division situations, in which individuals may not agree on the value of an object, the value of a seat in the U.S. House is not subjective. Therefore, the definition of fairness is a simple one that is mandated by the Constitution: that the seats be distributed among the states according to population.

Although the definition of fairness used to apportion seats in the U.S. House is not controversial, the method of apportionment can be. The first veto by U.S. president occurred in 1792 when George Washington rejected an apportionment bill advocated by Alexander Hamilton in favor of a method championed by Thomas Jefferson. This lesson considers the two

Alexander Hamilton.

methods of apportioning seats in a governmental body that were at the center of the Hamilton-Jefferson dispute.

Since the apportionment problem as applied to the U.S. House of Representatives involves 435 seats and 50 states, you will feel more comfortable starting with a simpler example. (Although they may seem artificial, many examples in this lesson are designed to reflect the large differences in populations among states.)

Central High School has sophomore, junior, and senior classes of 464, 240, and 196 students, respectively. The 20 seats on the school's student council are divided among the classes according to population. Since there are 900 students in the school and since 900 ÷ 20 = 45, ideally each representative in the council would represent 45 students. In other words, the **ideal ratio** of students to seats is 45.

Thomas Jefferson.

$$\text{Ideal ratio} = \frac{\text{Total population}}{\text{Number of seats}}$$

In cases of political representation, the ideal ratio is often called the *ideal district size.* If, for example, the population of the United States is 250 million, then the ideal district size is 250,000,000 ÷ 435, or about 575,000. Ideally, each member of the U.S. House would represent 575,000 people. This ideal cannot be achieved because district boundaries cannot cross state lines.

Because Central High's sophomore class has 464 members, it deserves 464 ÷ 45 = 10.31 seats. Accordingly, 10.31 is called the sophomore class **quota**. Similarly, the junior and senior quotas are 5.33 and 4.36 seats, respectively.

The Quotas
....
Sophomores:	10.31	
Juniors:	5.33	$\text{Quota} = \dfrac{\text{Class Size}}{\text{Ideal Ratio}}$
Seniors:	4.36	

In the case of the U.S. House, a state's quota is determined by dividing its population by the ideal district size. For example, a state with 2 million people deserves 2,000,000 ÷ 575,000, or about 3.5, seats.

It isn't possible to split a seat in Central High's council and give 0.36 of it to the seniors, 0.33 of it to the juniors, and 0.31 of it to the sophomores. The school must decide a fair way to award this seat to one of the classes.

The methods favored by Hamilton and Jefferson have one thing in common: Each begins by ignoring the decimal part of each quota and assigning a number of seats equal to the whole number part of the quota. Regardless of whether the quota is 10.31 or 10.91, both Hamilton and Jefferson begin by awarding 10 seats. Ignoring the decimal part of a number in this way is called *truncating*.

The total of the truncated quotas is 10 + 5 + 4 = 19 seats. The difference in the Hamilton and Jefferson methods lies in the way they award the remaining seat.

The **Hamilton method** awards the remaining seat to the class whose quota has the largest decimal part. Since the decimal part of the senior quota, 0.36, is larger than either of the other two decimal parts, the senior class gets the extra seat. The results of the Hamilton method are summarized in the following table.

Class Size	Quota	Hamilton Apportionment
464	10.31	10
240	5.33	5
196	4.36	5

The Hamilton method seems reasonable to most people. Perhaps some members of your class proposed a similar method in Lesson 2.1. However, the Hamilton method has fallen out of favor in the United States for reasons you will consider in this lesson's exercises, after you have examined Jefferson's approach.

You might think of the **Jefferson method** as conducting a race to see whether the sophomore quota can increase to 11 or the junior quota can increase to 6 or the senior quota can increase to 5 first. Here is how it conducts this race.

Since a quota is found by dividing the class size by the ideal ratio, the quota becomes larger when the ideal ratio becomes smaller. For example, consider what happens if the ideal ratio is decreased from 45 students per seat to 40 students per seat. The results are summarized in the following table.

Class Size	Quota with Ideal Ratio of 45	Quota with Ideal Ratio of 40
464	464 ÷ 45 = 10.31	464 ÷ 40 = 11.6
240	240 ÷ 45 = 5.33	240 ÷ 40 = 6.0
196	196 ÷ 45 = 4.36	196 ÷ 40 = 4.9
Seats	10 + 5 + 4 = 19	11 + 6 + 4 = 21

For example, the sophomore class receives 10 seats when the ideal ratio is 45 and 11 seats when the ratio is 40. Therefore, there must be some ratio between 45 and 40 that causes the sophomore apportionment to be exactly 11. It can be found by dividing the sophomore class size by 11: $464 \div 11 \approx$ 42.18. The number 42.18 is called the **Jefferson adjusted ratio**.

$$\text{Jefferson adjusted ratio} = \frac{\text{Class size}}{\text{Truncated quota} + 1}$$

Similarly, the junior quota passes 6 when the ratio drops below $240 \div 6 = 40$, and the senior quota passes 5 when the ratio drops below $196 \div 5 = 39.2$.

Proponents of the Jefferson method argue that since the ideal ratio does not produce a complete apportionment of the seats, it should be abandoned for a new ratio, as close to the ideal as possible, that does give a complete apportionment. If the ideal ratio is gradually decreased from 45, it will reach a value at which the sophomores receive another seat before it reaches a value at which either of the other classes receives another seat, as shown in the following table.

Ideal Ratio	Sophomore Seats	Junior Seats	Senior Seats
45	10	5	4
↓			
42.18	11	5	4
↓			
40	11	6	4
↓			
39.2	11	6	5

Greater detail can be seen in the following table, in which the adjusted ratio is decreased in steps of 0.2. Note that the sophomore class quota passes the next integer before either the junior or the senior quota does.

Adjusted Ratio	Sophomore	Junior	Senior
45.00	10.31	5.33	4.36
44.80	10.36	5.36	4.38
44.60	10.40	5.38	4.39
44.40	10.45	5.41	4.41
44.20	10.50	5.43	4.43
44.00	10.55	5.45	4.45
43.80	10.59	5.48	4.47
43.60	10.64	5.50	4.50
43.40	10.69	5.53	4.52
43.20	10.74	5.56	4.54
43.00	10.79	5.58	4.56
42.80	10.84	5.61	4.58
42.60	10.89	5.63	4.60
42.40	10.94	5.66	4.62
42.20	**11.00**	5.69	4.64
42.00	11.05	5.71	4.67
41.80	11.10	5.74	4.69
41.60	11.15	5.77	4.71
41.40	11.21	5.80	4.73
41.20	11.26	5.83	4.76
41.00	11.32	5.85	4.78
40.80	11.37	5.88	4.80
40.60	11.43	5.91	4.83
40.40	11.49	5.94	4.85
40.20	11.54	5.97	4.88
40.00	11.60	**6.00**	4.90
39.80	11.66	6.03	4.92
39.60	11.72	6.06	4.95
39.40	11.78	6.09	4.97
39.20	11.84	6.12	**5.00**
39.00	11.90	6.15	5.03

The Jefferson method can be summarized in algorithmic form:

1. Divide the total population by the number of seats to obtain the ideal ratio.

2. Divide the population of each class (state, district, etc.) by the ideal ratio to obtain the class quota.

3. Assign a number of seats to each class equal to its truncated quota.

4. If the number of seats assigned matches the total number of seats to be apportioned, then stop.

5. If the number of seats assigned is smaller than the total number of seats to be apportioned, then divide the size of each class by one more than the number of seats assigned to it in step 3, to obtain an adjusted ratio.

6. Give an extra seat to the class with the largest adjusted ratio. (In other words, to the class for which the adjusted ratio is closest to the ideal ratio.)

This algorithm applies only to situations in which the total of the truncated quotas falls one short of the number of seats available. In some cases, there is more than a one-seat shortfall after truncation. This lesson's exercises consider what to do in such cases.

Exercises

1. The student council at Central High has had difficulty deciding a number of issues because of conflicts between the sophomore representatives and the representatives of the other two classes. The vote has been a 10–10 tie. The council has decided to add a seat in order to prevent frequent ties.
   a. On the basis of the data in this lesson, which class do you think should have the extra seat? Why?
   b. Find the new ideal ratio of students per seat for the 21-seat council.
   c. Use the new ratio to determine the quota for each of the three classes.
   d. Use the Hamilton method to allocate the 21 seats on the new council to the three classes.
   e. Compare the Hamilton apportionment for the 21-seat council to that of the 20-seat council and explain why the results contitute a paradox.

2. A senior council member who recently studied apportionment in the school's discrete mathematics course is unhappy over the loss of one of the senior seats and proposes the apportionment be made by a different method.

a. Find an adjusted ratio for each class as described in the Jefferson algorithm of this lesson.

b. Decrease the 21-seat ideal ratio until all 21 seats are allocated. State the number of seats given each class by the Jefferson method.

c. Compare the 21-seat Jefferson apportionment with the 20-seat Jefferson apportionment. Does the Jefferson method produce a paradox similar to the one described in Exercise 1?

3. Revise the Jefferson apportionment algorithm given in this lesson to account for situations in which more than one seat remains after truncation.

4. The paradox you observed in Exercise 1 occurs because increases in a divisor do not produce equal changes in quotients. When the size of a representative assembly increases and the total population remains the same, the ideal ratio decreases. As an example, consider two classes (states, districts, etc.) with populations of 100 and 230; an increase in the size of the council has caused the ideal ratio to decrease from 22 to 21.

a. Complete the following table and explain why it could result in the shift of a council seat from one class to the other.

Class Size	Quota with Ideal Ratio of 22	Quota with Ideal Ratio of 21
100		
230		

b. Will the paradox observed in Exercise 1 result in the loss of a seat for a small class or a large one? Why?

c. Why do you think Thomas Jefferson opposed the Hamilton method? (If you're not sure, look up the 1790 census results for Virginia, Jefferson's home state.)

5. The student council members at Central High, aware of the strange results that slight differences can make, decide to monitor the council's apportionment. At the end of the first quarter of the school year, the class numbers have changed somewhat:

Sophomores	459
Juniors	244
Seniors	197

a. Use the Hamilton method to divide the council's 21 seats among the classes.

At the end of the first semester the classes have changed again:

Sophomores	460
Juniors	274
Seniors	196

At the council's first meeting of the new semester, the members are amazed when one of the representatives of the senior class, the only class that has decreased in size, demands that the council be reapportioned.

b. Use the Hamilton method to re-apportion the council.

c. Explain why the results constitute a paradox.

d. Use an analysis similar to that of Exercise 4 to explain why this type of paradox occurs. Will it have an adverse effect on small classes or large classes?

6. The cartoon on page 62 has been described as "an indiscreet attempt to apply continuous division procedure to a discrete problem." Explain what this means.

7. The U.S. population determined by the 1990 census was 248,709,873. The populations of Texas, New York, and California were 16,986,510, 17,990,455, and 29,760,021, respectively.

a. Find the quota for each of these three states.

b. Compare the quotas to the actual apportionment mentioned in the news article.

c. Do you think the apportionment treated any of these three states unfairly? Explain.

Texas Could Get Two More House Positions Following Next Census

LUBBOCK AVALANCHE-JOURNAL,
April 12, 1997

When Texas' robust population gains are tallied in the next census, the Lone Star state in all likelihood will surpass New York to become the second-largest presence in the House of Representatives.

Texas currently boasts the third-largest House delegation with 30 seats. California is first with 52, New York second with 31.

House seats are apportioned on the basis of states' populations. When the 2000 Census is completed, some seats in the Northeast and Midwest are expected to migrate to Texas and other booming Sunbelt states.

On the basis of population projections for 2000, the Congressional Research Services estimates that 11 seats will shift.

The House is limited to 435 seats. Each state is guaranteed one seat and the others are then distributed on the basis of the population count mandated by the Constitution.

d. The 1990 census total for Montana was 799,065. The apportionment that followed decreased Montana's representation from seats 2 to 1. Montana went to court to challenge the apportionment method but lost its case. Do you think it would have been fairer to give Montana 1 seat or 2? Explain.

Computer/Calculator Explorations

8. Develop a spreadsheet to do the Jefferson apportionment for the three classes in this lesson's example. When finished, the spreadsheet should be similar to the following.

	A	B	C	D
		Population	Quota	Seats
1		Population	Quota	Seats
2	Sophomore	464	10.311111	10
3	Junior	240	5.333333	5
4	Senior	196	4.355556	4
5	Total	900		19
6	Seats	20		
7	Ideal ratio	45		

The values in columns C and D and in cells B5 and B7 should be calculated with formulas. The formulas in column D require the use of your spreadsheet's truncation function. On many spreadsheets, this function is abbreviated TRUNC. For example, TRUNC(C2,0) truncates the value in cell C2 so it has 0 decimal places.

When your spreadsheet is done, show how the proper apportionment can be found by changing the value in cell B7 (the ideal ratio).

Projects

9. Research and report on methods that have been used to apportion the U.S. House of Representatives and controversies that have arisen. Why has the apportionment method been changed? Why has the size of the House been changed? Did paradoxes occur with any of the methods?

10. The president of the U.S. is chosen by the electoral college. The number of electoral votes a state has is determined by the size of its congressional delegation. Thus, apportionment affects the electoral college vote. Research and report on the affect of apportionment on the election of the president. Has the apportionment method ever made a difference in whom the electoral college elected president?

More Apportionment Algorithms and Paradoxes

Dissatisfaction with paradoxes that sometimes occur with the Hamilton method led to its abandonment as a method of apportioning the U.S. House of Representatives. The Jefferson method was attacked by small states because it favors large states. This lesson considers alternatives to the Jefferson and Hamilton methods and some recent developments in the debate over which method of apportionment is fairest.

The Jefferson method is one of several *divisor methods* of apportionment. The term *divisor* is used because these methods determine quotas by dividing the population by an ideal ratio or an adjusted ratio. This ratio is the divisor. The Hamilton method is not a divisor method.

Two divisor methods given considerable attention today are a method named for Daniel Webster and another named for Joseph Hill, U.S. statistician. The Webster and Hill methods differ from the Jefferson method in the way that they round quotas. Recall that the

The United States House of Representatives is apportioned by the Hill method, although this has not always been the case. At various times it has been apportioned by the Jefferson method, the Hamilton method, and the Webster method.

Jefferson method truncates a quota and apportions a number of seats equal to the integer part of the quota. Quotas of 11.06 and 11.92 both receive 11 seats under the Jefferson method.

Judge Pessimistic on Reapportionment for Chichester Soon

PHILADELPHIA INQUIRER, January 13, 1998, by Bill Ordine

A federal judge in Philadelphia had bad news for anyone thinking an end might be near in the protracted dispute between the Chichester School District and a group seeking reapportionment of the school board seats.

The lengthy federal court battle is the result of a lawsuit filed by Upper Chichester residents who contend that they are underrepresented.

The Chichester School District covers Upper Chichester, Lower Chichester, Marcus Hook, and Trainer. Until now, school board members have been elected under a system in which Upper

Chichester—which now has about two-thirds of the district's population—gets four seats, with the remaining five divide among the three other towns.

Terry Elizabeth Silva, attorney for Upper Chichester residents, asked for a new election for March 10. She argued for a tri-regional plan that would likely guarantee Upper Chichester at least 6 seats.

Michael Levin, attorney for the school district, presented two slightly different plans, depending on which population figures were used. In each case, part of Upper Chichester would make up one region, and in the two other regions, portions of Upper Chichester would be mixed with other municipalities.

The **Webster method** uses the rounding method with which you are familiar: A quota above or equal to 11.5 receives 12 seats, and a quota below 11.5 receives 11 seats. The number 11.5 is sometimes called the **arithmetic mean** of 11 and 12. The arithmetic mean of two numbers is the number halfway between them. It can be calculated by dividing the sum of the two numbers by 2.

The **Hill method** rounds by using the geometric mean instead of the arithmetic mean. The **geometric mean** of two numbers is the square root of their product. If the quota exceeds the geometric mean of the integers directly above and below the quota, the quota is rounded up; otherwise it is rounded down. For example, a quota between 11 and 12 must exceed $\sqrt{11 \times 12} \approx 11.4891$ to receive 12 seats (see Figure 2.1).

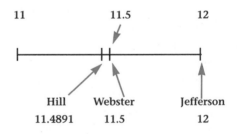

Figure 2.1 The Hill, Webster, and Jefferson roundup points for quotas between 11 and 12.

The Hill method can cause confusion because the quota for which it awards an extra seat must be calculated. For example, if the quota is 7.4903 and you are using the Webster method, you know immediately that 7 seats are awarded because the quota is slightly below 7.5. However, you do not know whether the Hill method awards 7 or 8 seats until you have calculated $\sqrt{7 \times 8} \approx 7.4833$. This calculation tells you that the Hill method awards 8 seats.

The following table summarizes the apportionment of the Central High student council example from Lesson 2.3 by the methods discussed in this and the previous lesson.

Class		Initial Apportionment			
Size	Quota	Hamilton	Jefferson	Webster	Hill
464	10.31	10	10	10	10
240	5.33	5	5	5	5
196	4.36	5	4	4	4

The Jefferson, Webster, and Hill methods all fail to assign one of the seats and so require an adjusted ratio. The following table lists the adjusted ratio necessary for each class to gain a seat by each method.

Class		Adjusted Ratio for	
Size	Jefferson	Webster	Hill
464	$464 \div 11 = 42.1818$	$464 \div 10.5 = 44.1905$	$464 \div \sqrt{10 \times 11} = 44.2407$
240	$240 \div 6 = 40.0000$	$240 \div 5.5 = 43.6364$	$240 \div \sqrt{5 \times 6} = 43.8178$
196	$195 \div 5 = 39.2000$	$196 \div 4.5 = 43.5556$	$196 \div \sqrt{4 \times 5} = 43.8269$

Recall that the adjusted Jefferson ratio for the sophomore class is found by dividing the class size by 11. The adjusted Webster ratio for the sophomore class is found by dividing the class size by 10.5. The adjusted Hill ratio for the sophomore class is found by dividing the sophomore class size by the geometric mean of 11 and 12, or 10.4881.

Each of these methods requires that the ideal ratio of 45 be decreased until it is smaller than exactly one of the adjusted ratios. To compare the methods, it can be helpful to list the adjusted ratios in decreasing order:

Jefferson: 42.1818 (sophomores), 40.0000 (juniors), 39.2000 (seniors)

Webster: 44.1905 (sophomores), 43.6364 (juniors), 43.5556 (seniors)

Hill: 44.2407 (sophomores), 43.8269 (seniors), 43.8178 (juniors)

All three methods award the extra seat to the sophomores. Note, however, that the Hill method lists the senior class second rather than third.

Because the Webster and Hill methods do not truncate quotas, they sometimes apportion too many seats rather than too few. In this lesson's exercises you will consider how to apply the Webster and Hill methods in such cases and learn of some surprising results.

Exercises

1. a. Complete the following apportionment table for the 21-seat Central High student council described in Exercise 1 of Lesson 2.3 (see page 65).

Class		Initial Apportionment			
Size	Quota	Hamilton	Jefferson	Webster	Hill
464	――――	――――	――――	――――	――――
240	――――	――――	――――	――――	――――
196	――――	――――	――――	――――	――――

   b. The Jefferson method distributes only 19 seats. You can use your results from Exercise 1 of Lesson 2.3 to complete the Jefferson apportionment. Give the final Jefferson apportionment.

   Both the Webster and Hill methods apportion 22 seats. Therefore, the ideal ratio must be increased until one of the classes loses a seat. The sophomore class, for example, will lose a seat under the Webster method if its quota drops below 10.5. This requires an adjusted ratio of $464 \div 10.5 \approx 44.1905$. For the sophomore class to lose a seat under the Hill method, its quota must drop below $464 \div \sqrt{10 \times 11} \approx 44.2407$.

   c. Complete the following table of adjusted ratios for the Webster and Hill methods.

Class	Adjusted Ratio For	
Size	Webster	Hill
464	$464 \div 10.5 = 44.1905$	$464 \div \sqrt{10 \times 11} = 44.2407$
240	――――	――――
196	――――	――――

   d. List the adjusted ratios for the Webster method in increasing order.
      The ideal ratio must be increased until it passes the first ratio in your
      list. This class loses one seat. Give the final Webster apportionment.
   e. List the adjusted ratios for the Hill method in increasing order. The
      ideal ratio must be increased until it passes the first number in your
      list. This class loses one seat. Give the final Hill apportionment.
   f. Which apportionment methods would be favored by each class?
      Explain.

2. Since the number of seats assigned to a class rarely equals its quota
   exactly, apportionments are not completely fair. This fact has led people
   to ask whether one apportionment is less fair than another. One way to
   measure the unfairness of an apportionment is to total the discrepancies
   between the quota and the number of seats assigned to each class. For
   example, if the quota is 11.25 seats and 11 seats are apportioned, the
   unfairness is 0.25 seats. If 12 seats are apportioned, the unfairness is
   0.75 seats.
   a. Use the apportionments for the 20-seat Central High student council
      to measure the discrepancy for each class by means of each method.
      Record the results in the following table.

| Class | | Amount of Discrepancy | | | |
Size	Quota	Hamilton	Jefferson	Webster	Hill
464	10.31	——	——	——	——
240	5.33	——	——	——	——
196	4.36	——	——	——	——
Total discrepancy		——	——	——	——

   b. Which method has the smallest total discrepancy?
   c. Do you think smallest total discrepancy is a good criterion for choos-
      ing an apportionment method? Explain.

3. Another way to measure the unfairness of an apportionment is to
   compare the representation of two classes (states, districts, etc.) as a
   percentage. For example, if a class with 250 students has 5 seats, then
   each seat represents $250 \div 5 = 50$ people. If another class has 270
   students and 6 seats, the representation is $270 \div 6 = 45$ people per
   seat. The representation is unfair to the first class by $50 - 45 = 5$ people
   per seat, which is $5 \div 50 = 0.10$, or 10% of its representation.
   a. The initial Hill apportionment in Exercise 1 apportioned one seat
      too many. Compare the representation of the junior and senior

classes if one seat is taken from the juniors. What is the unfairness percentage to the juniors?

b. Compare the representation of the junior and senior classes if one seat is taken from the seniors. What is the unfairness percentage to the seniors?

c. By this percentage measure, is it less unfair to take a seat from the juniors or to take a seat from the seniors? Which apportionment in Exercise 1 agrees with your answer?

4. None of the divisor methods is plagued by the paradoxes that caused the demise of the Hamilton method. Divisor methods, however, can cause their own peculiar problems. As an illustration, consider the case of South High School. The freshman class has 1,105 members, the sophomore class has 185, the junior class 130, and the senior class 80. The 30 members of the student council are apportioned among the classes by the Webster method.

a. What is the ideal ratio?

b. Complete the following Webster apportionment table.

Class Size	Quota	Initial Webster Apportionment
1,105	22.1	22
185	___	___
130	___	___
80	___	___

c. Because the Webster method apportions too many seats, the ideal ratio must be decreased. Calculate the adjusted ratio necessary for each class to lose a seat and enter the results in the following table.

Class Size	Adjusted Ratio
1,105	$1105 \div 21.5 = 51.3953$
185	___
130	___
80	___

d. Determine the final apportionment. Explain your method.

e. Explain why the freshman class would consider the final apportionment unfair.

The Webster apportionment in Exercise 3 demonstrates a **violation of quota**. This occurs whenever a class (district, state, etc.) is given a number of seats that does not equal either the integer part of its quota or one more than that. It can occur with any divisor method and is considered a flaw of divisor methods.

The following table shows how the quotas for each class change as the ideal ratio is gradually increased. The freshman quota drops much more quickly than the others do.

Adjusted Ratio	Freshman	Sophomore	Junior	Senior
50.00	22.10	3.70	2.60	1.60
50.20	22.01	3.69	2.59	1.59
50.40	21.92	3.67	2.58	1.59
50.60	21.84	3.66	2.57	1.58
50.80	21.75	3.64	2.56	1.57
51.00	21.67	3.63	2.55	1.57
51.20	21.58	3.61	2.54	1.56
51.40	21.50	3.60	2.53	1.56
51.60	21.41	3.59	2.52	1.55
51.80	21.33	3.57	2.51	1.54
52.00	21.25	3.56	2.50	1.54

Around 1980 Michel L. Balinski and H. Peyton Young proved an important impossibility theorem: Any apportionment method sometimes produces at least one of these undesirable results: violation of quota and the two paradoxes that occur with the Hamilton method (see Exercises 1 and 5 of Lesson 2.3).

Mathematician of Note

Michel L. Balinski is a director of the Laboratoire d'Econometrie and the Ecole Polytechnique in Paris.

Mathematician of Note

H. Peyton Young is professor of economics at Johns Hopkins University.

5. Every person in a small community belongs to exactly one of the community's four political parties. Membership is distributed as shown in the following table.

Party	Membership
A	561
B	200
C	100
D	139

The 20 seats in the community's council are apportioned by the Jefferson method.

a. Determine the Jefferson apportionment.

b. Parties C and D decide to join together to form a single party. Determine the new apportionment.

c. Would the apportionment that results from the merger of parties C and D have occurred with any of the other methods you have studied? Explain.

6. Compare the current quota for Upper Chichester with the one proposed by its residents in the news article on page 70. Do you feel the Upper Chichester residents are justified in their demands? How does this situation compare with weighted voting situations you studied in Lesson 1.5?

7. Compare the Hill roundoff point with the Webster roundoff point for quotas of various sizes. How, for example, do they compare for quotas between 2 and 3, between 10 and 11, between 100 and 101? Can you prove any relationship between the two?

Computer/Calculator Explorations

8. Extend the spreadsheet you made in Exercise 8 of Lesson 2.3 to include columns for the Webster and the Hill methods. You will need to use your spreadsheet's rounding function to do the Webster apportionment. For the Hill apportionment you will need to use the spreadsheet's logical functions. Logical functions on a spreadsheet are similar to those on a calculator. For example, the calculation $A(B < 5) + (A + 1)(B > 5)$ produces the value of A when $B < 5$ and the value of $A + 1$ when $B > 5$ because an equation or inequality has a value of 0 when it is false and 1 when it is true.

Projects

9. Research and report on the work of Balinski and Young. How did they prove their result? What method of apportionment did they recommend for the U.S. House of Representatives?

10. Research divisor methods for other ways in which violation of quota can occur. For example, can you find four classes for which the number of apportioned seats falls two short of the size of the council and results in awarding both seats to the same class?

11. Investigate measures of fairness such as those given in Exercises 2 and 3. For each apportionment method discussed in this and the previous lesson, which measure of fairness produces the same apportionment?

12. Research and report on the process used by the U.S. Census Bureau to apportion U.S. House seats after each census. How does the bureau implement the Hill method?

13. Research and report on apportionment methods used in other countries. Which methods are used and why?

14. Research and report on the way presidential primary results are used to apportion convention delegates among the candidates.

Fair Division Algorithms: The Continuous Case

The problem of fairly dividing a cake is in some ways similar to estate division. Like the cash in an estate, a cake can be divided in any number of ways. Moreover, the participants in a cake division situation may not agree on the value of a particular slice of cake, just as the heirs to an estate may place different values on a house. However, unlike estate division and legislative apportionment, a cake division problem is strictly continuous and involves no items such as cash and legislative seats that are of indisputable value.

The problem of fairly dividing a continuous medium is not limited to the activities of children. States, for example, must agree on how to share water resources.

Before beginning your exploration of cake division, you may want to review the work that you and others in your class did in Lesson 2.1. In that lesson, you concluded that the resolution of a cake division problem by an outside authority may not yield a solution that is fair in the eyes of both individuals. You also concluded that the use of a random event such as a coin toss may not result in a division that is fair in the eyes of both parties. In order for two people to feel that a piece of cake has been divided fairly, each must feel that he or she received at least half the cake.

An acceptable solution to the problem of fairly dividing a piece of cake is not possible without a *definition of fairness*. Therefore, a division among n people is called fair if each person feels that he or she received at least 1/nth of the cake (or other object).

Lesson 2.1 also showed that an appeal to a measurement scheme such as weighing may not be adequate because an individual's evaluation of a piece of cake may be based on more than just size. Cake icing, for example, could mean more to one person than to another. Along with a definition of fairness, a good solution to the cake division problem requires a few realistic *assumptions* about the participants.

1. Each individual is capable of dividing a portion of the cake into several portions that the individual feels are equal.

2. Each individual is capable of placing a value on any portion of the cake. The total of the values a person places on all portions of a cake is 1, or 100%.

3. The value that each individual places on a portion of the cake may be based on more than just the size of the portion.

The two-person cake division problem is usually solved by having one person cut the cake into two pieces and letting the other person choose one of

Water Fears on Babbitt's Shoulders

LAS VEGAS REVIEW JOURNAL,
December 15, 1997,
by Susan Greene

While politicians have squabbled over how to fund expansions of Southern Nevada's water delivery system, a quieter discussion has progressed this year about how to ensure there's water to deliver.

"It's clear Nevada will need a regional approach to grow into the future," Interior Secretary Bruce Babbitt said in December 1996 during the annual meeting of the Colorado River Water Users Association. One year later, as the group prepares to reconvene, local negotiators will count on Babbitt to approve a deal they say would help keep lawns watered and showers flowing for the next 30 years.

Federal law has granted Nevada 300,000 acre-feet of river water annually; about 250,000 of that is in use, and local water agencies have committed nearly all of the remainder to developers. The river, which flows into Lake Mead, is Southern Nevada's main source of water.

Arizona tapped 2.6 million of its 2.8 million acre-feet allotment in 1996, but this year plans to use or store for future use its full share.

California is entitled to 4.4 million acre-feet, but is expected to use 5.3 million this year; the excess comes from water not used by upstream states, including some 50,000 acre-feet of Nevada's unused entitlement.

An acre-foot is enough water for the typical needs of one to two families each year.

them. It is instructive to examine the roll of the above definition and assumptions in this solution.

The requirement that one person (the cutter) cut the cake into two portions that the person feels are equal is the first assumption. The second assumption ensures that the second person (the chooser) will place values on the two pieces of the cake that total 1. However, because of the third assumption, the chooser need not place the same value on both pieces. Even if the chooser feels that his or her piece is more than half the cake, the division is fair because the definition of fairness requires only that each person feel that his or her piece is at least half the cake.

No solution to the problem of fairly dividing a cake among more than two people is adequate unless it adheres to the definition of a fair cake division. For example, a fair division among three people must result in each person receiving a piece the person feels is at least one-third of the cake.

In mathematics, there is often more than one way to solve a problem. That is the case with the problem of dividing a cake fairly among three or more people. Often a solution to a problem can be based on the solution to a simpler problem. The following solution to the three-person problem is based on the two-person solution.

Call the three individuals Ann, Bart, and Carl. The solution is described in algorithmic form:

1. Ann cuts the cake into two pieces that she feels are equal.

2. Bart chooses one of the pieces; Ann gets the other.

3. Ann cuts her piece into three pieces that she considers equal; Bart does the same with his.

4. Carl chooses one of Ann's three pieces and one of Bart's.

To see that this is indeed a solution to the three-person problem, you must be satisfied that it adheres to the definition of fairness. That is, you must believe that each person has received a portion that the person feels is at least 1/3 of the cake.

Consider Ann. In step 1, Ann feels that each piece is one-half of the cake. She therefore feels that the piece she received in step 2 is half the cake. She feels that each piece she cut in step 3 is one-third of half the cake, or one-sixth of the cake. She therefore feels that she receives two-sixths, or one-third, of the cake.

Bart's case is similar except that he may feel that the portion he chose in step 2 is more than half the cake. Thus, he may feel that each piece he cut in step 3 is more than one-sixth of the cake and that his final share is more than two-sixths, or one-third.

Carl's case is quite different from either Ann's or Bart's. He may feel that the two pieces Ann cut in step 1 are unequal. He may, for example, consider one piece 0.6 of the cake and the other 0.4. Likewise, he may not feel that the cuts made in step 3 resulted in equal pieces. He could, for example, decide that the piece he valued at 0.6 was divided into pieces he values at 0.3, 0.2, and 0.1. Similarly, he could decide that the piece he valued at 0.4 was divided into pieces he values at 0.2, 0.1, and 0.1. However, because he chooses first, he will pick the largest piece from each: 0.3 from the piece he valued at 0.6 and 0.2 from the piece he valued at 0.4. Thus, he feels that the value of the portion he receives is $0.3 + 0.2 = 0.5$.

The previous example gave Carl a portion he valued at more than one-third, but it was only one example. To see that Carl will always get a portion he values as at least one-third, use the variable x to represent the value Carl places on one of the pieces cut by Ann in step 1. His value for the other piece cut by Ann must be $1 - x$. Although Carl may not feel that the piece he values at x has been divided into equal thirds, he will choose the largest of the three pieces and value it as at least one-third of x, or $\frac{1}{3}x$. Similarly, he will value the piece he chooses from the part he values as $1 - x$ as at least $\frac{1}{3}(1 - x)$. Thus, the total value of his two pieces is at least $\frac{1}{3}x + \frac{1}{3}(1 - x) = \frac{1}{3}x + \frac{1}{3} - \frac{1}{3}x = \frac{1}{3}$. This method is referred to as the **cut-and-choose method**.

This lesson's exercises consider several issues related to cake division and several ways of solving problems with three or more participants.

Exercises

1. In the division among Ann, Bart, and Carl, who will value his or her share as exactly one-third? Who might feel that he or she received more than one-third? Explain.

2. Does the division among Ann, Bart, and Carl result in three portions or three pieces? Does your answer violate the definition of fairness or any of the three assumptions?

3. For each of the four steps of the division among Ann, Bart, and Carl, state which of the three fairness assumptions is applied.

   In Exercises 4 and 5, suppose Carl feels that Ann's initial division in step 1 is even, that Ann's subdivision in step 3 is also even, but that

Bart's subdivision in step 3 is not. (Give your answers as fractions or as decimals rounded to the nearest 0.1.)

4. What value will Carl place on the piece he takes from Ann? Explain.

5. Although Carl feels that the piece Bart divided is half the cake, he does not feel that Bart subdivided it equally. He could, for example, place values of 0.3, 0.1, and 0.1 or values of 0.4, 0.06, and 0.04 on the three pieces.
    a. Explain why the largest value Carl can place on any of Bart's three subdivisions is 0.5.
    b. What is the smallest value Carl could place on the piece he takes from Bart? Explain.
    c. What is the largest total value Carl could place on the two pieces he takes from Amy and Bart?
    d. What is the smallest value Carl could place on the two pieces he takes from Amy and Bart?

6. Is the problem of fairly dividing cookies among three children discrete or continuous? Explain.

Reprinted by permission of United Feature Syndicate.

7. In mathematics, a fundamental principle of counting is that if there are *m* ways of performing one task and *n* ways of performing another, then there are $m \times n$ ways of performing both. For example, a tossed coin may fall in two ways, and a rolled die may fall in six ways. Together they may fall in a total of $2 \times 6 = 12$ ways.

    a. If two people each have a piece of cake and each person cuts his or her piece into three pieces, show how to use the fundamental counting principle to determine the number of pieces that result.

    b. If *k* people each have a piece of cake and each cuts his or her piece into $k + 1$ pieces, what are two equivalent expressions for the total number of pieces that result?

    c. If $k + 1$ boxes each contain $k + 5$ toothpicks, what are two equivalent expressions for the total number of toothpicks?

    d. Two offices are being filled in an election: mayor and governor. If there are three candidates for governor and four for mayor and conventional voting procedures are used, in how many ways can a person vote?

8. Consider the following division of a cake among three people: Arnold, Betty, and Charlie. Arnold cuts the cake into three pieces. Betty chooses one of the pieces, and Charlie chooses either of the remaining two. Arnold gets the third piece.

    a. Will Arnold feel he has received at least one-third of the cake? Might he feel he has received more than one-third?

    b. Will Betty feel she has received at least one-third of the cake? Might she feel she has received more than one-third?

    c. Will Charlie feel he has received at least one-third of the cake? Might he feel he has received more than one-third?

9. Arnold, Betty, and Charlie decide to divide a cake in the following way: Arnold slices a piece he considers one-third of the cake. Betty inspects the piece. If she feels it is more than one-third of the cake, she must cut enough from the cake so that she feels it is one-third of the cake. The removed portion is returned to the cake. Charlie now inspects Betty's piece and has the option of doing similarly if he thinks it is still more than one-third of the cake. The piece of cake is given to the last person who cut from it.

One of the remaining two people slices a piece that he or she feels is half of what remains of the cake. The other person inspects the piece with the option of removing some of the cake if he or she feels it is more than half the remainder.

   a. Will the person who receives the first piece feel that it is at least one-third of the cake? Could he or she feel it is more than one-third? Explain your answers.

   b. Will the person who receives the second piece feel that it is at least one-third of the cake? Could he or she feel it is more than one-third? Explain your answers.

   c. Will the person who receives the third piece feel that it is at least one-third of the cake? Could he or she feel it is more than one-third? Explain your answers.

   This method is called the **inspection method.**

10. The definition of a fair cake division used in this lesson does not state that each person should feel his or her portion of the cake is at least as large as the portion received by each of the other participants.

   a. Construct an example for the three-person cut-and-choose division among Ann, Bart, and Carl to show that Carl may feel one of the others got a piece bigger than his.

   b. Could the definition of fair cake division result in jealously? Explain.

Computer/Calculator Exploration

11. Use the moving knife program that accompanies this book to divide a cake of any shape among yourself and two other people in your class by means of the **moving knife method**.

   a. Explain why each of the people in your group feels that he or she received at least one-third of the cake.

   b. Might any of the people feel jealous of the share received by another? Explain.

12. A procedure for fairly dividing a cake can be considered "optimal" if it fulfills some criterion. For example, a method might be called optimal if it results in exactly the same number of pieces as participants. By this criterion, are the cut-and-choose method, the inspection method, or the moving knife method optimal? Explain.

13. Could the cut-and-choose, the inspection, or the moving knife method be extended to divide a cake among four people? Explain how this could be done.

Mathematical Induction

The cut-and-choose method, the inspection method, and the moving knife method all produce a fair division of a cake among two or three people. It appears that all these methods can be extended to larger groups, but you or others in your class may not be convinced. This lesson considers a method mathematicians use to prove that a discrete process will work indefinitely. The method is called **mathematical induction**.

For example, consider the cut-and-choose method. Since mathematical induction is used to prove that a result known to work in a few cases can

be extended indefinitely, it is important to review the situations for which the cut-and-choose method has been demonstrated. It was first offered as a means by which two people could fairly divide a cake. It was extended to three people by requiring that two of them first apply the two-person method. Then each of the two cut his or her piece into three pieces that he or she considered equal. The third person selected a piece from each. Therefore, the cut-and-choose method is known to work for groups of two and three people.

Before applying mathematical induction, consider how to extend the cut-and-choose method to four people, Ann, Bart, Carl, and Daiva. Begin by having Ann, Bart, and Carl apply the three-person method. Then each of them divides his or her portion into four portions that he or she considers equal. Daiva chooses one from each. To be sure this division is fair, you must agree that each of the three cutters and the chooser feel that the received share is one-fourth of the cake.

Consider one of the cutters, Ann. The three-person solution guarantees that she feels she has at least one-third of the cake. Therefore, Ann divides what she considers a third into four equal pieces. Since she retains three of these pieces, the value she attaches to her portion is at least $\frac{3}{4} \times \frac{1}{3} = \frac{1}{4}$.

Daiva may not feel that the three-person solution produced three equal portions, but must feel that the total value of the three portions is 1. Suppose Daiva attaches values of p_1, p_2, and p_3 to the three portions, where $p_1 + p_2 + p_3 = 1$. Because Daiva is given first choice of a portion from each of the cutters, she will place a value of at least $\frac{1}{4}p_1 + \frac{1}{4}p_2 + \frac{1}{4}p_3 = \frac{1}{4}(p_1 + p_2 + p_3) = \frac{1}{4}(1)$ on the resulting portion. Therefore, Daiva considers the division fair.

It appears that this division can be extended to a 5-person situation, to a 6-person situation, and, in general, to an n-person situation. If, for example, a cake is to be divided fairly among 10 people, the division can be based on a 9-person division, which, in turn, is based on an 8-person division, which is based on . . . .

Mathematical induction generalizes this pattern of solutions by proving it is always possible to extend the solution to a group that is one larger than the previous. The generalization is achieved by using a variable instead of a specific number. In other words, assuming you know how to divide a cake fairly among k people, you must show that it is possible to use that knowledge to divide a cake fairly among $k + 1$ people.

The proof begins by applying the assumption that k people can fairly divide the cake. When they have finished the division, each of them divides his or her share into $k + 1$ portions that he or she feels are equal. The $(k + 1)$th person then selects one portion from each.

It is now necessary to show that each of the $k + 1$ people has a share valued as at least $\frac{1}{k+1}$.

In the case of the k cutters, the k-person solution guarantees that each has a portion valued as at least $\frac{1}{k}$ of the cake. This portion is divided into $k + 1$ equal portions, so the value of each is at least $\frac{1}{k(k + 1)}$. Since k of the $k + 1$ portions are retained by the cutter, the total value attached to them is

$$k\left(\frac{1}{k(k + 1)}\right) = \frac{1}{k + 1}$$

Although the chooser may not feel that each of the k portions that result from the k-person solution is $\frac{1}{k}$ of the cake, he or she must feel that the total value is 1. Suppose this person assigns values of $p_1, p_2, . . ., p_k$ to the k portions. Then $p_1 + p_2 + \cdots + p_k = 1$. Because the chooser is given first choice of a portion from each of the cutters, the chooser will place a value of at least

$$\frac{1}{k}p_1 + \frac{1}{k}p_2 + \cdots + \frac{1}{k}p_k = \frac{1}{k}(p_1 + p_2 + \cdots + p_k) = \frac{1}{k}(1)$$

on the resulting portion. Therefore, the chooser considers the division fair.

The proof is complete because it demonstrates that whenever a cake can be divided fairly among k people, it can also be divided fairly among $k + 1$ people. In other words, the cut-and-choose method can be extended indefinitely. Note that the mathematical induction argument is merely a generalization of the one given to justify the extension to four people.

Keep in mind that it is senseless to attempt to show that a process can be extended unless it is known to work in at least one instance. In the case of the cut-and-choose method, previous work had shown it applied to groups of two and three.

Verifying a Formula with Mathematical Induction

One of the most common uses of mathematical induction is to verify that a formula that describes a numerical pattern will do so indefinitely. The identification and verification of numerical patterns helps people analyze and predict trends in science, business, economics, world affairs, and many other areas. By confirming that a formula accurately describes a pattern, mathematical induction helps avoid erroneous predictions that can waste time and money.

For example, consider a situation in which Luis and Britt are investigating the number of handshakes that are made in a group of people if every person shakes hands with every other person. Luis notes that if there is only one person in the group, no handshakes are made and that if there are two people, one handshake is made. Britt draws a graph (see Figure 2.2) in which the vertices represent people, and the segments, or edges, represent

Figure 2.2 Graph representing handshakes among three people.

handshakes. She concludes that a group of three people requires a total of three handshakes.

Luis suggests organizing the data into a table.

Number of People in the Group	Number of Handshakes
1	0
2	1
3	3

Britt and Luis must now determine a formula that predicts the number of handshakes in a group from the number of people in the group. They will need mathematical induction to prove that any conjectured formula will work for a group of any size. In this lesson's exercises you will help Britt and Luis complete their work and then try some similar problems on your own.

Exercises

1. To use mathematical induction, you must be able to express numerical patterns in symbols. Some of the expressions you write in this exercise will be used in the mathematical induction proof of a formula for the handshake problem.
   a. If there are three people in a group and another person joins the group, there will be four people in the group. If a person leaves the original group, there will be two. Write an expression for the number of people if there are k people in the group and another person joins. Do the same if a person leaves the group of k people.
   b. Repeat part a if the original group has $k + 1$ people and if the original group has $2k$ people.

2. Draw a graph like Britt's and make a table like Luis's.
   a. Add another vertex to the graph to represent a fourth person, and draw segments to represent the additional handshakes that result from the addition of the fourth person. In your table, write the total number of handshakes in a group of four people.
   b. Add a fifth vertex to represent a fifth person and draw segments to represent the additional handshakes. In your table, write the total number of handshakes in a group of five people.

3. a. Suppose that there are seven people in a group and each of them has shaken hands with each of the others. If an eighth person enters the group, how many additional handshakes must be made? Explain.

  b. Suppose that there are k people in a group and each of them has shaken hands with every other person. If a new person enters the group, how many additional handshakes must be made? Explain.

  c. If H_n represents the number of handshakes in a group of n people, write a recurrence relation that expresses the relationship between H_n and H_{n-1}. Write a recurrence relation that expresses the relationship between H_{n+1} and H_n.

4. After studying the data, Britt wonders whether the number of handshakes in a group can be found by multiplying the number of people by the number that is 1 less than that and then dividing by 2.

  a. If her guess is correct, how many handshakes would there be in a group of 10 people?

  b. On the basis of Britt's guess, write an expression for the number of handshakes if there are k people in a group. Do the same for a group of $2k$ people. Do the same for a group of $k + 1$ people.

  Britt's formula, if correct, is sometimes called a solution of the recurrence relation you wrote in part c of Exercise 3. One of the solution's advantages is that it allows you to determine the number of handshakes in a group without knowing the number of handshakes in a smaller group.

  To prove that Britt's guess is correct, you must show that whenever the solution is known to work, it is possible to extend it to a group that is one larger. In other words, whenever the conjecture works for a group of k people, it will also work for a group of $k + 1$ people.

  c. Assume that Britt's formula works for a group of k people. Write her formula for the number of handshakes in a group of k people.

  d. You must show that Britt's formula works for a group of $k + 1$ people. Write her formula for the number of handshakes in a group of $k + 1$ people.

  e. If an additional person enters a group of k people, how many new handshakes are necessary? Explain.

  An expression for the total number of handshakes in a group of $k + 1$ people can be found by adding the expression for the number of handshakes in a group of k people (part c) to the number of additional handshakes when another person enters the group (part e): $\frac{k(k-1)}{2} + k$.

f. You can conclude that Britt's formula will work indefinitely if this expression is equivalent to the one you wrote in part d. Use algebraic procedures to transform the expression $\frac{k(k-1)}{2} + k$ until it matches the one you wrote in part d.

5. Although Britt's formula finds the number of handshakes in a group of people, it could also be used to find the number of potential two-party conflicts in a group.
   a. Use the formula to compare the number of potential conflicts when the size of a group doubles. Does the number of potential conflicts also double? Explain.
   b. Why do the results of Exercise 4 suggest that some of the costs associated with government, such as that of maintaining a police force, may outpace the growth of population?

In Exercises 1 through 4, you supplied many of the steps of the mathematical induction proof. In Exercise 6, which considers a formula for the relationship between the number of candidates on a ballot and the number of ways of casting an approval vote, you will again supply many of the steps of the induction process. However, before the proof can begin, you must do several preliminary steps leading to the conjecture of a formula. The preliminary steps are summarized here.

Preliminary Steps

Do the following before using mathematical induction to prove that a formula describes a relationship.

1. Organize a table of data for several small values. For example, how many ways of voting are there with 1, 2, 3, or 4 choices on the ballot?

2. Study the data and attempt to describe the data with a recurrence relation. For example, how many additional ways of voting are there when another choice is added to the ballot?

3. Conjecture a formula that predicts the outcome for a collection of k items. For example, what is a formula that predicts the number of ways of voting when there are k choices on the ballot?

4. Verify that your formula works for the values in your table.

6. In Exercise 12 of Lesson 1.4 (page 29), you considered the number of ways of voting under the approval system. The data you gathered in Lesson 1.4 are reproduced below. You will now use mathematical induction to verify a formula for the number of ways of voting under the approval system when there are n choices on the ballot.

Number of Choices on the Ballot	Number of Possible Ways of Voting
1	2
2	4
3	8
4	16

a. Collecting these data completes the first of the preliminary steps. The second step requires that you determine a recurrence relation that describes the relationship between the number of ways of voting when there are $k + 1$ choices on the ballot (V_{k+1}) and the number of ways of voting when there are k choices on the ballot (V_k). Do so, but be careful; if you do not establish the recurrence relation properly, the proof that comes later will fail. Here's a hint.

With Three Choices There are Eight Ways		The New Ways When a Choice Is Added
{ }		{D}
{A}		{A, D}
{B}	Append D	{B, D}
{C}	to each	{C, D}
{A, B}	$\rightarrow$	{A, B, D}
{A, C}		{A, C, D}
{B, C}		{B, C, D}
{A, B, C}		{A, B, C, D}

b. You are ready for the third and fourth preliminary steps. Study the data carefully. Notice that the values in the second column are all powers of 2. What formula does this suggest for the number of ways of voting when there are n choices? Check the formula with each pair of values in the table.

Parts a and b complete the preliminary process. You are now ready to prove that the formula you conjectured in part b of Exercise 6 will work indefinitely.

The Proof

After you have conjectured a formula and shown that it works for a few cases, you can prove that it is correct by using mathematical induction:

1. State the meaning of your formula for a collection of size k. This is the assumption or hypothesis that you will use in your proof. (It is similar to the "given" in a geometric proof.)

2. State the meaning of your formula for a collection of size $k + 1$. You can do this by replacing k with $k + 1$ in the formula you wrote in step 1. This is the goal. (It is similar to the "prove" in a geometric proof.)

3. Use your recurrence relation to describe the effect of an additional object on the formula you stated in step 1.

4. Use algebraic procedures to transform the expression you wrote in step 3 until it matches the one you stated in step 2.

c. Assume that the formula you conjectured in part b works for a ballot with k choices: $V_k = 2^k$. You must show that the formula works for a ballot with $k + 1$ choices. Write the formula for a ballot with $k + 1$ choices to complete the first two steps of the proof process.

d. Write an expression for the total number of ways of voting on a ballot with $k + 1$ choices by applying the recurrence relation you gave in part a to the formula you stated in part b. Use algebraic procedures to show that the result is equivalent to your answer in part c. This completes steps 3 and 4 of the proof process, and your induction proof is finished.

In Exercises 7 to 12, collect and organize data into a table, examine the data, conjecture a formula, and use mathematical induction to prove that the formula is correct. If you need help with any of the steps, use the summaries of the preliminary steps and the proof in Exercise 6.

7. Dominoes come in sets of different sizes. A double-six set, for example, contains dominoes that pair every number of spots from 0 through 6 with itself and with every other number of spots. Find a formula for the number of dominoes in a double-k set.

8. Bowling pins are normally set in a triangular configuration. Find a formula for the number of pins in a triangular configuration of k rows.

9. An ancient legend says that the inventor of the game of chess was offered a reward of his own choosing for the delight the game gave the king. The inventor asked for enough grains of wheat to be able to place one grain on the first square of the chessboard, two on the second, four on the third, and so forth, doubling the number of grains each time. Find a formula for the total number of grains on a chessboard after the kth square has been filled.

10. It takes four toothpicks to make a 1×1 square, and it takes 12 toothpicks to make a 2×2 square that is subdivided into 1×1 squares. Find a formula for the number of toothpicks needed to make a $k \times k$ square that is subdivided into 1×1 squares.

11. In Exercises 16 and 17 of Lesson 1.4 (page 31), you found a recurrence relation for the number of ways of selecting exactly two items when there are several choices on a ballot. Find a formula for the number of ways of selecting exactly two items when there are k choices on the ballot.

12. In the popular song "The Twelve Days of Christmas," one gift is given on the first day, one plus two on the second, and so on. Find a formula for the total number of gifts given on the kth day.

13. Find a formula for the number of building blocks in a set of $k \times k$ steps.

A 2×2 set of steps has 3 building blocks

A 3×3 set of steps has 6 building blocks.

14. Mathematicians often use mathematical induction to establish facts about numbers.
    a. Use mathematical induction to prove that a formula for the kth odd integer is $2k - 1$.
    b. Find and prove a formula for the sum of the first k odd integers. (Hint: The formula in part a may be of help.)
15. a. Find and prove a formula for the kth even integer.
    b. Find and prove a formula for the sum of the first k even integers.

Projects

16. Research and report on the use of mathematical induction in computer science.

Envy-Free Division

The first mathematician to take a serious interest in the problem of fairly dividing a cake may have been Hugo Steinhaus, a Polish mathematician who studied the problem during World War II. Steinhaus appears to have been motivated only by the challenge the problem posed. In the last half-century the world has served up an increasing number of problems that resemble cake division. Mathematicians and social scientists have risen to the occasion.

Mathematician of Note

Hugo Steinhaus (1887–1972) did important mathematical research at universities in Lvov and Wroclaw.

Steinhaus used the same definition of fairness that you used in this chapter: a cake is fairly divided among n people if each person feels that he or she has received at least $1/n$th of the cake. The cleverness of the people who have tackled fair division problems is demonstrated by the existence of an estate division algorithm that can result in everyone's getting more than a fair share by this definition.

However, the fairness definition is deficient in at least one important way. Fair division situations are sometimes charged with emotion. Not all participants behave as rationally as an impartial observer might expect. For example, according to this definition, a participant in a four-person estate division should be happy if the value placed on his or her received share is 0.3. But what if this individual feels someone else's share has a value of 0.35? Feeling that your share is fair is not the same as feeling it is larger than everyone else's.

For example, consider three individuals, A, B, and C, who are dividing a cake by the cut-and-choose method. Suppose that A and B are the cutters and that C places a value of 0.9 on A's piece and 0.1 on B's. Following the subsequent division of A's and B's pieces, C values A's three pieces at 0.3 each. C will now choose a piece valued at 0.3 from A, leaving A with two pieces C values at 0.3 each. Even if C gets all of B's pieces, C can do no better than 0.4 against A's 0.6.

The existence of situations in which envy arises suggests an alternative definition of fairness: Each person should feel that the received portion is at least as big as every other person's. The two-person cut-and-choose solution is free of envy, but the three-person cut-and-choose solution is not.

A three-person envy-free solution was devised independently by several mathematicians in the 1960s. It begins by having A cut the cake into three pieces. If B feels the division is not fair, B trims the best piece to make it the same value as the second-best. C gets first choice of one of the three pieces, B second choice, and A third. However, when B chooses, B must take the trimmed piece if C did not choose it. The trimmings can be disposed of by, say, feeding them to the dog. If the trimmings are substantial, they can be divided in a second round of cutting and choosing (although the roles of A, B, and C change).

Since C chooses first, C's piece will seem at least as good to C as either of the others. After C chooses, B is left at least one of the two pieces B considers better. Since B is forced to pick the trimmed piece if it is available, A is left one of the two pieces A cut on the first round.

None of the mathematicians who discovered the three-person solution succeeded in extending it to four or more. Then, in 1992, Steven Brams and Alan Taylor

Mathematician of Note

Steven Brams is a political scientist at New York University.

Mathematician of Note

Alan Taylor is a mathematician at Union College in Schenectady, New York.

The Fairness Equation: Solomon's Wisdom + Math

SAN JOSE MERCURY NEWS,
April 27, 1996

Mathematics now may be used to solve a problem that has tormented people since the dawn of humanity: how to divide things fairly.

Mathematician Alan Taylor and political scientist Steven Brams say they have devised a system based on "preference points" that can split just about anything—from the spoils of war to a child's birthday cake—into "envy-free pieces."

Not only do all parties get what they think is fair, they say, each thinks it got the better of the other guys.

Taylor and Brams work is only a small part of a rapidly surfacing trend. Mathematics is invading political science in attempts to find rational approaches to complex, often highly emotional questions.

discovered an algorithm that produces an envy-free solution for any number of people.

Although the method of Brams and Taylor applies to a divisible object such as a cake, the pair quickly adapted it to situations in which only a portion of the goods is divisible (i.e., the cash in an estate). In a recent book on fair division, Brams and Taylor apply the method to a wide range of real-world fair division problems, including situations, such as the division of chores, in which the "goods" are really "bads."

1. Write a summary of what you think are the important points of this chapter.

2. Joan, Henry, and Sam are heirs to an estate that includes a vacant lot, a boat, a computer, a stereo, and $10,000 in cash. Each heir has submitted bids as summarized in the following table.

	Joan	Henry	Sam
Vacant lot	$8,000	$7,500	$6,200
Boat	$6,500	$5,700	$6,700
Computer	$1,340	$1,500	$1,400
Stereo	$800	$1,100	$1,000

   For each heir, find the fair share, the items received, the amount of cash, and the final settlement. Summarize your results in a matrix.

3. Anne, Beth, and Jay are heirs to an estate that includes a computer, a used car, and a stereo. Each heir has submitted bids for the items in the estate as summarized in the following table.

	Anne	Beth	Jay
Computer	$1,800	$1,500	$1,650
Car	$2,600	$2,400	$2,000
Stereo	$1,000	$800	$1,200

   For each heir, find the fair share, the items received, the amount of cash, and the final settlement. Summarize your results in a matrix.

4. States A, B, and C have populations of 647, 247, and 106, respectively. There are 100 seats to be apportioned among them.

    a. What is the ideal ratio?

    b. Find the quota for each state.

    c. Apportion the 100 seats among the three states by the Hamilton method.

    d. What is the initial Jefferson apportionment?

    e. Find the Jefferson adjusted ratio for each state.

    f. Apportion the 100 seats by the Jefferson method.

    g. What is the initial Webster apportionment?

    h. Find the Webster adjusted ratio for each state.

    i. Apportion the 100 seats by the Webster method.

    j. What is the initial Hill apportionment?

    k. Find the Hill adjusted ratio for each state.

    l. Apportion the 100 seats by the Hill method.

    m. Suppose the populations of the states change to 650, 255, and 105, respectively. Reapportion the 100 seats by the Hamilton method.

    n. Explain why the results in part m constitute a paradox.

5. Discuss the theorem proved by Michel Balinski and H. Peyton Young. That is, what did they prove?

6. Arnold, Betty, and Charlie are dividing a cake in the following way.

Arnold divides the cake into what he considers six equal pieces.

The pieces are then chosen in this order: Betty, Charlie, Betty, Charlie, Arnold, Arnold.

Who is guaranteed a fair share by his or her own assessment?

7. Four people have divided a cake into four pieces that each considers fair, and then a fifth person arrives. Describe a method of dividing the four existing pieces so that each of the five people receives a fair share.

8. For the following situation, collect and organize data into a table, examine the data and conjecture a formula, and use mathematical induction to prove that your formula is correct.

   In a set of concentric circles, a ring is any region that lies between any two of the circles. Find a formula for the number of rings in a set of k concentric circles.

9. On the basis of the enrollment in each of a high school's courses, the administration must decide the number of sections that are offered. A number of factors affect the decision. For example, financial considerations require about 25 students in each section and a maximum of 300 sections for all courses. Recommend a procedure the school might use to divide 300 sections fairly among all courses on the basis of the enrollment in those courses.

10. Discuss how fair division methods you studied in this chapter might change to accommodate a situation in which the objects being divided were undesirable (i.e., the division of household chores among children).

Bibliography

Balinski, Michel, and H. Peyton Young. 1982. *Fair Representation: Meeting the Ideal of One Man, One Vote.* New Haven, CT: Yale University Press.

Bradberry, Brent A. 1992. "A Geometric View of Some Apportionment Paradoxes." *Mathematics Magazine* 65(1): 3–17.

Brams, Steven J., and Alan D. Taylor. 1996. *Fair Division: From Cake-Cutting to Dispute Resolution.* Cambridge/New York: Cambridge University Press.

Bunch, Brian. 1982. *Mathematical Fallacies and Paradoxes.* New York: Van Nostrand-Reinhold.

Burrows, Herbert, et al. 1989. *Mathematical Induction.* Lexington, MA: COMAP.

Cole, K. C. 1998. *The Universe and the Teacup: The Mathematics of Truth and Beauty.* New York: Harcourt, Brace.

COMAP. 1997. *For All Practical Purposes: Introduction to Contemporary Mathematics.* 4th ed. New York: W. H. Freeman.

Eisner, Milton P. 1982. *Methods of Congressional Apportionment.* Lexington, MA: COMAP.

Falletta, Nicholas. 1983. *The Paradoxican.* Garden City, NY: Doubleday.

Hively, Will. 1995. "Dividing the Spoils." *Discover* (March): 49.

Lambert, J. P. 1988. *Voting Games, Power Indices, and Presidential Elections.* Lexington, MA: COMAP.

Meyer, Rochelle Wilson, and Walter Meyer. 1990. *Play It Again Sam: Recurrence Equations in Mathematics and Computer Science.* Lexington, MA: COMAP.

Peterson, Ivars. 1996. "Formulas for Fairness." *Science News* 149(18).

Steinhaus, Hugo. 1969. *Mathematical Snapshots.* 3rd ed. New York: Oxford University Press.

Matrix Operations and Applications

I t has been said that sports fans are the nation's foremost consumers of statistics and that baseball fans are the most prominent among them. Whether it's in the information conveyed by a scoreboard, as on this one that commemorates the 1941 National League pennant race, or in the current information on league leaders, baseball records are full of numbers.

How can large collections of data be organized and managed in an efficient way? What calculations provide meaningful information to people who use the data? How can computers and calculators assist them? Baseball statisticians, business executives, and wildlife biologists are among the diverse groups of people who turn to the mathematics of matrices for answers to these questions.

Addition and Subtraction of Matrices

In Lesson 2.2, you were introduced to matrices as a natural way to organize, manipulate, and display information. As you have seen, matrices provide a very handy device for representing sets of discrete data that can be described with two characteristics, one characteristic being represented by the rows of the matrix and the other by the columns. In this lesson you will be introduced to some of the terminology and notation used in working with matrices. Matrix addition and subtraction will also be developed.

Matrix Terminology

As a first example, suppose that you and a few of your friends are planning a pizza and video party and some decisions have to be made about ordering food. You decide to call the pizza houses that deliver in your neighborhood to ask about prices for large single-topping pizzas, liter containers of cold drinks, and family-sized salads with house dressing. You could record the information you get in a table such as the following.

	Gina's	Vin's	Toni's	Sal's
Pizza	$12.16	$10.10	$10.86	$10.65
Drinks	$1.15	$1.09	$0.89	$1.05
Salad	$4.05	$3.69	$3.89	$3.85

Or you might choose to write your data in **matrix form,** which simply means writing the numbers in a rectangular array and enclosing them in brackets or parentheses.

$$
\begin{array}{c}
 \quad \text{Gina's} \quad\ \text{Vin's} \quad\ \text{Toni's} \quad\ \text{Sal's} \\
\begin{array}{c} \text{Pizza} \\ \text{Drinks} \\ \text{Salad} \end{array}
\left[\begin{array}{cccc}
\$12.16 & \$10.10 & \$10.86 & \$10.65 \\
\$1.15 & \$1.09 & \$0.89 & \$1.05 \\
\$4.05 & \$3.69 & \$3.89 & \$3.85
\end{array}\right]
\end{array}
$$

For the sake of simplicity, you could omit the row and column labels and dollar signs in your matrix and write only the values. If you delete the labels, however, you will have to remember that the rows represent the prices for pizzas, drinks, and salads, while the columns represent the various pizza houses.

$$
\left[\begin{array}{cccc}
12.16 & 10.10 & 10.86 & 10.65 \\
1.15 & 1.09 & 0.89 & 1.05 \\
4.05 & 3.69 & 3.89 & 3.85
\end{array}\right]
$$

Each of the individual entries in the matrix is called an **element** or a **component** of the matrix. A matrix, such as this one, with three rows and four columns is described as a matrix with **order** or **dimension** 3 by 4 (written as 3 × 4). It is also customary to refer to this matrix as simply a 3 × 4 matrix.

> In general, if a matrix has m rows and n columns, in which m and n represent counting numbers, it is called an m by n matrix.

After looking over your data, you might decide to drop Gina's options from the possible choices since they are more expensive item by item than any of the others. If you do this, you will be left with a 3 × 3 **square matrix.** Notice that in a square matrix the number of rows equals the number of columns or $m = n$.

	Vin's	Toni's	Sal's
Pizza	$10.10	$10.86	$10.65
Drinks	$1.09	$0.89	$1.05
Salad	$3.69	$3.89	$3.85

Matrices do not always have multiple rows and columns. For example, if your group thinks about which pizza most people prefer, regardless of the price, you might decide that it would be Sal's. If you list only the prices for Sal's offerings, the result will be a **column matrix** of order 3×1. Matrices that have only one column are sometimes referred to as **column vectors.**

	Sal's
Pizza	10.65
Drinks	1.05
Salad	3.85

If you choose to look at the pizza prices alone, they can be represented with a 1×3 **row matrix,** or **row vector.**

	Vin's	Toni's	Sal's
Pizza	[$10.10	$10.86	$10.65]

Notice that when you give the order of a matrix, you write the number of rows followed by the number of columns. The simplest order would be that of a **1 × 1 matrix** such as [10.65] containing a single element.

Exercises

1. Write a definition for a matrix in your own words. Justify the validity of your definition through discussion with other students in your class.

2. Bring to class at least two matrices from newspapers or magazines (such as the one in the newsclip on page 112).
   a. What are the dimensions of each of your matrices?
   b. What is represented by the rows and columns of your matrices?
   c. Be prepared to share your matrices with other class members.

3. A trendy garment company receives orders from three boutiques. The first boutique orders 25 jackets, 75 shirts, and 75 pairs of pants. The

second boutique orders 30 jackets, 50 shirts, and 50 pairs of pants. The third boutique orders 20 jackets, 40 shirts, and 35 pairs of pants. Display this information in a matrix whose rows represent the boutiques and whose columns represent the type of garment ordered. Label the rows and columns of your matrix accordingly.

4. Although matrices contain many data values, they can also be thought of as single entities. This feature allows us to refer to a matrix with a single capital letter:

$$
\begin{array}{c}
\begin{array}{ccc} \text{Vin's} & \text{Toni's} & \text{Sal's} \end{array} \\
A = \begin{array}{c} \text{Pizza} \\ \text{Drinks} \\ \text{Salad} \end{array}
\begin{bmatrix} \$10.10 & \$10.86 & \$10.65 \\ \$1.09 & \$0.89 & \$1.05 \\ \$3.69 & \$3.89 & \$3.85 \end{bmatrix}
\end{array}
$$

or
$$
\begin{array}{c}
\begin{array}{ccc} \text{Vin's} & \text{Toni's} & \text{Sal's} \end{array} \\
S = \text{Pizza} \begin{bmatrix} \$10.10 & \$10.86 & \$10.65 \end{bmatrix}
\end{array}
$$

Individual entries in a matrix are identified by row number and column number, in that order. For example, the value 10.65 is the entry in row 1 and column 3 of matrix A and is referenced as A_{13}. Entry A_{13} represents or *is interpreted as* the cost of a pizza at Sal's. Notice that A_{31} is not the same as A_{13}. Entry A_{31} has the value $3.69 and represents the cost of a salad at Vin's. In the row matrix S, the entries are referenced as S_1, S_2, and S_3.
a. What is the value of A_{21}? Of A_{12}? Of A_{32}?
b. Write an interpretation of each of the entries in part a.
c. Write an interpretation of S_3.

5. For breakfast Patty had cereal, a medium banana, a cup of 2% milk, and a slice of buttered toast. She recorded the following information in her food journal. Cereal: 165 calories, 3 g fat, 33 g carbohydrate, and no cholesterol. Banana: 120 calories, no fat, 26 g carbohydrate, and no cholesterol. Milk: 120 calories, 5 g fat, 11 g carbohydrate, and 15 mg cholesterol. Buttered toast: 125 calories, 6 g fat, 14 g carbohydrate, and 18 mg cholesterol.
a. Write this information in a matrix N whose rows represent the foods. Label the rows and columns of your matrix.
b. State the values of N_{23}, N_{32}, and N_{42}.
c. Write an interpretation of N_{23}, N_{32}, and N_{42}.

6. Suppose that as you continue to plan your pizza party, you discover that the local supermarket has a sale on 2-liter bottles of soft drinks and you decide not to order drinks from a pizza house after all. Write and label a 2 × 3 matrix that represents the prices for just pizza and salad at Vin's, Toni's, and Sal's.

7. Suppose further that when you were calling the pizza houses about prices, you also collected the following information about the cost of additional toppings and salad dressings.

	Vin's	Toni's	Sal's
Additional toppings	$1.15	$1.10	$1.25
Additional dressings	$0.00	$0.45	$0.50

Represent the information from this table in another 2 × 3 matrix whose rows represent the additional toppings and dressings and whose columns represent the three pizza houses. Label the rows and columns of your matrix.

8. Your next step is to compute what it would cost to order pizzas with two toppings and to allow a choice of two salad dressings. This can be done by simply adding corresponding components of your two price matrices. If you let A represent the basic price matrix and B represent the matrix of additional costs, then you can add A and B to get a third matrix C, which will represent the total prices for pizza and salads at each pizza house.

$$A = \begin{bmatrix} 10.10 & 10.86 & 10.65 \\ 3.69 & 3.89 & 3.85 \end{bmatrix}$$

and

$$B = \begin{bmatrix} 1.15 & 1.10 & 1.25 \\ 0.00 & 0.45 & 0.50 \end{bmatrix}$$

then

$$A + B = \begin{bmatrix} 10.10 + 1.15 & 10.86 + 1.10 & \underline{\qquad} \\ \underline{\qquad} & \underline{\qquad} & \underline{\qquad} \end{bmatrix}$$

$$= \begin{bmatrix} \underline{\qquad} & \underline{\qquad} & \underline{\qquad} \\ \underline{\qquad} & \underline{\qquad} & \underline{\qquad} \end{bmatrix} = C.$$

Complete the addition and label the rows and columns of matrix C.

9. In Exercise 8, the entries of matrix C represent the sum of the corresponding entries in matrices A and B. For example, C_{13}, which represents the cost of a pizza with an extra topping at Sal's, equals the sum of A_{13} and B_{13}.
   a. What is the value of A_{21}? Of B_{21}? Of C_{21}?
   b. Write an interpretation of A_{21}, B_{21}, and C_{21}.

   > In general, if A and B are m by n matrices, then
   > $C = A + B$ is a matrix whose entries represent the
   > sum of the corresponding entries in matrices A and
   > B. This matrix sum is represented with symbols as
   >
   > $$C_{ij} = A_{ij} + B_{ij} \text{ where } 1 \le i \le m \text{ and } 1 \le j \le n.$$

   It is clear from this discussion that matrices of unlike dimensions can not be added. It also does not make sense to add matrices whose row and column labels represent unlike quantities.

10. Suppose that a physician associates with each patient a row matrix whose components represent that person's age, weight, and height. Would it be appropriate to add together the matrices associated with two different patients? Explain your answer.

11. Suppose that the manager of a convenience store associates with each customer a column vector whose components represent the customer's purchases. Would it make sense to add together the vectors representing the purchases of two or more customers? Explain your answer.

12. Through July 20, 1997, the three baseball players with the highest batting averages in the National League had the following batting statistics.

	AB	R	H	HR	RBI	Avg
L. Walker (Colorado)	343	88	138	27	79	.402
Gwynn (San Diego)	372	64	147	15	84	.395
Piazza (Los Angeles)	332	56	118	19	62	.355

The following statistics for the same three players were published on September 30, 1997.

	AB	R	H	HR	RBI	Avg
L. Walker (Colorado)	568	143	208	49	130	.366
Gwynn (San Diego)	592	97	220	17	119	.372
Piazza (Los Angeles)	556	104	201	40	124	.362

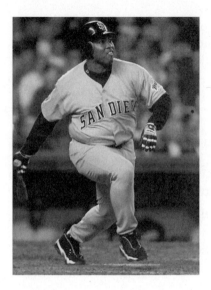

At the end of the 1997 season only Ty Cobb had won more batting titles (12) than the Padres' Tony Gwynn (8).

Find and label a matrix that displays the changes in these statistics over the course of the season. Notice that the batting averages for two of the three player's decreased from July through September. How will you show this in your matrix?

13. Write a definition for **matrix subtraction** based on your calculations in Exercise 12. Justify the validity of your definition through discussion with your classmates.

14. The matrices that follow give the average times in minutes for physical fitness test endurance runs for U.S. youth in 1980 and 1989 (U.S. Department of Education, 1991). The runs were 3/4 mile for ages 10 and 11 and 1 mile for ages 12 through 17.

	1980		1989	
	Boys	Girls	Boys	Girls
10- and 11-year-olds	6.5	7.4	7.3	8.0
12- and 13-year-olds	8.4	9.8	9.1	10.5
14- and 17-year-olds	7.5	9.6	8.6	10.7

a. Find and label the matrix that represents the change in times in minutes for each age group from 1980 to 1989.
b. In which age group and sex was there the greatest increase in average time for the endurance runs? The smallest increase?
c. How would you explain this increase in times across age groups?

15. In statistics, a correlation matrix is a matrix whose entries represent the degree of relationship between variables. The values in a correlation matrix range from −1 to 1, where 0 indicates that there is no relationship, a negative value indicates that as one variable increases the other decreases, and a positive value indicates that as one variable increases the other one also increases. In a study of the relationship between ACT test scores, high school class rank, and college grade point average,

the following correlation matrix was generated. Notice that the row labels and the column labels are the same in a correlation matrix.

	ACT Comp.	ACT Eng.	ACT Math	ACT Soc. St.	ACT Sci.	H.S. Rank	Coll. GPA
ACT Composite	1.00	0.80	0.79	0.81	0.82	0.59	0.51
ACT English	0.80	1.00	0.54	0.58	0.55	0.53	0.48
ACT Math	0.79	0.54	1.00	0.42	0.52	0.57	0.42
ACT Social studies	0.81	0.58	0.42	1.00	0.61	0.39	0.39
ACT Science	0.82	0.55	0.52	0.61	1.00	0.44	0.36
High school rank	0.59	0.53	0.57	0.39	0.44	1.00	0.45
College GPA	0.51	0.48	0.42	0.39	0.36	0.45	1.00

Source: Aksamit, Mitchell, and Pozehl, 1986.

The ones in this matrix are located along what is called the **main diagonal**, in which the row and column numbers are the same. If we call this matrix R, then these diagonal elements are referenced as R_{ii} ($i = 1, 2, . . . , 7$). We say that this matrix is **symmetric**, since $R_{ij} = R_{ji}$ ($i, j = 1, 2, . . . , 7$).

In a symmetric matrix you need only to know the values along the main diagonal and either the triangle above the main diagonal (the upper triangle) or below it (the lower triangle). Because of this feature, correlation matrices are often written with blanks in either the upper or lower triangle.

a. Why do you think the values along the main diagonal of a correlation matrix are all 1s?

b. Could a matrix that is not square be symmetric? Why?

c. Why are the values in a correlation matrix symmetrical about the main diagonal?

d. Which variable had the highest correlation with college GPA?

e. Which subject area test had the highest correlation with high school rank?

16. In your study of algebra, you learned that the commutative and associative properties hold for addition over the set of real numbers (That is, for all real numbers a, b, and c, $a + b = b + a$ and $a + (b + c) = (a + b) + c$.

Campus Administrators Announce New Parking Plans

SCARLET,
March 11, 1999

Campus planners are working overtime to finalize plans for three new parking garages and a beefed up shuttle system to replace parking stalls that will be lost in the next four to five years due to construction. Administrators acknowledge that the new parking and transit plan has a downside: parking fees will increase significantly. It will cost about $50 million to build the new garages and support the shuttle system. These costs will be borne by the users. Typical faculty/staff permits will escalate in $5 or $6 monthly increments annually through 2004. Student fees will also increase, although in $4 or $5 increments. Administrators admit that this will be painful and that the rate of escalation will leave some folks breathless.

Permit Fees

Year	Structure Completed	F/S/S Monthly Perimeter Permits	Monthly Student Permits	Monthly F/S Permits	Monthly Student Reserved	Monthly F/S Reserved
1998		$4	$9	$11	$26	$31
1999		$5	$13	$17	$35	$40
2000		$6	$17	$23	$45	$50
2001	South	$7	$21	$29	$50	$60
2002	East	$8	$25	$34	$55	$70
2003		$9	$29	$39	$60	$80
2004	North–1	$10	$32	$44	$65	$90

a. Do you think that the commutative and associative properties hold for addition of matrices? Why?

b. Use the following matrices to test your conjecture in part a.

$$A = \begin{bmatrix} 4 & -2 \\ 3 & 1 \end{bmatrix},$$

$$B = \begin{bmatrix} 1 & 3 \\ -2 & 5 \end{bmatrix}, \quad C = \begin{bmatrix} 2 & 4 \\ 1 & -1 \end{bmatrix}.$$

17. Do you think that the commutative and associative properties hold for subtraction of matrices? Why? Test your conjecture using matrices A, B, and C in Exercise 16b.

18. A matrix all of whose entries are the number zero is called a **zero matrix** and is denoted using a capital letter O alone or with subscripts $O_{m \times n}$.

a. Show that $A + O = O + A = A$ and that $A - A = O$.

b. Show that $A + (-A) = (-A) + A = O$, where the matrix $-A$, called the negative of A, is obtained by negating each entry in A.

Multiplication of Matrices, Part 1

In the previous lesson, matrix addition and subtraction were defined by looking at some matrix models of real-world situations. In this lesson, matrix multiplication is approached in much the same manner.

Multiplication by a Scalar

Examine again the data in the pizza problem posed in Lesson 3.1. Suppose that a decision that the group must make is how many of each type of pizza and salad it can afford to order. You can start this decision-making process by computing the cost of ordering four of each of the pizzas and salads represented by matrix C in Exercise 8 of Lesson 3.1. To do this, multiply each element in matrix C by 4 to get a new matrix T, equal to $4C$. An operation of this type is called **multiplication of a matrix by a scalar**. Multiplication of a matrix by a scalar is analogous to multiplication of integers in that $4C$ could also be interpreted as repeated addition or $4C = C + C + C + C$.

$$4C = 4 \times \begin{bmatrix} 11.25 & 11.96 & 11.90 \\ 3.69 & 4.34 & 4.35 \end{bmatrix} = \begin{bmatrix} 4(11.25) & 4(11.96) & 4(11.90) \\ 4(3.69) & 4(4.34) & 4(4.35) \end{bmatrix}$$

$$= \begin{bmatrix} 45.00 & 47.84 & 47.60 \\ 14.76 & 17.36 & 17.40 \end{bmatrix} = T.$$

Labeling the rows and columns of the matrix, you have

$$
T = \begin{array}{c} \\ \text{Pizza} \\ \text{Salad} \end{array}
\begin{array}{ccc} \text{Vin's} & \text{Toni's} & \text{Sal's} \\ \left[\begin{array}{ccc} \$45.00 & \$47.84 & \$47.60 \\ \$14.76 & \$17.36 & \$17.40 \end{array} \right] \end{array}
$$

Multiplication by a Row Matrix

Multiplication of a matrix by a scalar is only one way in which multiplication can be applied in matrix situations. **Multiplication of a column matrix by a row matrix** is illustrated in the following problem. Suppose that Jon, a student at Washington High, runs out to the nearby Super X to buy some junk food to stock up his locker for between-class snacks. He chooses four small bags of chips, five candy bars, a box of cheese crax, three packs of sour drops, and two bags of cookies. Jon's purchases can be represented by a row matrix Q.

$$
\begin{array}{cccccc} & \text{Chips} & \text{Candy} & \text{Crax} & \text{Drops} & \text{Cookies} \\ Q = [& 4 & 5 & 1 & 3 & 2 \quad]. \end{array}
$$

Suppose further that chips cost 30 cents a bag, candy bars cost 35 cents each, crax cost 50 cents a box, sour drops cost 20 cents a pack, and cookies sell for 75 cents a bag. These prices can be represented in column matrix P.

$$
P = \begin{array}{c} \\ \text{Chips} \\ \text{Candy} \\ \text{Crax} \\ \text{Drops} \\ \text{Cookies} \end{array}
\begin{array}{c} \text{Cents} \\ \left[\begin{array}{c} 30 \\ 35 \\ 50 \\ 20 \\ 75 \end{array} \right] \end{array}.
$$

The obvious question to ask now is, "How much did Jon pay for all these goodies?" To find the answer, multiply the price vector P by the quantity vector Q.

$$Q \times P = [4 \quad 5 \quad 1 \quad 3 \quad 2] \begin{bmatrix} 30 \\ 35 \\ 50 \\ 20 \\ 75 \end{bmatrix}$$

$$= [4(30) + 5(35) + 1(50) + 3(20) + 2(75)]$$
$$= [120 + 175 + 50 + 60 + 150]$$
$$= [555] \text{ cents} = [\$5.55].$$

This matrix computation is, of course, exactly what the clerk at Super X would do in figuring Jon's bill. The price of each item is multiplied by the number purchased and the products are summed. In order to do this computation, it is obvious that the number of items and the number of prices must be the same. Items and prices must also correspond.

> In general, if Q is a row matrix and P is a column matrix, each having the same number of components, then the product QP is defined. The matrix product Q times P is found by multiplying the corresponding components and summing the results.

As a second example, suppose another student, Trilby, goes along with Jon to the Super X. Her purchases are a bag of chips, two candy bars, two packs of gum that cost 25 cents each, and a medium drink for 75 cents.

1. Write and label a row matrix Q that represents the quantity of each item that Trilby purchased and a column matrix P that represents the price for each item.

2. Perform the multiplication Q times P to find the total cost of Trilby's purchases.

3. ·Compare your work with that of other students in your class. Was your final matrix [$2.25]?

As you can see from these examples, when a column matrix is multiplied by a row matrix the result is a single value. In other words, when a $(k \times 1)$ column matrix P is multiplied by a $(1 \times k)$ row matrix

Q, the result is a (1×1) single-value matrix C. Schematically this product looks like

$$\underset{(1 \times k)}{Q} \times \underset{(k \times 1)}{P} = \underset{(1 \times 1)}{C.}$$

Same

Dimensions of the product

In the previous examples you saw how multiplication of a matrix by a scalar and multiplication of a column matrix by a row matrix are defined. The next step is to define **multiplication of a matrix with more than one column by a row matrix.** This type of matrix multiplication is illustrated by examining some additional options in the pizza problem.

Suppose your group decides to order five pizzas and three salads and you want to know the total cost at each of the pizza houses for this combination. If you do the calculations involved here without using matrices, you proceed by multiplying the pizza price by 5 and adding the result to 3 times the salad price for each pizza house as follows.

Cost at Vin's: $5(\$11.25) + 3(\$3.69) = \$56.25 + \$11.07 = \$67.32.$

Cost at Toni's: $5(\$11.96) + 3(\$4.34) = \$59.80 + \$13.02 = \$72.82.$

Cost at Sal's: $5(\$11.90) + 3(\$4.35) = \$59.50 + \$13.05 = \$72.55.$

To use matrix multiplication to solve this problem, the number of pizzas and salads you plan to order are modeled by the 1×2 row matrix A.

$$\begin{array}{cc} \text{Pizzas} & \text{Salads} \\ A = [\quad 5 & 3 \quad]. \end{array}$$

The prices for pizzas and salads at each of the three pizza houses are modeled by the 2×3 matrix C.

$$C = \begin{array}{c} \\ \text{Pizzas} \\ \text{Salads} \end{array} \begin{array}{ccc} \text{Vin's} & \text{Toni's} & \text{Sal's} \\ \begin{bmatrix} 11.25 & 11.96 & 11.90 \\ 3.69 & 4.34 & 4.35 \end{bmatrix}. \end{array}$$

Now when matrix C is multiplied by the row matrix A, the expected result is another matrix whose components will give the total cost for five pizzas

and three salads at each pizza house. To accomplish this, it makes sense
to multiply each of the columns in matrix C by the row matrix A:

$$A \times C = [5 \quad 3] \times \begin{bmatrix} 11.25 & 11.96 & 11.90 \\ 3.69 & 4.34 & 4.35 \end{bmatrix}$$

$$= [5(11.25) + 3(3.69) \quad 5(11.96) + 3(4.34) \quad 5(11.90) + 3(4.35)]$$
$$= [56.25 + 11.07 \quad 59.80 + 13.02 \quad 59.50 + 13.05]$$
$$= [67.32 \quad 72.82 \quad 72.55].$$

The last step is to label the entries in the final product:

$$\begin{array}{ccc} \text{Vin's} & \text{Toni's} & \text{Sal's} \\ [\$67.32 & \$72.82 & \$72.55]. \end{array}$$

Notice that in carrying out this matrix computation, the process was
exactly the same as that of figuring the costs without the use of matrices.
In the matrix multiplication, each of the components in the columns of
matrix C was multiplied by the corresponding component the row matrix
A. These products were then summed to give the components of the final
product matrix.

You can see from this model that matrix multiplication of this sort can
only be defined when the number of entries in the row matrix equals the
number of rows in the multidimensional matrix. In addition, the product
will be a row matrix with the same number of entries as there are columns
in the second matrix.

> In general, the product of a $(1 \times k)$ row matrix A
> and a $(k \times n)$ matrix C is a $(1 \times n)$ row matrix P.

This result can be represented schematically as follows.

$$\begin{array}{ccccc} A & \times & C & = & P \\ (1 \times k) & & (k \times n) & & (1 \times n) \end{array}$$

Same

Dimensions of the product

Another example: Suppose your group wants to look at a couple of other combinations of pizzas and salads before it makes its final decision about how many to order. Use matrix multiplication to calculate the totals for (1) four pizzas and three salads and (2) four pizzas and four salads at each of the three pizza houses. Be sure to label your matrices. Check the steps in your work and discuss the interpretations of each of your answers with other students in your class. Did you get (1) [$56.07 $60.86 $60.65] and (2) [$59.76 $65.20 $65.00]?

Exercises

1. Refer to matrix T in the first example of this lesson.
   a. What does matrix T represent?
   b. What is the cost of four pizzas at Sal's?
   c. Interpret T_{12} and T_{21}.

2. Nancy has a small shop in the Oldmarket where she makes and sells four different kinds of jewelry: earrings (e), pins (p), necklaces (n), and bracelets (b). She fashions each item out of either cultured pearls or jade beads. The following matrix represents Nancy's sales for May.

$$M = \begin{array}{c} \text{Pearl} \\ \text{Jade} \end{array} \begin{matrix} e & p & n & b \\ \begin{bmatrix} 8 & .4 & 6 & 5 \\ 20 & 10 & 12 & 9 \end{bmatrix} \end{matrix}.$$

   Nancy hopes to sell twice as many of each piece in June.
   a. Calculate a matrix J, where $J = 2M$ to represent the number of each item Nancy will sell in June if she reaches her goal.
   b. Label the rows and columns of matrix J.
   c. How many jade necklaces does Nancy expect to sell in June?
   d. Interpret J_{21} and J_{12}.

3. Matt reads on the side of his cereal box that each ounce of cereal contains the following percentages of the minimum daily requirements of:

Vitamin A	25%
Vitamin C	25%
Vitamin D	10%

If Matt eats 3 ounces of cereal for breakfast, what percentages of each vitamin will he get? Show the matrices and matrix operation involved in your calculation. Label your matrices.

4. The regents at a state university recently announced a 7% raise of tuition rates per semester hour. The current rates per semester hour are shown in the following table.

	Undergraduate	Graduate
Resident	$75.00	$99.25
Nonresident	$204.00	$245.25

   a. Write and label a matrix that represents this information.
   b. Find a new matrix that represents the tuition rates per semester hour after the 7% raise goes into effect. Label your matrix.
   c. Find a matrix that represents the dollar increase for each of the categories. Label your matrix.
   d. Which matrix multiplication operation did you use in part b?

5. As you saw in the second example in this lesson (page 115), if Q is a (1×5) row matrix and P is a (5×1) column matrix, then the product Q times P is a single-value matrix with dimension (1×1). Suppose you multiply the column matrix P times the row matrix Q.
   a. What will be the dimension of the product P times Q? Justify your answer using a schematic diagram similar to the ones following the examples in this lesson.
   b. Multiply matrix P times matrix Q. (Refer to the second example in this lesson for the values in these matrices.)
   c. Does the product P times Q have a meaningful interpretation in this situation? Explain your answer.

6. Teresita's credit union has investments in three states—Massachusetts, Nebraska, and California. The deposits in each state are divided between consumer loans and bonds.

   The amount of money (in thousands of dollars) invested in each category is displayed in the following table.

	Mass.	Neb.	Cal.
Loans	230	440	680
Bonds	780	860	940

The current yields on these investments are 6.5% for consumer loans and 7.2% for bonds. Use matrix multiplication to find the total earnings for each state. Label your matrices.

7. Nancy is doing some remodeling of her home. She makes a trip to the lumber company to pick up ten 2 × 6s, four 4 × 6s, and two 5 × 5s. In 8-foot lengths, 2 × 6s cost $3.00, 4 × 6s cost $8.50, and 5 × 5s cost $9.50.

   a. Write and label a row matrix and a column matrix to represent the information in this problem.

   b. Will everyone necessarily write the same row and column matrices? Explain your answer.

   c. Perform a matrix multiplication to find the total cost of Nancy's purchases.

8. Chiu has $10,000 in a 12-month CD at 7.3% (annual yield), $17,000 in a credit union at 6.5%, and $12,000 in bonds at 7.5%. Use vector multiplication to find Chiu's earnings for a year. Label your vectors.

9. The **transpose** (A^T) of a matrix A is the matrix obtained by interchanging the rows and columns of matrix A.

   a. Describe the transpose of a row matrix and of a column matrix.

   b. Write and label the transpose (M^T) of matrix M in Exercise 2.

   c. What might be a reason for wanting to know the transpose of a matrix?

10. Refer to Exercise 2. Suppose it takes Nancy 2 hours to make a pair of earrings, 1 hour to make a pin, 2.5 hours to make a necklace, and 1.5 hours to make a bracelet.

   a. Write and label a row matrix that represents this information.

   b. Use matrix multiplication to find a matrix that represents the total hours Nancy spends making each type of jewelry (cultured pearls or jade) for the month of May. (Hint: Use the transpose of matrix M that you found in the previous exercise.)

   c. Label your product matrix.

   d. Interpret each of the entries in the product matrix.

11. Suppose Nancy has expanded her jewelry business and now has shops in the Westmarket and Eastmarket plazas as well as in the Oldmarket. Her sales of cultured pearl sets for July are shown in the following table.

	Old	West	East
Earrings	10	8	12
Pins	6	5	4
Necklaces	3	2	2
Bracelets	4	3	2

Earrings sell for $40 a pair, pins for $35 each, necklaces for $80, and bracelets for $45. Use matrix multiplication to find Nancy's total sales at each location. Label your matrices.

12. During the first week of a recent fund-raiser for the math club at Washington High, Anne sold the following number of candy canes.

	Mon	Tues	Wed	Thurs	Fri
Canes	10	15	20	30	50

a. Write this information in a column matrix C. Label your matrix.
b. Find a row matrix N such that the product N times C gives the total number of candy canes that Anne sold for the week.
c. Find a row matrix A such that the product A times C gives the average number of candy canes that Anne sold each day. (Hint: What fraction would you multiply the total number of canes by to find the average?)

Multiplication of Matrices, Part 2

In this lesson, you will continue to explore matrix multiplication by looking at the products of multidimensional matrices. There was an example of this type of matrix multiplication in the exercises of Lesson 2.2. Now, if you look at the three multiplications done in Lesson 3.2 to compare the cost of different combinations of pizza and salad, it seems to make sense to combine all three options into a single matrix B and to perform a single matrix multiplication.

$$
B = \begin{matrix} & \text{No. pizzas} & \text{No. salads} \\ \text{Option 1} \\ \text{Option 2} \\ \text{Option 3} \end{matrix} \begin{bmatrix} 4 & 3 \\ 4 & 4 \\ 5 & 3 \end{bmatrix}
$$

If you multiply matrix B times matrix C (see Lesson 3.2), the product (call it D) is a 3×3 matrix whose rows represent the three options and whose columns represent the three pizza houses. The components of this matrix give the total cost for each of the three options at each of the three pizza houses.

Notice as you follow the steps of this matrix multiplication that the computations are exactly the same as shown in the three separate calculations in the previous lesson. You expect, then, that row 1 of the product represents the cost of four pizzas and three salads, that row 2 of the product represents the cost of four pizzas and four salads, and that row 3 of the product represents the cost of five pizzas and three salads at each of the pizza houses.

Then, matrix B times matrix C =

$$
\begin{array}{c}
\text{Pizzas} \quad \text{Salads}\\
\begin{array}{c}\text{Option 1}\\\text{Option 2}\\\text{Option 3}\end{array}
\begin{bmatrix} 4 & 3 \\ 4 & 4 \\ 5 & 3 \end{bmatrix}
\end{array}
\times
\begin{array}{c}
\text{Vin's} \quad \text{Toni's} \quad \text{Sal's}\\
\begin{array}{c}\text{Pizzas}\\\text{Salads}\end{array}
\begin{bmatrix} 11.25 & 11.96 & 11.90 \\ 3.69 & 4.34 & 4.35 \end{bmatrix}
\end{array}
$$

or

$$
\begin{bmatrix} 4 & 3 \\ 4 & 4 \\ 5 & 3 \end{bmatrix}
\times
\begin{bmatrix} 11.25 & 11.96 & 11.90 \\ 3.69 & 4.34 & 4.35 \end{bmatrix}
$$

$$
= \begin{bmatrix}
4(11.25)+3(3.69) & 4(11.96)+3(4.34) & 4(11.90)+3(4.35) \\
4(11.25)+4(3.69) & 4(11.96)+4(4.34) & 4(11.90)+4(4.35) \\
5(11.25)+3(3.69) & 5(11.96)+3(4.34) & 5(11.90)+3(4.35)
\end{bmatrix}
$$

$$
= \begin{bmatrix}
45.00+11.07 & 47.84+13.02 & 47.60+13.05 \\
45.00+14.76 & 47.84+17.36 & 47.60+17.40 \\
56.25+11.07 & 59.80+13.02 & 59.50+13.05
\end{bmatrix}
$$

$$
= \begin{bmatrix}
56.07 & 60.86 & 60.65 \\
59.76 & 65.20 & 65.00 \\
67.32 & 72.82 & 72.55
\end{bmatrix} = D.
$$

If you label the product matrix for clarity's sake, the result is

$$
D = \begin{array}{c}
\qquad\qquad \text{Vin's} \quad \text{Toni's} \quad \text{Sal's}\\
\begin{array}{c}\text{Option 1}\\\text{Option 2}\\\text{Option 3}\end{array}
\begin{bmatrix} 56.07 & 60.86 & 60.65 \\ 59.76 & 65.20 & 65.00 \\ 67.32 & 72.82 & 72.55 \end{bmatrix}.
\end{array}
$$

In this matrix, D_{11} represents the cost of four pizzas and three salads at Vin's. How would you interpret D_{23} and D_{33}?

> In order for the multiplication of two matrices to be defined, the matrices must be **conformable**, which means that the number of columns in the first matrix must equal the number of rows in the second matrix.

Notice, also, that the order of the product matrix is the number of rows of the first matrix by the number of columns of the second. This can be shown schematically as follows.

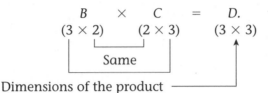

In general, if we multiply a matrix *P* with *m* rows and *k* columns times a matrix *Q* with *k* rows and *n* columns, the products will be a matrix *R* with *m* rows and *n* columns.

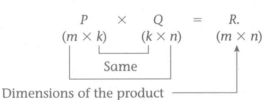

The dimensions of these matrices can also be described using the row and column labels. Matrix *B* classifies the data according to Options (rows) and Foods (columns). Hence you can refer to matrix *B* as an Options by Foods matrix. Likewise you can describe *C* as a Foods by Houses matrix. The product *B* times *C*, in turn, results in a matrix of dimension Options by Houses, which is what you wanted to know. Notice, also, that when the multiplication was performed, the common label (Foods) was eliminated, so the product matrix was left with the row label of the first factor and the column label of the second. Schematically, the dimension of each of the matrices involved in computing the product can be described as follows.

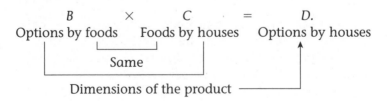

Using row and column labels in this manner helps determine whether a matrix multiplication will result in a meaningful interpretation or, indeed, whether it will give you the results that you wish.

Exercises

1. Mike, Liz, and Kate are heirs to an estate that consists of a condominium, a customized BMW, and choice season tickets to the Nebraska Cornhusker football games, and for the purposes of fair division, they have submitted the bids shown in matrix E.

$$
E = \begin{array}{c} \\ \text{Mike} \\ \text{Liz} \\ \text{Kate} \end{array}
\begin{array}{ccc} \text{Condo} & \text{BMW} & \text{Tickets} \\ \left[\begin{array}{ccc} \$185,000 & \$76,000 & \$250 \\ \$175,000 & \$60,000 & \$215 \\ \$180,000 & \$75,000 & \$325 \end{array} \right] \end{array}.
$$

The awarding of the items in the estate is indicated by matrix A.

$$
A = \begin{array}{c} \\ \text{Condo} \\ \text{BMW} \\ \text{Tickets} \end{array}
\begin{array}{ccc} \text{Mike} & \text{Liz} & \text{Kate} \\ \left[\begin{array}{ccc} 1 & 0 & 0 \\ 1 & 0 & 0 \\ 0 & 0 & 1 \end{array} \right] \end{array}.
$$

a. Find the matrix product $P = EA$. Label the rows and columns of P.

b. Write an interpretation of the entries in matrix P. (Refer to Exercise 6 in Lesson 2.2, pages 56 and 57.)

2. Emma and Ken go out to eat at Sammy's Drive Inn. Ken has a small cheeseburger, a baked potato with sour cream, and a shake. Emma orders a Sammy's special, medium fries, and a shake. The approximate numbers of calories, grams of fat, and milligrams of

Industries such as the fast-food business have changed dramatically since the advent of the computer in using matrices as a natural tool for storing and manipulating data.

cholesterol in each of these foods are represented in the following table.

	Calories	Fat (g)	Cholesterol (mg)
Cheeseburger	450	40	50
Sammy's special	570	48	90
Potato/sour cream	500	45	25
French fries	300	30	0
Shake	400	22	50

a. Write a matrix Q that describes Emma's and Ken's choices, with the columns' representing the foods. Label the rows and columns of this matrix.

b. Write a matrix C that represents the information in the preceding table with the rows' representing the foods. Label the rows and columns of this matrix.

c. What is the dimension of matrix Q and of matrix C?

d. What is the dimension of the product Q times C? Show why your answer is correct by using a schematic diagram.

e. The dimension of matrix Q could be described as Persons by Foods. Describe the dimensions of matrices C and Q times C in a similar manner. Justify your answer for matrix Q times C with a schematic diagram.

f. Multiply matrix Q times matrix C to get a matrix R. Label the rows and columns of matrix R.

g. Interpret R_{12}, R_{21}, and R_{23}.

3. a. What must be true about the dimensions of matrices A and B if the product $C = AB$ is defined?

b. If the products AB and BA are both defined, what must be true about the dimensions of matrices A and B? Why?

c. Find two nonsquare matrices A and B, where AB and BA are both defined. Compute AB and BA. Does $AB = BA$? Why?

d. As illustrated by your answer in part b, if AB and BA are both defined, it does not necessarily follow that $AB = BA$ (i.e., *in general, matrix multiplication is not commutative*). Using 2×2 matrices, find examples in which $AB = BA$ and in which AB is not equal to BA.

4. Let A be any 3×3 matrix and let

$$I = \begin{bmatrix} 1 & 0 & 0 \\ 0 & 1 & 0 \\ 0 & 0 & 1 \end{bmatrix}.$$

Show that $IA = AI = A$.

The matrix *I* is called an **identity matrix**. An identity matrix is any matrix in which each of the entries along the main diagonal are 1s and all other entries are 0s. Identity matrices act in the same way for matrix products as the number 1 does for number products.

5. Given the matrices *A*, *B*, and *C*.

$$A = \begin{bmatrix} 1 & 0 & 1 \\ 0 & 1 & 1 \end{bmatrix} \quad B = \begin{bmatrix} 3 & 1 \\ 2 & 2 \\ -1 & 1 \end{bmatrix} \quad C = \begin{bmatrix} 1 & -1 & 0 \\ 2 & 1 & 1 \end{bmatrix}$$

   a. Do you think that $A(BC) = (AB)C$?
   b. Test your conjecture by computing the products $A(BC)$ and $(AB)C$.
   c. The computations in part b show one case in which **matrix multiplication is associative.** Do you think this property holds for all matrices *A*, *B*, and *C* for which the product $A(BC)$ is defined? Why or why not?

6. Find two (2 × 2) matrices *A* and *B* to demonstrate that $(A + B)(A - B)$ is not necessarily equal to $A^2 - B^2$.

7. In algebra you learned that two numbers whose product is 1 (the identity element for multiplication) are called inverses of each other. For example, 5 and $\frac{1}{5}$ (or 5^{-1}) are inverses of each other since $5\left(\frac{1}{5}\right) = \left(\frac{1}{5}\right)5 = 1$.

   Similarly, if *A* and *B* are two square matrices such that $AB = BA = I$, then *A* and *B* are called **inverses** of each other. The inverse of *A* is denoted A^{-1}.

   a. Verify that the matrices *A* and *B* are inverses of each other by computing *AB* and *BA*.

   $$A = \begin{bmatrix} 2 & 3 \\ 1 & 2 \end{bmatrix} \quad B = \begin{bmatrix} 2 & -3 \\ -1 & 2 \end{bmatrix}.$$

   b. Not all square matrices will have an inverse. Use algebra to show that matrix *C* does not have an inverse.

   $$C = \begin{bmatrix} 2 & 4 \\ 3 & 6 \end{bmatrix}.$$

8. Carefully plot the points $A(0, 0)$, $B(6, 2)$, $C(8, 6)$, and $D(2, 4)$ on graph paper. Connect the points to form a polygon $ABCD$. This polygon can be represented with a matrix P as follows.

$$P = \begin{matrix} & A & B & C & D \\ & \begin{bmatrix} 0 & 6 & 8 & 2 \\ 0 & 2 & 6 & 4 \end{bmatrix} \end{matrix}.$$

a. Multiply the matrix representing polygon $ABCD$ by the matrix

$$T_1 = \begin{bmatrix} -1 & 0 \\ 0 & 1 \end{bmatrix}.$$

b. Plot and label the four points represented in your new matrix as A', B', C', and D'. Connect the points to form polygon $A'B'C'D'$.

c. Describe the relationship between polygon $A'B'C'D'$ and polygon $ABCD$.

d. Multiply the matrix representing polygon $A'B'C'D'$ by the matrix

$$T_2 = \begin{bmatrix} 1 & 0 \\ 0 & -1 \end{bmatrix}.$$

e. Plot and label the four points represented in your new matrix as A'', B'', C'', and D''. Describe the relationship between polygon $A''B''C''D''$, and polygon $A'B'C'D'$.

f. Multiply T_2T_1 to get a new matrix R. Multiply R times the matrix P, that represents the original polygon $ABCD$, and plot the resulting points. What effect does multiplication by R have on $ABCD$? Do the following to test your conjecture: Use a blank sheet of unlined paper and trace both your axes and polygon $ABCD$. Leave your copy on top of the original polygon and place the point of your pencil on the origin. Now, holding the original paper in place, rotate the top sheet until your copy of $ABCD$ rests on top of polygon $A''B''C''D''$. Describe what happened to polygon $ABCD$.

g. Find a matrix T_3 that reflects polygon $A''B''C''D''$ about the y-axis into quadrant IV of your graph.

h. Find a matrix T_4 that rotates polygon $A'B'C'D'$ about the origin into quadrant IV. How does T_4 relate to T_2 and T_3?

For Exercises 9, 10, and 11, you need either a graphing calculator or access to computer software that performs matrix operations.

9. The matrix A is called an upper-triangular matrix.

$$A = \begin{bmatrix} 1 & 1 & 1 \\ 0 & 1 & 1 \\ 0 & 0 & 1 \end{bmatrix}.$$

   a. Calculate A^2, A^3, and A^4.
   b. Make a conjecture about the form of A^k.
   c. Test your conjecture by computing additional powers of A.
   d. Challenge: Prove your conjecture using mathematical induction.

10. The Fancy Bag manufacturing company that makes and markets fine leather bags has three factories—one in New York, one in Nebraska, and one in California. One of the bags they make comes in three styles—handbag, standard shoulder bag, and roomy shoulder bag. The production of each bag requires three kinds of work—cutting the leather, stitching the bag, and finishing the bag.

   Matrix T gives the time (in hours) of each type of work required to make each type bag.

	Cutting	Stitching	Finishing
Handbag	0.4	0.6	0.4
$T =$ Standard	0.5	0.8	0.5
Roomy	0.6	1.0	0.6

   Matrix P gives daily production capacity at each of the factories.

	Handbag	Standard	Roomy
New York	10	15	20
$P =$ Nebraska	25	15	12
California	20	12	10

   Matrix W provides the hourly wages of the different workers at each factory.

	Cutting	Stitching	Finishing
New York	7.50	8.50	9.00
$W =$ Nebraska	7.00	8.00	8.50
California	8.40	9.60	10.10

Matrix D contains the total orders received at each factory for the months of May and June.

$$D = \begin{array}{c} \\ \text{Handbag} \\ \text{Standard} \\ \text{Roomy} \end{array} \begin{array}{c} \text{May} \\ \left[\begin{array}{cc} 600 \\ 800 \\ 400 \end{array} \right. \end{array} \begin{array}{c} \text{June} \\ \left. \begin{array}{c} 800 \\ 1{,}000 \\ 600 \end{array} \right] \end{array}.$$

a. Matrix T can be described as a Bag by Work matrix. Describe matrices P, W, and D in a similar manner.

Use the matrices above (or their transposes) to compute the following. Label the rows and columns of the matrix in each answer. Hint: The label dimensions from part a will help you decide what your matrix products should look like.

b. The hours of each type of work needed each month to fill all orders.

c. The production cost per bag at each factory.

d. The cost of filling all May orders at the Nebraska factory. (Hint: In this example the answer, a single value, is the product of a row vector and a column vector).

e. The daily hours of each type of work needed at each factory if production levels are at capacity.

11. (For those students who have studied trigonometry.)

a. Plot the polygon $ABCD$ represented in Exercise 8.

b. Multiply the matrix P by the following transformation matrix.

$$T_1 = \begin{bmatrix} \cos 30° & -\sin 30° \\ \sin 30° & \cos 30° \end{bmatrix}.$$

c. Plot the resulting polygon and label it $A'B'C'D'$. How does polygon $A'B'C'D'$ relate to polygon $ABCD$? Try repeating the transformation using 180° to test your conjecture.

d. Write a matrix that will rotate a polygon through 60°. Does this transformation matrix have the same effect as applying T_1 twice? Test your conjecture.

e. Find a matrix that rotates polygon $ABCD$ through 90° and another that rotates it through $-90°$. Find the product of these two transformation matrices. What is the relationship between these two matrices? Test your conjecture by finding the product of the matrices that will rotate the polygon through 60° and $-60°$.

12. Challenge: Refer to Exercise 9 and explore the following.

a. Replace the 1s in the upper-triangular matrix A with 2s, 3s, and 4s and repeat parts a to d of Exercise 9 for each of your new upper-triangular matrices.

b. Use the results of part a to make a conjecture for A^k when the 1s in A are replaced by any natural number m.

c. Prove your conjecture in part b using mathematical induction.

Computer/Calculator Exploration

13. Write a program for the graphing calculator based on the method of Exercise 11 that will allow you to enter the coordinates of the vertices of a polygon and the angle of rotation. Design your program so that both the original polygon and the rotation will be displayed.

Project

14. Research and write a short report on additional applications of matrices in trigonometry. Possible topics include the representation vectors and complex numbers as matrices.

Population Growth: The Leslie Model, Part 1

Age-specific population growth is a topic that is of great concern to people in fields as diverse as urban planning and wildlife management. Urban policymakers are interested in knowing how many people there will be in various age groups after certain periods of time have elapsed. Those in wildlife management worry about maintaining animal populations at levels that can be supported in their natural habitats without damage to the environment.

If the age distribution of a population at a certain date is known, along with birth and survival rates for age-specific groups, a model can be created to determine the age distributions of the survivors and descendants of the original population at successive intervals of time. The problem used to illustrate this model was posed in 1945 by P. H. Leslie of the Bureau of Animal Population at Oxford University in Oxford, England. In this problem, the growth rate of a population of an imaginary species of small brown rats, *Rattus norvegicus,* is examined. The lifespan of these rodents is 15–18 months. They have their first litter at approximately 3 months and continue to reproduce every 3 months until they reach the age of 15 months. Birth-rates and age-specific survival rates for 3-month periods are summarized in the following table. In order to simplify the situation as much as possible, birth rates and survival rates are held constant over time and only the female population is considered.

Age (months)	Birthrate	Survival Rate
0–3	0	0.6
3–6	0.3	0.9
6–9	0.8	0.9
9–12	0.7	0.8
12–15	0.4	0.6
15–18	0	0

The actual number of female births in a particular age group can be found by multiplying the birth rate by the number of females currently in the age group. The survival rate is the probability that a rat will survive and move into the next age group.

Suppose the original female rat population is 42 animals with the age distribution shown in the following table.

Age (months)	0–3	3–6	6–9	9–12	12–15	15–18
Number	15	9	13	5	0	0

In examining this model, one question that might be asked is how many female rats will there be after 3 months have passed. Another question might be asked about the age distribution of this new group. To answer these questions, it is necessary to find the number of new female babies introduced into the population and the number of female rats that survive in each group and move up to the next age group.

The number of new births after 3 months (1 cycle) can be found by multiplying the number of female rats in each age group times the corresponding birth rates and then finding the sum:

$$15(0) + 9(0.3) + 13(0.8) + 5(0.7) + 0(0.4) + 0(0)$$

$$= 0 + 2.7 + 10.4 + 3.5 + 0 + 0 = 16.6.$$

The number of female rats in the 0–3 age group after 3 months is about 17. The number of female rats who survive in each age group and move up to the next can be found as follows (SR stands for survival rate).

Age	No.	SR	Number moving up to the next age group
0–3	15	0.6	(15)(0.6) = 9.0 move up to the 3–6 age group.
3–6	9	0.9	(9)(0.9) = 8.1 move up to the 6–9 age group.
6–9	13	0.9	(13)(0.9) = 11.7 move up to the 9–12 age group.
9–12	5	0.8	(5)(0.8) = 4.0 move up to the 12–15 age group.
12–15	0	0.6	(0)(0.6) = 0 move up to the 15–18 age group.
15–18	0	0	No rodent lives beyond 18 months.

The sum of the number of female rats in each age group results in a total population of female rats equal to 16.6 + 9.0 + 8.1 + 11.7 + 4.0 + 0, or 49.4.

After 3 months (1 cycle) the female rat population has grown from 42 to approximately 50. The distribution of female rats is shown in the following table. Notice that in the table the number of female rats in each age group is not rounded to the nearest integer. This is because when the values are to be used for further analysis, rounding off can mean a significant difference in calculations over time even though it doesn't make sense to have a fractional part of a rat.

Age	0–3	3–6	6–9	9–12	12–15	15–18
Number	16.6	9.0	8.1	11.7	4.0	0

Exercises

1. Use the preceding table (the distribution of the female rat population after 3 months) and the process introduced in this lesson to compute the following. Note that each of the calculations refers to female rats only.
   a. Calculate the number of newborn rats (aged 0–3) after 6 months (2 cycles).
   b. Calculate the number of rats that survive in each age group after 6 months and move up to the next age group.
   c. Use the results to parts a and b to show the distribution of the rat population after 6 months. Approximately how many rats will there be after 6 months?
   d. Use your population distribution from part c to calculate the number of rats and the approximate number in each age group after 9 months (3 cycles). Continue this process to find the number of rats after 12 months (4 cycles).

e. Compare the original number of rats with the numbers of rats after 3, 6, 9, and 12 months. What do you observe?

f. What do you think might happen to this population if you extended the calculations to 15, 18, 21, . . . months?

2. Suppose that a species of deer has the following birth and survival rates.

Age (years)	Birthrate	Survival Rate
0–2	0	0.6
2–4	0.8	0.8
4–6	1.7	0.9
6–8	1.7	0.9
8–10	0.8	0.7
10–12	0.4	0

a. Given that the initial population for this species is 148 deer with the following distribution,

Age (years)	0–2	2–4	4–6	6–8	8–10	10–12
Number	50	30	24	24	12	8

find the number of newborn female deer after 2 years (1 cycle).

b. Arrange the initial population distribution in a row matrix and the birth rates in a column matrix. Multiply the row matrix times the column matrix. Interpret this result.

c. Calculate the number of deer that survive in each age group after 2 years and move up to the next age group.

d. Explore the possibility of multiplying the initial population distribution in a row matrix times some column matrix to find the number of deer after 2 years that move from:
   i. The 0–2 group to the 2–4 group. (Hint: the column matrix that you use will need to contain several zeros in order to produce the desired product.)
   ii. The 2–4 to the 4–6 group.
   iii. The 4–6 group to the 6–8 group.
   iv. The 6–8 group to the 8–10 group.
   v. The 8–10 group to the 10–12 group.

3. Using the birth and survival rate information for *Rattus norvegicus* from this lesson (see the table on page 133), find the population total and

distribution after 3 months (1 cycle) for the following initial populations.
a. [35 0 0 0 0 0].
b. [5 5 5 5 5 5].

4. Using the birth and survival rate information for the deer population
in Exercise 2, find the population total and distribution after 2 years
(1 cycle), 4 years (2 cycles), 6 years (3 cycles), 8 years (4 cycles), and
10 years (5 cycles) if the initial population is [25 0 0 0 0 0].

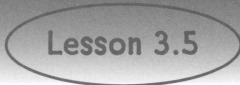
Population Growth: The Leslie Model, Part 2

In your beginning explorations of the Leslie model for population growth, you found that it was possible to use an initial population distribution, birth rates, and survival rates to predict population figures at future times. Looking 2, 3, or even 4 cycles into the future is not impossible, but the arithmetic soon becomes cumbersome. What do the wildlife manager and the urban planner do if they want to look 10, 20, or even more cycles into the future?

In Lesson 3.4, Exercise 2, you began to get a glimpse of the model that P. H. Leslie proposed. The use of matrices seems to hold the key, and with the aid of computer software or a calculator, looking ahead many cycles is not difficult. In fact, some very fascinating results are produced.

Let's return to our rat model. If the original population distribution (P_0) and a matrix that we will call L are multiplied, the population distribution at the end of cycle 1 (P_1) can be calculated.

$$P_0 L = [15 \ 9 \ 13 \ 5 \ 0 \ 0] \begin{bmatrix} 0 & 0.6 & 0 & 0 & 0 & 0 \\ 0.3 & 0 & 0.9 & 0 & 0 & 0 \\ 0.8 & 0 & 0 & 0.9 & 0 & 0 \\ 0.7 & 0 & 0 & 0 & 0.8 & 0 \\ 0.4 & 0 & 0 & 0 & 0 & 0.6 \\ 0 & 0 & 0 & 0 & 0 & 0 \end{bmatrix}$$

$$= [15(0) + 9(0.3) + 13(0.8) + 5(0.7) + 0(0.4) + 0(0)$$
$$15(0.6) \quad 9(0.9) \quad 13(0.9) \quad 5(0.8) \quad 0(0.6)]$$
$$= [16.6 \quad 9.0 \quad 8.1 \quad 11.7 \quad 4.0 \quad 0] = P_1.$$

The matrix L (called the **Leslie matrix**) is formed by augmenting or joining the column vector containing the birth rates of each age group and a series of column vectors that contain a survival rate as one entry and zeros everywhere else. Notice that the survival rates (of which there are $n - 1$ since no animal survives beyond the 15–18 age group) lie along the **super diagonal** that is immediately above the main diagonal of the matrix.

When the matrix L is multiplied by a population distribution P_k, a new population distribution P_{k+1} results. To find population distributions at the end of other cycles, the process can be continued.

$$P_1 = P_0L$$

$$P_2 = P_1L = (P_0L)L = P_0(LL) = P_0L^2.$$

In general, $P_k = P_0L^k$.

Using this formula to find the population distribution for the rats after 24 months (8 cycles) and the total population of the rats, you have

$$P^8 = P_0L^8 = [15 \quad 9 \quad 13 \quad 5 \quad 0 \quad 0] \begin{bmatrix} 0 & 0.6 & 0 & 0 & 0 & 0 \\ 0.3 & 0 & 0.9 & 0 & 0 & 0 \\ 0.8 & 0 & 0 & 0.9 & 0 & 0 \\ 0.7 & 0 & 0 & 0 & 0.8 & 0 \\ 0.4 & 0 & 0 & 0 & 0 & 0.6 \\ 0 & 0 & 0 & 0 & 0 & 0 \end{bmatrix}^8$$

$$= [21.03 \quad 12.28 \quad 10.90 \quad 9.46 \quad 7.01 \quad 4.27].$$

Total population $= 21.03 + 12.28 + 10.90 + 9.46 + 7.01 + 4.27 = 64.95$, or approximately 65 rats.

Point of Interest

```
[A][B]^8
[[21.026 12.283...
```

You can perform the calculation P_0L^8 on a calculator with matrix features.

Exercises

Note: For the following exercises, you need to have access to either a graphing calculator or computer software that performs matrix operations.

1. Using the original population distribution, [15 9 13 5 0 0], and Leslie matrix from the *Rattus norvegicus* example,
    a. Find the population distribution after 15 months (5 cycles).
    b. Find the total population after 15 months. (Hint: Multiply $P_0 L^5$ times a column matrix consisting of six 1s.)
    c. Find the population distribution and the total population after 21 months.

2. Suppose the *Rattus norvegicus* start dying off from overcrowding when the total female population for a colony reaches 250. Find how long it will take for this to happen when the initial population distribution is:
    a. [18 9 7 0 0 0].
    b. [35 0 0 0 0 0].
    c. [5 5 5 5 5 5].
    d. [25 15 10 11 7 13].

3. a. Complete the table for the given cycles of *Rattus norvegicus* using the original population distribution of [15 9 13 5 0 0].

Cycle	Total Population	Growth Rate
Original	42	
1	49.4	17.6%
2	56.08	13.5%
3	57.40	2.4%
4		
5		
6		

    b. What do you observe about the growth rates?
    c. Calculate the total populations for P_{25}, P_{26}, and P_{27}. What is the growth rate between these successive years? Hint: To find the growth

rate from P_{25} to P_{26}, subtract the total population for P_{25} from the total population for P_{26} and divide the result by the total population for P_{25}.

4. One characteristic of the Leslie model is that growth does stabilize at a rate called the **long-term growth rate** of the population. As you observed in Exercise 4, the growth rate of *Rattus norvegicus* converges to about 3.04%. This means that for a large enough k, the total population in cycle k will equal about 1.0304 times the total population in the previous cycle.
   a. Find the long-term growth rate of the total population for each of the initial population distributions in Exercise 2.
   b. How does the initial population distribution seem to affect the long-term growth rate?

5. Consider once again the deer species from Exercise 2 in Lesson 3.4. The birth and survival rates follow.

Age (years)	Birthrate	Survival Rate
0–2	0	0.6
2–4	0.8	0.8
4–6	1.7	0.9
6–8	1.7	0.9
8–10	0.8	0.7
10–12	0.4	0

   a. Construct the Leslie matrix for this animal.
   b. Given that P_0 = [50 30 24 24 12 8], find the long-term growth rate.
   c. Suppose the natural range for this animal can sustain a herd that contains a maximum of 1,250 females. How long before this herd size is reached?
   d. Once the long-term growth rate of the deer population is reached, how might the population of the herd be kept constant?

6. In his study of the application of matrices to population growth, P. H. Leslie was particularly interested in the special case in which the birthrate vector has only one nonzero element. The following example falls into this special case. Suppose there is a certain kind of bug that lives at most 3 weeks and reproduces only in the third week of life. Fifty percent of the bugs born in one week survive into the second week, and 70% of the bugs who survive into their second week also survive into

their third week. On the average, six new bugs are produced for each bug that survives into its third week. A group of five 3-week-old female bugs decide to make their home in a storage box in your basement.

a. Construct the Leslie matrix for this bug.

b. What is P_0?

c. How long will it be before there are at least 1,000 female bugs living in your basement?

7. Exercise 6 is an example of a population that grows in waves. Will the population growth for this population stabilize in any way over the long run? To explore this question, make a table of the population distributions P_{22} through P_{30}.

   a. Examine the population change from one cycle to the next. Can you find a pattern in the population growth?

   b. Examine the population change from P_{22} to P_{25}, P_{23} to P_{26}, P_{24} to P_{27}, P_{25} to P_{28}, P_{26} to P_{29}, and P_{27} to P_{30}. Are you surprised at the results? Why?

8. Change the initial population in Exercise 6 to $P_0 = [4\ 4\ 4]$ and repeat the instructions in Exercise 7 looking at the total population growth for each cycle.

9. Examine the changes in successive age groups from P_{22} to P_{25}, P_{23} to P_{26}, P_{24} to P_{27}, P_{25} to P_{28}, P_{26} to P_{29}, and P_{27} to P_{30}. Make a conjecture based on your results.

10. Using mathematical induction, prove that $P_k = P_0 L^k$ for any original population P_0 and Leslie matrix L, where k is a natural number.

Projects

11. Search the Web for applications of the Leslie matrix model in managing wildlife or domestic herds.

12. Research and report on the life and work of P. H. Leslie.

Harvesting Animal Populations

The following application of the Leslie matrix to a population of domestic sheep in New Zealand was originally published in 1967. (See G. Caughley, "Parameters for Seasonally Breeding Populations," *Ecology* 48(1967):834–839. Anton and Rorres (1987) developed the problem in their text, *Elementary Linear Algebra with Applications*. More recent references can be found by searching the Web for Leslie matrix applications.

This application involves what is referred to as the *harvesting* of a population. The term *harvesting* is defined as removal of the animals from the population. This could entail slaughtering some of the animals as is the case of wild deer or caribou herds that grow too large to be supported by their habitat. It could also mean selling or relocating some of the animals to start a new herd or colony as is the case of domestic herds or wild colonies of animals such as beaver. The ultimate goal is to find a stable distribution from which the population growth can be harvested at regular intervals so that the population can be held constant.

This model begins with an initial population that undergoes a growth period that is described by the Leslie matrix. At the end of the growth period, however, a certain percentage of each age group in the distribution is harvested. This is usually done in such a way that the unharvested population has the same age distribution as the initial population. A plan for harvesting the same percentage of each age group on a regular basis so the population remains the same after each harvesting is called a *sustainable harvesting policy*.

To describe this situation using matrices, suppose a Leslie matrix and a population distribution vector for n age groups of females in a population have been defined. If h_i, for $i = 1, 2, 3, . . ., n$, represents the fraction of females harvested for each of the age groups, an $n \times n$ diagonal matrix H called the harvesting matrix can be formed as follows.

$$H = \begin{bmatrix} h_1 & 0 & 0 & \cdots & 0 \\ 0 & h_2 & 0 & \cdots & 0 \\ 0 & 0 & h_3 & \cdots & 0 \\ \vdots & \vdots & \vdots & & \vdots \\ 0 & 0 & 0 & \cdots & h_n \end{bmatrix}.$$

If all the h_i's are the same value, then the harvesting is called uniform. From the definition of a sustainable harvesting policy, it follows that

$$\begin{bmatrix} \text{Age distribution} \\ \text{at end of} \\ \text{growth period} \end{bmatrix} - [\text{harvest}] = \begin{bmatrix} \text{age distribution} \\ \text{at beginning of} \\ \text{growth period} \end{bmatrix}.$$

Using matrix notation, where P represents the distribution at the beginning of the growth period, L represents the Leslie matrix, and H represents the harvesting matrix, this translates to

$$PL - PLH = P.$$

This model was applied in 1967 to a species of domestic sheep in New Zealand which has a lifespan of 12 years. The sheep were divided into 12 age groups with a growth period of 1 year. Birth and survival rates for each age group of sheep were found using demographics, and the following Leslie matrix was developed.

$$\begin{bmatrix} 0 & 0.845 & 0 & 0 & 0 & 0 & 0 & 0 & 0 & 0 & 0 & 0 \\ 0.045 & 0 & 0.975 & 0 & 0 & 0 & 0 & 0 & 0 & 0 & 0 & 0 \\ 0.391 & 0 & 0 & 0.965 & 0 & 0 & 0 & 0 & 0 & 0 & 0 & 0 \\ 0.472 & 0 & 0 & 0 & 0.950 & 0 & 0 & 0 & 0 & 0 & 0 & 0 \\ 0.484 & 0 & 0 & 0 & 0 & 0.926 & 0 & 0 & 0 & 0 & 0 & 0 \\ 0.546 & 0 & 0 & 0 & 0 & 0 & 0.895 & 0 & 0 & 0 & 0 & 0 \\ 0.543 & 0 & 0 & 0 & 0 & 0 & 0 & 0.850 & 0 & 0 & 0 & 0 \\ 0.502 & 0 & 0 & 0 & 0 & 0 & 0 & 0 & 0.786 & 0 & 0 & 0 \\ 0.468 & 0 & 0 & 0 & 0 & 0 & 0 & 0 & 0 & 0.691 & 0 & 0 \\ 0.459 & 0 & 0 & 0 & 0 & 0 & 0 & 0 & 0 & 0 & 0.561 & 0 \\ 0.433 & 0 & 0 & 0 & 0 & 0 & 0 & 0 & 0 & 0 & 0 & 0.370 \\ 0.421 & 0 & 0 & 0 & 0 & 0 & 0 & 0 & 0 & 0 & 0 & 0 \end{bmatrix}$$

Snow Goose Glut: Hunters Will Get Every Advantage

LINCOLN JOURNAL STAR
January 23, 1999

When snow geese arrive in south central Nebraska in the coming weeks, waterfowl hunters will be ready for them. They'll be allowed to kill 20 snow geese per day and no conservation officer will ever ask to check their possession limits, because there are no possession limits.

Wildlife managers are trying to address a staggering problem of "white" or "light" geese population. Biologists estimate that 3 million light geese will migrate through Nebraska and they say the number should be reduced to about 1 million. As many as a million birds have been reported on a single 2,000-acre wetland in Nebraska's Rainwater Basin.

By grubbing their beaks into fragile tundra soils, the huge numbers of geese are causing long-term damage to the summer arctic habitat used by many other waterfowl and shorebirds. This is the reason the Nebraska Game and Parks Commission has taken such dramatic steps to help hunters reduce the population.

If left to reproduce without a harvesting policy in place, the sheep would eventually approach a stable growth rate. If the shepard allows this to happen without a harvesting policy, the income from selling the wool would not cover the cost of feeding the flock. The stable growth rate that can be found using the Leslie matrix can be used to approximate a uniform harvesting policy. In this case the uniform harvesting policy is one in which roughly 18% of the sheep from each of the 12 age groups is harvested each year.

1. Write a summary of what you think are the important points of this chapter.

2. Can a matrix with dimension 3 by 5 be added to a matrix with dimension 5 by 3? Explain your answer.

3. The math club is planning a Saturday practice session for an upcoming math contest. For lunch they have ordered 35 Mexican special combination lunches, 6 large bags of corn chips, 6 containers of salsa sauce, and 12 six-packs of assorted cold drinks.
   a. Write this information in a row matrix L. Label your matrix.
   b. Interpret L_2 and L_4.
   c. The math club pays $4.50 per lunch, $1.97 per bag of corn chips, $2.10 for each container of salsa, and $2.89 for each six-pack of cold drinks. Use multiplication of a row and column vector to find the total cost to the math club. Label your vectors.

4. A youth fellowship group is planning a spring retreat. They have contacted three lodges in the vicinity to inquire about rates. They found that Crystal Lodge charges $13.00 per person per day for lodging, $20.00 per day for food, and $5.00 per person for use of the recreational facilities. Springs Lodge charges $12.50 for lodging, $19.50 for meals, and $7.50 for use of the recreational facilities. Bear Lodge charges $20.00 per night for lodging, $18.00 a day for meals, and there is no extra charge for using the recreational facilities. Beaver Lodge charges a flat rate of $40.00 a day for lodging (meals included) and no additional fee for use of the recreational facilities.
   a. Display this information in a matrix C. Label the rows and columns.
   b. State the values of C_{22} and C_{43}
   c. Interpret C_{13} and C_{31}.

5. Mrs. Jones has been bothered by flies, spiders, and a variety of beetles on her summer porch. She has been shopping for a vacuum-powered insect disposal system. She found one at Z-Mart and another model at Base Hardware. The Z-Mart system cost $39.50, disposal cartridges were 6 for $24.50, and storage cases were $8.50 each. At Base Hardware the system cost $49.90, cartridges were 6 for $29.95, and cases were $12.50 each.

   a. Write and label a matrix showing the prices for the three parts at the two stores.

   b. Mrs. Jones decided to wait and see if the prices for insect disposal systems would be reduced during the upcoming end-of-summer sales. When she went back during the sales, the Z-Mart prices were reduced by 10% and the Base Hardware prices were reduced by 20%. Construct a matrix showing the sale prices for each of the three parts at the two stores.

   c. Use matrix subtraction to compute how much Mrs. Jones could save for each part at the two stores.

   d. Suppose Mrs. Jones is interested in purchasing the vacuum-powered insect disposal systems for herself and three of her friends. Use multiplication of a matrix by a scalar to show how much she would pay for four of each part of the system at the two stores at the sale prices.

6. An artist fashions plates and bowls from small pieces of colored woods. The artist currently has orders for five 10-inch plates, three large bowls, and seven small bowls. Each plate requires 100 pieces of ebony, 800 pieces of walnut, 600 pieces of rosewood, and 400 pieces of maple. It takes 200 ebony pieces, 1,200 walnut pieces, 1,000 rosewood pieces, and 800 pieces of maple to make a large bowl. A small bowl takes 50 pieces of ebony, 500 walnut pieces, 450 rosewood pieces, and 400 pieces of maple.

   a. Write a row matrix showing the current orders for this artist's work.

   b. Construct a matrix showing the number of pieces of wood used in an individual plate or bowl.

   c. Use matrix multiplication to compute the number of pieces of each type of wood the artist will need for the plates and bowls that are on order.

   d. Suppose it takes the artist 3 weeks to fashion a plate, 4 weeks to make a large bowl, and 2 weeks to complete a small bowl. Use matrix multiplication to show how long it will take the artist to fill all the orders for plates and bowls.

7. Jon has money invested in three sports complexes in Smith City. His return (annual) from a $50,000 investment in a tennis club is 8.2%. He receives 6.5% from a $100,000 investment in a golf club and 7.5% on a $75,000 investment in a soccer club. Use vector multiplication to find Jon's income from his investments for one year. Label your vectors.

8. Three music classes at Central High are selling candy as a fund-raiser. The number of each kind of candy sold by each of the three classes so far is shown in the following table.

	Jazz Band	Symphonic Band	Orchestra
Chocolate delights	300	220	250
Chocolate overdose	240	330	400
Chocolate chewies	150	200	180
Sour balls	175	150	160

The profit for each type of candy is sour balls, 30 cents; chocolate overdose, 50 cents; chocolate delights, 25 cents; and chocolate chewies, 35 cents. Use matrix multiplication to compute the profit made by each class on its candy sales.

9. Write the transpose (A^T) of matrix A, where

$$A = \begin{bmatrix} 4 & 2 & 6 \\ 5 & 1 & 3 \end{bmatrix}.$$

10. The dimensions of matrices P, Q, R, and S are 3×2, 3×3, 4×3, and 2×3, respectively. If matrix multiplication is possible, find the dimensions of the following matrix products. If it is not possible, tell why.
    a. QP.
    b. RQ.
    c. QS.
    d. RPS.

11. Let matrix

$$M = \begin{bmatrix} 1 & 1 \\ 1 & 1 \end{bmatrix}.$$

    a. Calculate M^2, M^3, and M^4.
    b. Predict the components of M^5 and check your prediction.

c. Generalize to M^n, where n is a natural number.

d. Prove your conjecture in part c using mathematical induction.

e. Repeat parts a, b, c, and d for the matrix

$$M = \begin{bmatrix} 1 & 0 \\ 2 & 3 \end{bmatrix}.$$

12. Complete the following statement: If a square matrix A has an inverse A^{-1}, then the product $AA^{-1} =$ the _____ matrix I, where I is a _____.

13. Which of the following matrices are inverses of each other? Explain your answers.

a. $\begin{bmatrix} -1 & 3 \\ 2 & -5 \end{bmatrix}$ and $\begin{bmatrix} 5 & 3 \\ 2 & 1 \end{bmatrix}$.

b. $\begin{bmatrix} 1 & 0 \\ 0 & 1 \end{bmatrix}$ and $\begin{bmatrix} 1 & 0 \\ 0 & 1 \end{bmatrix}$.

c. $\begin{bmatrix} 2 & 1 & 0 \\ 3 & 2 & 1 \end{bmatrix}$ and $\begin{bmatrix} 1 & -1 \\ -1 & 2 \\ -1 & 2 \end{bmatrix}$.

14. The students at Central High are planning to hire a band for the Senior Prom. Their choices are bands A, B, and C. They survey the Sophomore, Junior, and Senior classes and find the following percentages of students (regardless of sex) prefer the bands,

	10th	11th	12th
A	20%	35%	40%
B	30%	30%	25%
C	50%	35%	35%

The student population by class and sex is:

	Male	Female
10th	235	225
11th	205	215
12th	175	190

Use matrix multiplication to find:

a. The number of males and females who prefer each band.

b. The total number of students who prefer each band.

15. The characteristics of the female population of a herd of small mammals is shown in the following table.

	Age Groups (months)					
	0–4	4–8	8–12	12–16	16–20	20–24
Birthrate	0	0.5	1.1	0.9	0.4	0
Survival Rate	0.6	0.8	0.9	0.8	0.6	0

Suppose the initial female population for the herd is given by

$$P_0 = [22 \quad 22 \quad 18 \quad 20 \quad 7 \quad 2].$$

a. What is the expected life span of this mammal?
b. Construct the Leslie matrix for this population.
c. Determine the long-term growth rate for the herd.
d. Suppose this mammal starts dying off from overcrowding when the total female population for the herd reaches 520. How long will it take for this to happen?

Bibliography

Aksamit, D. L., J. V. Mitchell, and B. J. Pozehl. 1987. "Relationships between PPST and ACT Scores and Their Implications for Basic Skills Testing of Prospective Testin." *Journal of Teacher Education* November–December: 48–52.

Anton, H., and C. Rorres. 1987. *Elementary Linear Algebra with Applications.* New York, NY: Wiley.

Cozzens, M. B., and R. D. Porter. 1987. *Mathematics and Its Applications.* Lexington, MA: Heath.

Kemeny, J. G., J. N. Snell, and G. L. Thompson. 1957. *Finite Mathematics.* Englewood Cliffs, NJ: Prentice-Hall.

Leslie, P. H. 1945. "On the Uses of Matrices in Certain Population Mathematics." *Biometrika* 33:183–212.

Lincoln Journal Star, June 9, 1991, p. 2E, and June 16, 1991, p. 4D.

Maurer, S. B., and A. Ralston. 1991. *Discrete Algorithmic Mathematics.* Reading, MA: Addison-Wesley.

North Carolina School of Science and Mathematics. 1988. *New Topics for Secondary School Mathematics: Matrices.* Reston, VA: National Council of Teachers of Mathematics.

Ross, K. A., and C. R. B. Wright. 1985. *Discrete mathematics.* Englewood Cliffs, NJ: Prentice-Hall.

Tuchinsky, Philip M. 1986. *Matrix Multiplication and DC Ladder Circuits.* Lexington, MA: COMAP.

U.S. Department of Education. 1991. *Youth Indicators 1991: Trends in the Well-being of American Youth.* Washington, DC: U.S. Government Printing Office.

Graphs and Their Applications

When the boundaries or names of countries change, cartographers have to be prepared to provide the public with new maps. For years, mapmakers and mathematicians alike have wondered about the number of colors it takes to color a map.

What is the minimum number of colors needed to color any map? How do you optimally color a map? What do map coloring and meeting time scheduling for your school organizations have in common? Whether you're a cartographer who must find a way to color a map, a businessperson who must determine whether a project can be completed on time, or a planner who wants to know the most efficient way to route a city's garbage trucks, you'll find the answer in an area of mathematics known as graph theory.

Modeling Projects

How does a building contractor organize all of the jobs needed to complete a project? How do your parents manage to get all the food for a Thanksgiving dinner done at the same time? Many people believe that planning is a simple activity. After all, everybody does it. Planning your day-to-day activities seems to be second nature. What most people fail to realize is that for people in the business world who must plan and control work on extensive projects, this haphazard manner of planning is not the most efficient way to complete a job. A more scientific, organized method must be used.

One way to model projects that consist of several different subprojects, or tasks, is through the use of a diagram, or **graph**, that is made up of points called **vertices** and connecting lines called **edges** (See Figure 4.1).

Figure 4.1 Graph with three vertices and two edges.

Explore This

The Central High yearbook staff has only 16 days left before the deadline for completing their book. They are running behind schedule and still have several tasks left to finish. The remaining tasks and time that it takes to complete each task are listed in the following table.

Task	Time (days)
Start	0
A Buy film	1
B Load cameras	1
C Take photos of clubs	3
D Take sports photos	2
E Take photos of teachers	1
F Develop film	2
G Design layout	5
H Print and mail pages	3

Is it possible for the yearbook to be completed on time if the tasks have to be done one after the other? If some tasks can be done at the same time as others, can the deadline be met?

As you may have noticed, some of the yearbook staff jobs can be done simultaneously, while several of them cannot be started until others have been completed. Assuming the following prerequisites, how soon can the project be completed?

Task	Time (days)	Prerequisite Task
Start	0	—
A Buy film	1	None
B Load cameras	1	A
C Take photos of clubs	3	B
D Take sports photos	2	C
E Take photos of teachers	1	B
F Develop film	2	D, E
G Design layout	5	D, E
H Print and mail pages	3	G, F

Drawing a graph of this information makes it easier to see the relationships among the tasks. In the graph in Figure 4.2 on page 154, the tasks are represented by points (vertices), and the arrows (directed edges) indicate which tasks must be finished before a new task can begin. Each edge also shows the number of days it takes to complete the preceding task. Note that tasks with the same prerequisites are aligned vertically. Although this is not necessary, it helps to make the graph easier to follow.

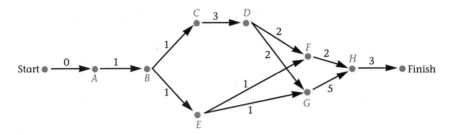

Figure 4.2 Diagram of the order of tasks necessary to complete the yearbook.

Exercises

In Exercises 1 through 4, use the task table to draw a graph with appropriately labeled vertices and edges.

1.

Task	Time	Prerequisites
Start	0	
A	5	None
B	6	A
C	4	A
D	4	B
E	8	B, C
F	4	C
G	10	D, E, F
Finish		

2.

Task	Time	Prerequisites
Start	0	
A	1	None
B	2	None
C	3	A, B
D	5	B
E	5	C
F	5	C, D
G	4	D, E
H	4	E, F
Finish		

3.

Task	Time	Prerequisites
Start	0	
A	5	None
B	7	A
C	4	A
D	3	B
E	7	B, C
F	5	C
G	8	D, E, F
Finish		

4.

Task	Time	Prerequisites
Start	0	
A	5	None
B	8	A, D
C	9	B, I
D	7	None
E	8	B
F	12	I
G	4	C, E, F
H	9	None
I	5	D, H
Finish		

5. To help organize the task of completing the family dinner, Mrs. Shu listed the following tasks.

Task	Time (min.)	Prerequisite Task
Start	0	—
A Wash hands		
B Defrost hamburger		
C Shape meat into patties		
D Cook hamburgers		
E Peel and slice potatoes		
F Fry potatoes		
G Make salad		
H Set table		
I Serve food		

   a. Complete the table by making reasonable time estimates in minutes for each of these tasks and indicating the prerequisites.
   b. Construct a graph using the information from your table.
   c. What is the least amount of time needed to prepare dinner?

6. Your best friend, Matt, has always been very disorganized. He is now preparing to leave for college and desperately needs your help.
   a. Create a table of at least six activities that will need to be completed before Matt can leave home. Give the times and prerequisites of these activities.
   b. Construct a corresponding graph.
   c. For your task list, what is the least amount of time it will take to get Matt off to school?

7. Consider the following graph.

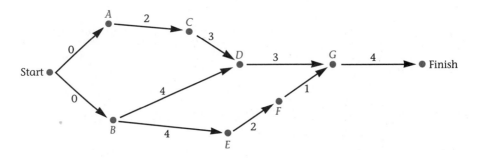

a. Complete the following task table for this graph.

Task	Time	Prerequisite Task
Start	0	—
A		
B		
C		
D		
E		
F		
G		
Finish		

b. What is the least amount of time it will take to complete all of the tasks in this graph? Explain why it cannot be completed in less time.

Critical Paths

It is relatively easy to find the shortest time needed to complete a project if the project consists of only a few activities. But as the tasks increase in number, the problem becomes more difficult to solve by inspection alone.

In the 1950s the U.S. government was faced with the need to complete very complex systems such as the U.S. Navy Polaris Submarine project. In order to do this efficiently, a method was developed called PERT (Program Evaluation and Review Technique) in which those tasks which were critical to the earliest completion of the project were targeted. This path of targeted tasks from the start to the finish of the project became known as the **critical path**.

Recall the graph in Lesson 4.1 that represented the Central High yearbook project. How might you go about finding a systematic way to identify the critical path for this project? To do this, an **earliest-start time** (EST) for each task must be found. The EST is the earliest that an activity can begin if all the activities preceding it begin as early as possible.

To calculate the EST for each task, begin at the start and label each vertex with the smallest possible time that is needed before the task can begin. The label for C in Figure 4.3 is found by adding the EST of B to the 1 day that it takes to complete task B ($1 + 1 = 2$). Task G cannot be completed until both predecessors, D and E, have been completed. Hence, G cannot begin until 7 days have passed.

In the case of the yearbook staff, the earliest time in which the project can be completed is 15 days. As paradoxical as it may seem, the least amount of time that it takes to complete all of the tasks in the project corresponds to the time it takes to complete the longest path through the graph from start to finish. A path with this longest time is the desired critical path. In Figure 4.3, the critical path is Start–*ABCDGH*–Finish.

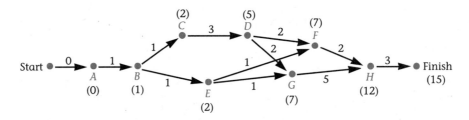

Figure 4.3 Yearbook diagram showing the earliest-start time for each task.

Example

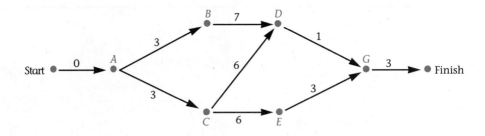

1. Copy the graph, label the vertices with the EST for each task, and determine the earliest completion time for the project. All times are in minutes.

2. Identify the critical path.

The solutions are:

1.

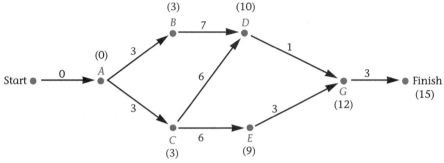

The earliest time in which the project can be completed is 15 minutes.

2. Since the critical path is the longest path from the start to the finish, the critical path is Start–*ACEG*–Finish.

If it is desirable to cut the completion time of a project, it can be done by shortening the length of the critical path once it is found. In the preceding example, one way to shorten the time it takes to complete the project is to cut the time it takes to complete task *E* to 2 minutes instead of 3 minutes. If that is done, the completion time for the project is cut to 14 minutes.

The efficient management of large projects like the construction of a building requires the use of critical path analysis.

Exercises

1.

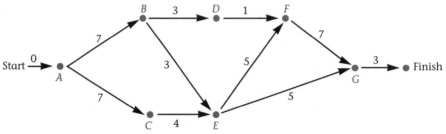

Complete the following.

Vertex	Earliest-Start Time
A	0
B	7
C	
D	
E	
F	
G	

Minimum project time =

Critical path(s) =

In Exercises 2 and 3, list the vertices of the graphs and give their earliest-start time, as in Exercise 1. Determine the minimum project time and all of the critical paths.

2.

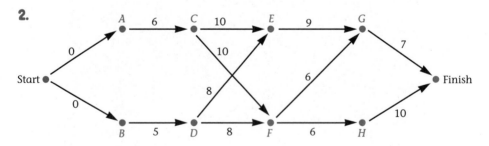

3.

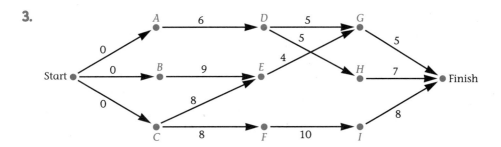

4. Using the information from the following table, construct a graph and label each of the vertices with its earliest-start time. Determine the minimum project time and critical path.

Task	Time	Prerequisites
Start	0	
A	2	None
B	4	None
C	3	A, B
D	1	A, B
E	5	C, D
F	6	C, D
G	7	E, F

5.

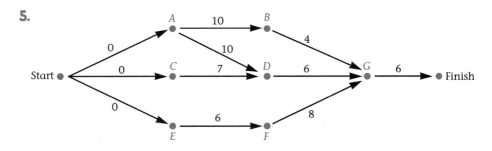

a. Copy the graph and label each vertex with its earliest-start time.
b. How quickly can the project be completed?
c. Determine the critical path.
d. What happens to the minimum project time if task A's time is reduced to 9 days? To 8 days?
e. Will the project time continue to be affected by reducing the time of task A? Explain why or why not.

6. Construct a graph with three critical paths.

7. Determine the minimum project time and the critical path for the following graph.

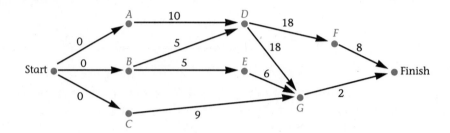

8. In the following graph, each vertex has been label with its EST, and the critical path is marked.

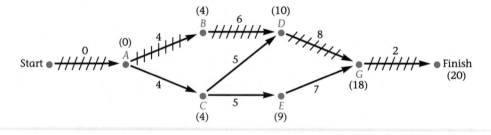

a. Task E can begin as early as day 9. If it begins on day 9, when will it be completed? If it begins on day 10? On day 11? What will happen if it begins on day 12?

b. What is the latest day on which task E can begin if task G is to begin on day 18?

If an activity is not on the critical path, it is possible for it to start later than its earliest-start time and not delay the project. The latest a task can begin without delaying the project's minimum completion time is known as the **latest-start time** (LST) for the task. For example, the LST for E is day 11.

c. In order to find the LST for vertex C, the times of the two vertices D and E need to be considered. Since vertex D is on the critical path, the latest it can start is day 10. For D to begin on time, what is the latest day on which C can begin? In part b, you found that the latest

E can start is day 11. In that case, what is the latest *C* can begin? From this information, what is the latest (LST) that *C* can begin without delaying *either* task *D* or *E*?

9. To find the LST for each task, it is necessary to begin with the Finish and work through the graph in reverse order to the Start. Each of the vertices in the following graph are labeled with their ESTs. The LSTs for several of the tasks have been calculated and are shown below the ESTs on the vertices. Find the LSTs for the remaining tasks.

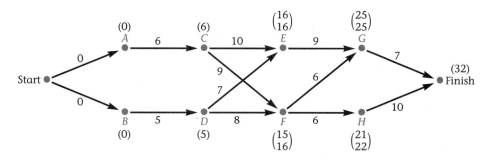

10. Write an algorithm to find the LSTs for the tasks in a graph. Test your algorithm on the graph in Exercise 1.

Projects

11. Interview the yearbook sponsors in your school to find out how they organize the publication of your school's yearbook. Create a task table that shows the approximate times and prerequisite tasks that must be completed before your yearbook can go to the publisher. Design a graph with the EST for each task, and identify the critical path.

12. Use the Internet or other sources to research and report on businesses or people who use PERT or similar evaluation techniques for project planning.

The Vocabulary and Representations of Graphs

Graphs have many applications in addition to critical path analysis. They are frequently used in social science, computer science, chemistry, biology, transportation, and communications. In the following lessons several of these applications are examined.

Recall that a graph is a set of points called vertices and a set of connecting lines called edges. Often graphs are used to model situations in which the vertices represent objects, and edges are drawn between the vertices on the basis of a particular relationship between the objects. Note that the important characteristics of a graph remain unchanged whether the edges are curved or straight.

Explore This

Case 1

Suppose the vertices of Figure 4.4 represent the starting five players on a high school basketball team, and the edges denote friendships. This graph

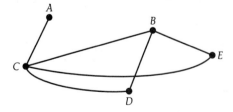

Figure 4.4 Graph showing five vertices and five edges.

indicates that player *C* is friends with all of the other players and that *E* has only two friends, *C* and *B*. Notice that edge *CE* and edge *DB* intersect in this graph, but the intersection does not create a new vertex.

1. Which player has only one friend?

2. How many friends does *E* have? Who are they?

3. Redraw the graph so that *A* has no friends.

Consider the following solutions.

1. Player *A*.

2. Two, *C* and *B*.

3. The graph shown in Figure 4.5.

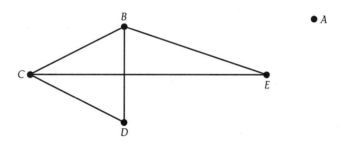

Figure 4.5 Graph that shows *A* with no friends.

A graph is **connected** if there is a path between each pair of vertices. The graph shown above in Figure 4.5 is not connected because there is no path from *A* to any of the other vertices.

Case 2

Figure 4.4 on page 164 could represent many different things other than basketball players and their friendships. For example, let the vertices in the figure represent rooms in your school. The vertices are connected if there are direct hallways between two rooms. Then, according to the graph in Figure 4.4, a student can get from room C directly to any of the other four rooms.

When two vertices are connected with an edge, they are said to be **adjacent**. In Figure 4.4, C is adjacent to A, B, D, and E. Although there is no single edge from D to A, it is possible to get from room D to room A by following a path that goes through room C. Although a path exists between D and A, they are not adjacent.

Try drawing a graph in which there is direct access from each room to every other room. Figure 4.6 shows two possible ways to represent this graph. Even though these graphs appear to be different, they are structurally the same, and so they are considered the same graph. It is important to note that there is no single correct way to draw a graph to represent a given situation. A good graph is one that clearly represents the information needed to solve some problem.

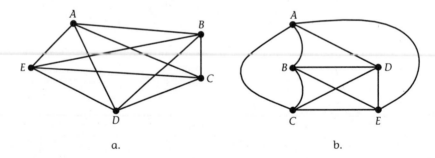

a.   b.

Figure 4.6 Two different representations of five rooms in a school, with each having direct access to every other room.

Graphs such as the ones in Figure 4.6, in which every pair of vertices is adjacent, are called **complete** graphs. Complete graphs are often denoted by K_n, where n is the number of vertices in the graph. Figure 4.6 depicts a K_5 graph.

Alternative Representations

Graphs can be represented in several different ways. A diagram, such as the one in Figure 4.4, is just one of these ways. Another way to represent the information is to list the set of vertices and the set of edges. For example, the graph in Figure 4.4 on page 164 can be described by

Vertices = {A, B, C, D, E}   Edges = {AC, CB, CE, CD, BD, BE}.

A third way to represent this information is with an adjacency matrix. This type of representation can be used to depict the vertices and edges of the graph in a computer or calculator.

A 5×5 matrix in which both the rows and columns correspond to the vertices A, B, C, D, and E can be used to represent Figure 4.4. If an edge exists between vertices, a 1 appears in the corresponding position in the matrix; otherwise a 0 appears.

$$
\begin{array}{c c c c c c}
 & A & B & C & D & E \\
A & 0 & 0 & 1 & 0 & 0 \\
B & 0 & 0 & 1 & 1 & 1 \\
C & 1 & 1 & 0 & 1 & 1 \\
D & 0 & 1 & 1 & 0 & 0 \\
E & 0 & 1 & 1 & 0 & 0
\end{array}
$$

The entry in row 2, column 4 is a 1. This indicates that vertices B and D are adjacent; that is, an edge exists between them.

Exercises

1. Mr. Butler bought six different types of fish. Some of the fish can live in the same aquarium, but others cannot. Guppies can live with Mollies, Swordtails can live with Guppies, Plecostomi can live with both Mollies and Guppies, Gold Rams can live only with Plecostomi, and Piranhas cannot live with any of the other fish. Draw a graph to illustrate this.

2. Construct a graph for each of the following sets of vertices and edges. Which of the graphs are connected? Which are complete?
   a. V = {A, B, C, D, E}
      E = (AB, AC, AD, AE, BE}.
   b. V = {M, N, O, P, Q, R, S}
      E = {MN, SR, QS, SP, OP}.

   c. V = {E, F, G, J, K, M}
      E = {EF, KM, FG, JM, EG, KJ}.
   d. V = {W, X, Y, Z}
      E = {WX, XZ, YZ, XY, WZ, WY}.

3. Draw a graph with vertices = {A, B, C, D, E, F} and edges = {AB, CD, DE, EC, EF}.
   a. Name two vertices that are not adjacent.
   b. F, E, C is one possible path from F to C. This path has a length of 2, since two edges were traveled to get from F to C. Name a path from F to C with a length of 3.
   c. Is this graph connected? Explain why or why not?
   d. Is this graph complete? Explain why or why not?

4. Draw a graph with five vertices in which vertex W is adjacent to Y, X is adjacent to Z, and V is adjacent to each of the other vertices.

5. Construct a graph for each adjacency matrix. Label the vertices A, B, C, ...

   a. $\begin{bmatrix} 0 & 1 & 0 & 0 \\ 1 & 0 & 1 & 1 \\ 0 & 1 & 0 & 1 \\ 0 & 1 & 1 & 0 \end{bmatrix}$
   b. $\begin{bmatrix} 0 & 1 & 0 & 0 & 1 \\ 1 & 0 & 1 & 0 & 1 \\ 0 & 1 & 0 & 1 & 1 \\ 0 & 0 & 1 & 0 & 1 \\ 1 & 1 & 1 & 1 & 0 \end{bmatrix}$
   c. $\begin{bmatrix} 0 & 1 & 0 & 0 & 0 \\ 1 & 0 & 0 & 0 & 0 \\ 0 & 0 & 0 & 1 & 1 \\ 0 & 0 & 1 & 0 & 1 \\ 0 & 0 & 1 & 1 & 0 \end{bmatrix}$

6. Create an adjacency matrix for each of the following graphs:

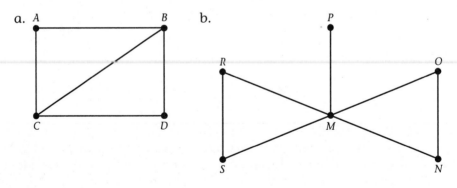

a.

b.

7. Give the adjacency matrix for the following graph.

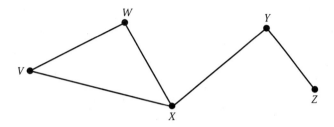

a. What do you notice about the main diagonal of the matrix?
b. Does your matrix possess symmetry? If so, where?
c. If an adjacency matrix has a 1 on the main diagonal, what would that indicate? What would a 2 in row 2, column 1 indicate?

8. Find the sum of each row of your matrix from Exercise 7. What do these sums tell you about the graph of the matrix?

9. In a graph, the number of edges that have a specific vertex as an endpoint is known as the **degree** or **valence** of that vertex. In the following graph, the degree of vertex W is 4. This is denoted by $\deg(W) = 4$. Find the degrees of each of the other vertices.

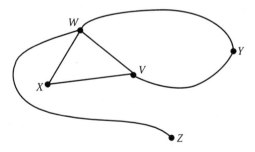

10. An edge that connects a vertex to itself is called a **loop**. If a graph contains a loop or **multiple edges** (more than one edge between two vertices), the graph is known as a **multigraph**. When finding the degree of a vertex on which there is a loop, the loop is counted twice. For example, deg(A) = 3.

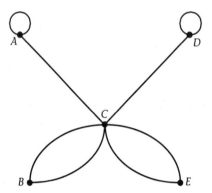

a. Find the degree of vertices B, C, D, and E.
b. Give the adjacency matrix for the above multigraph.
c. Compare an adjacency matrix for a graph and one for a multigraph. Without seeing the graph, can you tell which belongs to the graph and which belongs to the multigraph? Explain how you know.

11. Complete the following table for the sum of the degrees of the vertices in a complete graph.

Graph	Number of Vertices	Sum of the Degrees of All of the Vertices	Recurrence Relation
K_1	1	0	
K_2	2	2	$T_2 = T_1 + 2$
K_3	3	6	$T_3 = T_2 + 4$
K_4	—	—	—
K_5	—	—	—
K_6	—	—	—

Write a recurrence relation that expresses the relationship between the sum of the degrees of all the vertices for K_n and the sum for K_{n-1}.

12. Having completed the table in Exercise 11, what did you notice about the sum of the degrees of the vertices for any complete graph? Do you think this is true for any graph? If so, explain why this is true; if not, give a counterexample.

13. a. Try to construct a graph with four vertices, two of the vertices with degree 3 and two with degree 2. No loops or multiple edges may be used.
   b. Try to construct a graph with five vertices, three of the vertices with an odd degree and two with even degree. No loops or multiple edges may be used.
   c. What do you think might be true about the number of vertices with even degree and the number of vertices with odd degree in any graph? (Hint: Try a few examples to check your hypothesis.)

14. Describe the adjacency matrix of a complete graph.

15. Complete the following table for the given complete graphs.

Graph	Number of Vertices	Number of Edges	Recurrence Relation
K_1	1	0	
K_2	2	1	$S_2 = S_1 + 1$
K_3	3	3	$S_3 = S_2 + 2$
K_4	—	—	—
K_5	—	—	—
K_6	—	—	—

Write a recurrence relation that expresses the relationship between the number of edges of K_n and the number of edges of K_{n-1}.

16. Central High School is a member of a five-team hockey league. Each team in the league plays exactly two games, which must be against different teams. Show that there is only one possible graph for this schedule.

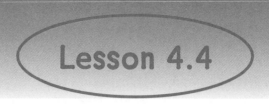
Euler Circuits and Paths

Now that you are familiar with some of the concepts of graphs and the way graphs convey connections and relationships, it's time to begin exploring how they can be used to model many different types of situations.

Explore This

Consider the graph in Figure 4.7. Try to draw this figure without lifting your pencil from the paper and without tracing any of the lines more than once. Is this possible?

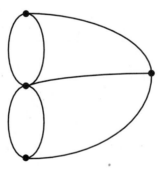

Figure 4.7 Graph.

The graph in Figure 4.7 represents an eighteenth-century problem that intrigued the famous Swiss mathematician Leonhard Euler (pronounced "oiler"). The problem was one that had been posed by the residents of Königsberg, a city in what was then Prussia but is now the Russian city of Kaliningrad. In the 1700s, seven bridges connected two islands in the Pregel River to the rest of the city (see Figure 4.8). The people of Königsberg wondered whether it would be possible to walk through the city by crossing each bridge exactly once and return to the original starting point.

Mathematician of Note

Leonard Euler (1707–1783). Euler was an extraordinary mathematician who published over 500 works during his lifetime. Even total blindness for the last 17 years of his life did not stop his effectiveness and genius. He is often referred to as "the father of graph theory."

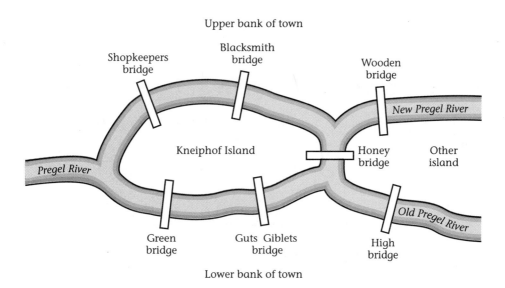

Figure 4.8 Representation of the seven bridges of Königsberg.

Using a graph like the one in Figure 4.7, in which the vertices represented the landmasses of the city and the edges represented the bridges, Euler found that it was not possible to make the desired walk through the city. In so doing, he also discovered a solution to problems of this general type.

What did Euler find? Try to reproduce the graphs in Figure 4.9 without lifting your pencil or tracing the lines more than once.

1. When can you draw the figures without retracing any edges and still end up at your starting point?

2. When can you draw the figure without retracing and end up at a point other than the one from which you began?

3. When can you not draw the figure without retracing?

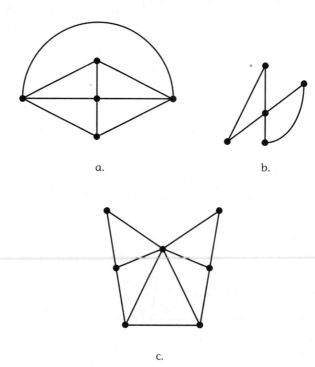

a.

b.

c.

Figure 4.9 Graphs to trace.

Euler found that the key to the solution was related to the degrees of the vertices. Recalling that the degree of a vertex of a graph is the number of edges that have that vertex as an endpoint, find the degree of each vertex of the graphs in Figure 4.9. Do you see what Euler noticed?

Euler hypothesized and later proved that in order to be able to traverse each edge of a connected graph exactly once and to end at the starting vertex, the degree of each vertex of the graph must be even, as in Figure 4.9b. In honor of Leonhard Euler, a path that uses each edge of a graph exactly once and ends at the starting vertex is called an **Euler circuit**.

Euler also noticed that if a connected graph had exactly two odd vertices, it was possible to use each edge of the graph exactly once but to end at a vertex different from the starting vertex. Such a path is called an **Euler path**. Figure 4.9a is an example of a graph that has an Euler path. Figure 4.9c has four odd vertices, and so it cannot be traced without lifting your pencil. It has neither an Euler circuit nor an Euler path.

An Euler circuit for a relatively small graph can usually be found by trial and error. However, as the number of vertices and edges increases, a systematic way of finding the circuit becomes necessary. The following algorithm gives a procedure for finding an Euler circuit for a connected graph with all vertices of even degree.

Euler Circuit Algorithm

1. Pick any vertex, and label it S.

2. Construct a circuit, C, that begins and ends at S.

3. If C is a circuit that includes all edges of the graph, go to step 8.

4. Choose a vertex, V, that is in C and has an edge that is not in C.

5. Construct a circuit C' that starts and ends at V using edges not in C.

6. Combine C and C' to form a new circuit. Call this new circuit C.

7. Go to step 3.

8. Stop. C is an Euler circuit for the graph.

Euler Circuit Algorithm

1. Pick any vertex, and label it S.

2. Construct a circuit, C, that begins and ends at S.

3. If C is a circuit that includes all edges of the graph, go to step 8.

4. Choose a vertex, V, that is in C and has an edge that is not in C.

5. Construct a circuit C' that starts and ends at V using edges not in C.

6. Combine C and C' to form a new circuit. Call this new circuit C.

7. Go to step 3.

8. Stop. C is an Euler circuit for the graph.

Example

Use the Euler circuit algorithm to find an Euler circuit for the following graph.

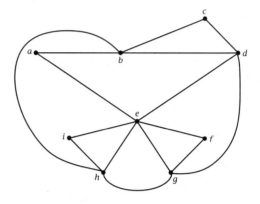

- Apply step 1 of the algorithm. Choose vertex b, and label it S.

- Let C be the circuit S, c, d, e, a, S.

- Circuit C does not contain all edges of the graph, so proceed to step 4 of the algorithm.

- Choose vertex d.

- Let C' be the circuit d, g, h, S, d.

- Combine C and C' by replacing vertex d in the circuit C with the circuit C'. Let C now be the circuit S, c, d, g, h, S, d, e, a, S.

- Go to step 3 of the algorithm.

- Circuit C does not contain all edges of the graph, so again proceed to step 4.

- Choose vertex g.

- Let C' be the circuit g, f, e, i, h, e, g

- Combine C and C' by replacing vertex g in the circuit C with the circuit C'. Let C now be the circuit S, c, d, g, f, e, i, h, e, g, h, S, d, e, a, S.

- Circuit *C* now contains all edges of the graph, so go to step 8 of the algorithm and stop. *C* is an Euler circuit for the graph.

Edges with Direction

Many applications of graphs require that the edges have direction. A city with one-way streets is one such example. A graph that has directed edges, edges that can be traversed in only one direction, is known as a **digraph** (see Figure 4.10). The number of edges coming into a vertex is known as the **indegree** of the vertex, and the number of edges going out of a vertex is known as the **outdegree**.

Examine Figure 4.10. This digraph can be described by

Vertices = {*A, B, C, D*}   Ordered edges = {*AB, BA, BC, CA, DB, AD*}.

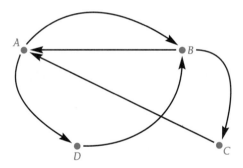

Figure 4.10 Digraph.

If you follow the indicated direction of each edge, is it possible to start at some vertex, draw the digraph, and end up at the vertex from which you started? That is, does this digraph have a directed Euler circuit?

Check the indegree and outdegree of each vertex. You will find that a connected digraph has an Euler circuit if the indegree and outdegree of each vertex are equal.

Exercises

1. State whether each graph has an Euler circuit, an Euler path, or neither. Explain why.

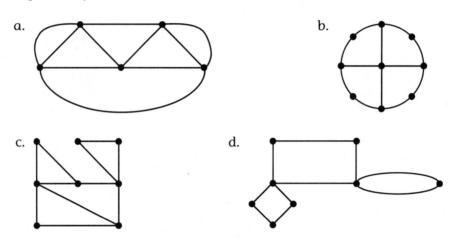

a.

b.

c.

d.

2. Sally began using the Euler circuit algorithm to find the Euler circuit for the following graph. She started at vertex *d* and labeled it *S*. The first circuit she found was *S, e, f, a, b, c, S*. Using Sally's start, continue the algorithm and find an Euler circuit for the graph.

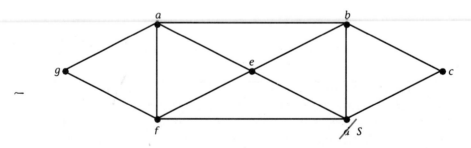

3. Use the Euler circuit algorithm to find an Euler circuit for the following graph.

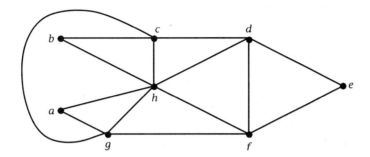

4. The text states that to apply the Euler circuit algorithm, the graph must be "connected with all vertices of even degree."
   a. Why is it necessary to state that the graph must be connected?
   b. Give an example of a graph with all vertices of even degree that does not have an Euler circuit.

5. Will a complete graph with 2 vertices have an Euler circuit? With 3 vertices? With 4 vertices? With 5 vertices? With *n* vertices?

6. The present-day Königsberg has two more bridges than it did in Euler's time. One bridge was added to connect the two banks on the river, *A* to *B* in the following figure. Another one was added to link the land to one of the islands, *B* to *D*. Is it now possible to make the famous walk and return to the starting point? Explain your reasoning.

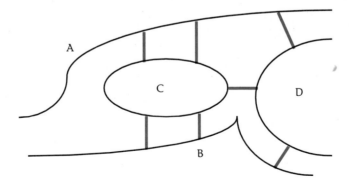

Königsberg's *original* seven bridges.

7. The street network of a city can be modeled with a graph in which the vertices represent the street corners, and the edges represent the streets. Suppose you are the city street inspector and it is desirable to minimize time and cost by not inspecting the same street more than once.

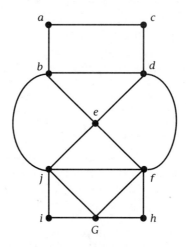

a. In this graph of the city, is it possible to begin at the garage (G) and inspect each street only once? Will you be back at the garage at the end of the inspection?

b. Find a route that inspects all streets, repeats the least number of edges possible, and returns to the garage.

8. Construct the following digraphs.

a. $V = \{A, B, C, D, E\}$
   $E = \{AB, CB, CE, DE, DA\}$.

b. $V = \{W, X, Y, Z\}$
   $E = \{WX, XZ, ZY, YW, XY, YX\}$.

9. Determine whether the digraph has a directed Euler circuit.

a.

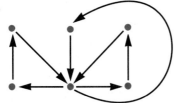

b.

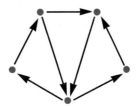

c.

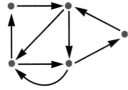

10. a. Does the following digraph have a directed Euler circuit? Explain why or why not.
   b. Does it have a directed Euler path? If it does, which vertices can be the starting vertex?
   c. Write a general statement explaining when a digraph has a directed Euler path.

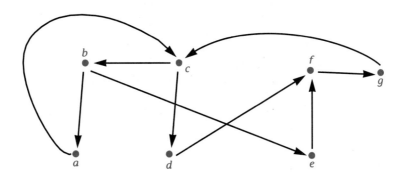

11. A digraph can be represented by an adjacency matrix. If there is a directed edge from vertex a to vertex b, then a 1 is placed in row a, column b of the matrix; otherwise a 0 is entered. Matrix M is the adjacency matrix for the following graph.

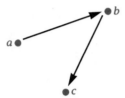

$$M = \begin{bmatrix} 0 & 1 & 0 \\ 0 & 0 & 1 \\ 0 & 0 & 0 \end{bmatrix}$$

Find the adjacency matrix for each of the following digraphs.

a.

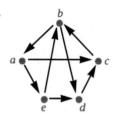

b.

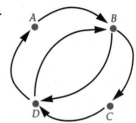

c.

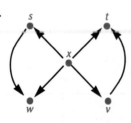

12. a. Construct a digraph for the following adjacency matrix.

$$\begin{bmatrix} 0 & 1 & 0 & 1 & 0 \\ 1 & 0 & 1 & 0 & 1 \\ 0 & 0 & 0 & 1 & 0 \\ 0 & 1 & 1 & 0 & 0 \\ 1 & 0 & 0 & 1 & 0 \end{bmatrix}$$

 b. Is there symmetry along the main diagonal of the adjacency matrix? Explain why or why not.
 c. Find the sum of the numbers in the second row. What does that total indicate?
 d. Find the sum of the numbers in the second column. What does that total indicate?

Computer/Calculator Explorations

13. Create a computer or calculator program that prompts the user to enter the adjacency matrix for a connected graph. The program should then tell the user whether or not the graph has an Euler circuit.

Projects

14. Leonhard Euler was known for many accomplishments in addition to his discoveries related to graph theory. After researching Euler's achievements, create a "biographic poster" that illustrates the important milestones of his life.

15. Research and report on algorithms that determine Euler circuits for graphs that have them.

Hamiltonian Circuits and Paths

Since its inception, graph theory has been closely tied to applications. In Lesson 4.4, you investigated situations in which you needed to traverse each edge of a graph. In this lesson, you will explore reasons to examine the vertices.

Explore This

Suppose once again that you are a city inspector, but instead of inspecting all of the streets in an efficient manner, you must inspect the fire hydrants that are located at each of the street intersections. This implies that you are searching for an optimal route that begins at the garage G, visits each intersection exactly once, and returns to the garage (see Figure 4.11).

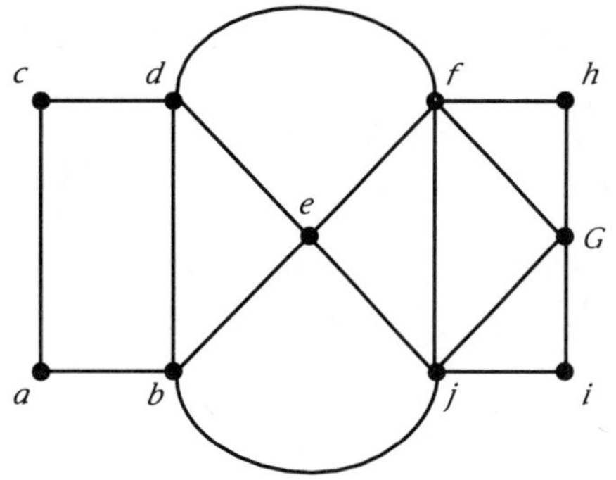

One path that meets these criteria is G, h, f, d, c, a, b, e, j, i, G. Notice that it is not necessary to traverse every edge of the graph when visiting each vertex exactly once.

In 1856, Sir William Rowan Hamilton used his mathematical knowledge to create a game called the Icosian game. The game consisted of a graph in which the vertices represented major

Figure 4.11 Street network.

cities in Europe, and the object of the game was to find a path that visited each of the 20 vertices exactly once. In honor of Hamilton and his game, a path that uses each vertex of a graph exactly once is known as a **Hamiltonian path**. If the path ends at the starting vertex, it is called a **Hamiltonian circuit**.

Try to find a Hamiltonian circuit for each of the graphs in Figure 4.12.

Mathematician of Note

Sir William Rowan Hamilton (1805–1865). Hamilton, a leading nineteenth-century Irish mathematician, was appointed Astronomer Royal of Ireland at the age of 22 and knighted at 30. His most notable discoveries were in algebra.

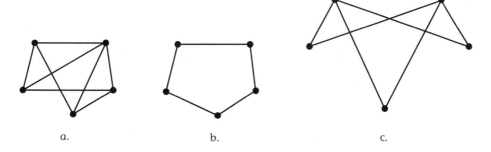

a. b. c.

Figure 4.12 Graphs with possible Hamiltonian circuits.

Mathematicians continue to be intrigued with this type of problem because a simple test for determining whether a graph has a Hamiltonian circuit has not been found, and it now appears that a general solution that applies to all graphs may be impossible. Fortunately, several theorems have been proved that help guarantee the existence of Hamiltonian circuits for certain kinds of graphs. The following is one such theorem.

> If a connected graph has *n* vertices, where *n* > 2 and each vertex has degree of a least *n*/2, then the graph has a Hamiltonian circuit.

To apply the Hamiltonian circuit theorem to Figure 4.12a on page 185, check the degree of each vertex. Since each of the five vertices of the graph has degree of at least 5/2, the theorem guarantees that the graph has a Hamiltonian circuit. Unfortunately, it does not tell you how to find it.

The theorem has another downside as well. If a graph has a vertex with degree less than $n/2$, then the theorem simply does not apply to that graph. It may or it may not have a Hamiltonian circuit. Each of the graphs in parts b and c of Figure 4.12 has some vertices of degree less than 5/2, so no conclusions can be drawn. By inspection, Figure 4.12b has a Hamiltonian circuit, but Figure 4.12c does not.

Tournaments

As with Euler circuits, it often is useful for the edges of the graph to have direction. Consider a competition in which each player must play every other player. By using directed edges, it is possible to indicate winners and losers. To illustrate this, draw a complete graph in which the vertices represent the players, and a directed edge from vertex A to vertex B indicates that player A defeats player B. This type of graph is known as a tournament. A **tournament** is a digraph that results from giving directions to the edges of a complete graph. Figure 4.13 shows a tournament in which A beats B, C beats B, and A beats C.

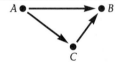

Figure 4.13 Tournament with three vertices.

One interesting property of such a digraph is that every tournament contains at least one Hamiltonian path. If there is exactly one such path, it can be used to rank the teams in order, from winner to loser.

Example

Suppose four teams play in the school soccer round-robin tournament. The results of the competition follow:

Game	AB	AC	AD	BC	BD	CD
Winner	B	A	D	B	D	D

Draw a digraph to represent the tournament. Find a Hamiltonian path and use it to rank the participants from winner to loser.

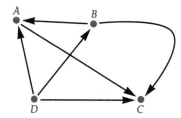

To find a solution, remember that a tournament results from a *complete* graph when direction is given to the edges. In this case, there is only one Hamiltonian path for the graph: *D, B, A, C*. Therefore, *D* finishes first, *B* is second, *A* is third, and *D* finishes fourth.

Exercises

1. Apply the theorem from page 185 to the graphs below. According to the theorem, which of the graphs have Hamiltonian circuits? Explain your reasoning.

a.

b.

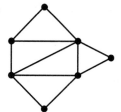

c.

2. Give two examples of situations that could be modeled by a graph in which finding a Hamiltonian path or circuit would be of benefit.

3. a. Construct a graph that has both an Euler and a Hamiltonian circuit.
   b. Construct a graph that has neither an Euler nor a Hamiltonian circuit.

4. Hamilton's Icosian game was played on a wooden regular dodecahedron (a solid figure with 12 sides). Here is a planar representation of the game.

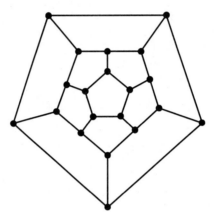

   a. Copy the graph onto your paper and find a Hamiltonian circuit for the graph.
   b. Is there only one Hamiltonian circuit for the graph?
   c. Can the circuit begin at any of the vertices or only some of them?

5. Draw a tournament with five players, in which player A defeats everyone, B defeats everyone but A, C is defeated by everyone, and D defeats E.

6. Find all the directed Hamiltonian paths for each of the following tournaments:

a.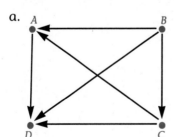

b.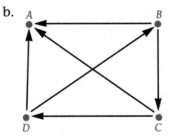

7. Draw a tournament with three vertices in which:
   a. One player wins all the games he or she plays.
   b. Each player wins exactly one game.
   c. Two players lose all of the games they play.

8. Draw a tournament with five vertices in which there is a three-way tie for first place.

9. When ties exist in a ranking for a tournament (e.g., more than one first place winner), there is more than one Hamiltonian path for the graph. Explain why this is so.

10. a. Write an algorithm that uses the outdegree of the vertices to find the Hamiltonian path for a tournament that has exactly one Hamiltonian path.
    b. Explain the difficulties that arise with your algorithm when the tournament has more than one Hamiltonian path.

11. Complete the following table for a tournament.

Number of Vertices	Sum of the Outdegrees of the Vertices
1	0
2	1
3	3
4	
5	
6	

    Write a recurrence relation that expresses the relationships between S_n, the sum of the outdegrees for a tournament with n vertices, and S_{n-1}.

12. In a tournament a **transmitter** is a vertex with a positive outdegree and a zero indegree. A **receiver** is a vertex with a positive indegree and a zero outdegree. Explain why a tournament can have at most one transmitter and at most one receiver.

13. Use mathematical induction on the number of vertices to prove that every tournament has a Hamiltonian path.
    a. Begin the mathematical induction process by showing that every tournament with one vertex has a Hamiltonian path.
    b. Assume that a tournament of k vertices has a Hamiltonian path and use this assumption to prove that a tournament of $k + 1$ vertices has a Hamiltonian path.

14. Consider the set of preference schedules from Lesson 1.3:

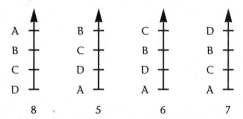

The first preference schedule could be represented by the following tournament.

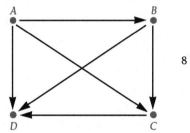

a. Construct tournaments for each of the three other preference schedules.

b. Construct a cumulative preference tournament that would show the overall results of the four individual preference schedules.

c. Is there a Condorcet winner in the election? (Recall from Lesson 1.3 that a Condorcet winner is one who is able to defeat each of the other choices in a one-on-one contest.)

d. Find a Hamiltonian path for the cumulative tournament. What does this path indicate?

15. a. Construct an adjacency matrix for the following digraph, and call the matrix *M*.

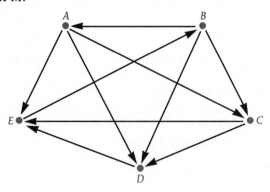

By summing the rows of M, you can see that a tie exists between A and B, each with three wins.

b. Square M. Notice that this new matrix M^2 gives the number of paths of length 2 between vertices. For example, the 3 in entry M_{25} indicates that 3 paths of length 2 exist between B and E. These paths are B, C, E; B, D, E; and B, A, E. In the case of a tournament, this means that B has three 2-stage wins, or **dominances,** over E.

c. Add M and M^2. Use the sum to determine the total number of ways that A and B can dominate in one and two stages. Who might now be considered the winner?

d. What would M^3 indicate? Find M^3 and see whether you are correct.

Projects

16. Design and build a Planar Icosian Game by enlarging the graph in Exercise 4 and copying it onto a piece of plywood or heavy cardboard. Use tacks for the vertices. The game is then played by tying a piece of string (approximately 12 inches or longer) on one of the tacks (vertices) and attempting to wind the string from tack to tack (following the lines on the graph) until all of the tacks are touched and the player is back to the initial tack (in other words, until a Hamiltonian circuit is found). Try the game with younger children and adults. Who seems to find a Hamiltonian circuit the quickest?

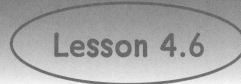

Graph Coloring

When it comes time to schedule meetings at school or register students into classes, problems often arise. Mathematicians have found that graphs are useful tools in helping to resolve conflicts of these types.

Explore This

Here is a table of clubs at Central High School and students who hold offices in these clubs.

	Math Club	Honor Club	Science Club	Art Club	Pep Club	Spanish Club
Matt	X	X	X	—	—	—
Marty	X	—	—	X	X	—
Kim	—	X	—	—	—	X
Lois	X	—	X	—	—	—
Dot	X	—	—	—	X	—

Each club at Central High wants to meet once a week. Since several students hold offices in more than one organization, it is necessary to arrange the meeting days so that no students are scheduled for more than one meeting on the same day. Is it possible to create such a schedule? What is the minimum number of days needed?

One possible solution to the problem is to use five days for the scheduling. The Math and Spanish Clubs could meet on Monday, and the remaining clubs could meet on the other four days. If the problem is to schedule the meetings in the fewest number of days, then the solution of using five days is not optimal. It is possible to create a schedule using only three days.

Finding such a schedule by trial and error is not difficult in this case, but a mathematical model would be helpful for more complicated problems. The first

Graphs can be of help when scheduling high school activities.

step in creating such a model might be to construct a graph in which the vertices represent clubs at Central High and the edges indicate conflict. If an edge joins two vertices, then those two clubs share an officer and cannot meet on the same night.

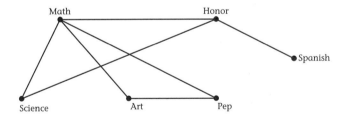

Figure 4.14 Graph showing the organizations of Central High that share an officer.

After completing the graph (see Figure 4.14), label the vertices with days of the week. When doing so, keep in mind that adjacent vertices must have different labels, since this is where the conflicts occur. One way of assigning days is to begin with the Math Club and label it with Monday. Look at the vertices not adjacent to the Math Club vertex. Since no one belongs to both the Math Club and the Spanish Club, also label the Spanish Club with Monday. Label the Honor Club with Tuesday. The Pep or Art

Club, but not both, can receive a Tuesday label. The one not labeled Tuesday is then labeled Wednesday, the third day. The resulting schedule is an optimal solution to the problem, but notice that it is not unique.

Problems of this type are called *coloring* problems because historically the labels placed on the vertices of the graphs were referred to as *colors*. The process of labeling the graph is called **coloring the graph**, and the minimum number of labels, or colors, that can be used is known as the **chromatic number** of the graph. The chromatic number for the graph in Figure 4.14 is 3.

Questions of this type first attracted interest in the nineteenth century when mathematicians such as Augustus de Morgan, William Rowan Hamilton, and Arthur Cayley became interested in a problem known as the four-color conjecture. The problem stated that any map that could be drawn on the surface of a sphere could be colored with, at most, four colors.

For over 100 years, this problem intrigued mathematicians. During that time, many claimed to have proved the conjecture, but flaws were always found in the proofs. It wasn't until 1976 that Kenneth Appel and Wolfgang Haken of the University of Illinois solved the famous problem, and the four-color conjecture became known as the four-color theorem.

The problem was proved in a way very different from earlier attempts. Appel and Haken used a high-speed computer in their verification. When the proof was finally complete, they had used over 1,200 hours of computer time and had examined approximately 1,936 basic forms of maps. Because of the unusual method of proof and the inability to emulate it by hand, criticism arose from many in the mathematical community. A recent simplification of the four-color theorem proof, by Neil Robertson, Daniel Sanders, Paul Seymour, and Robin Thomas, has removed the cloud of doubt hanging over the complex original proof of Appel and Haken.

Are you wondering what scheduling classes and coloring maps have in common? As it turns out, they are both problems that can be solved with the help of graph coloring techniques. One way to approach the problem of coloring a map is to represent each region of the map with a

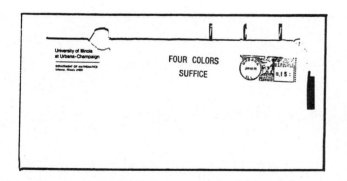

Postage meter stamp used by the University of Illinois to commemorate the proof of the four-color theorem.

vertex of a graph. Two vertices are then connected by an edge if the regions they represent have a common border. The process of coloring the graph resulting from a map is the same as when you colored the scheduling graph model.

Example

Color the following map using four or fewer colors.

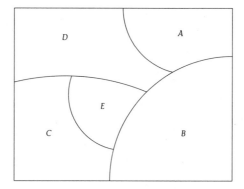

To find a solution, represent the map with a graph in which each vertex represents a region of the map, and draw edges between vertices if the regions on the map have a common border. Then label the graph using a minimum number of colors.

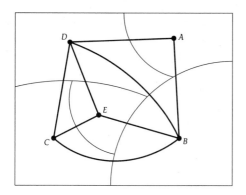 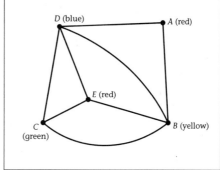

Four colors are necessary to color this map.

Exercises

1. Find the chromatic number for each of the following graphs.

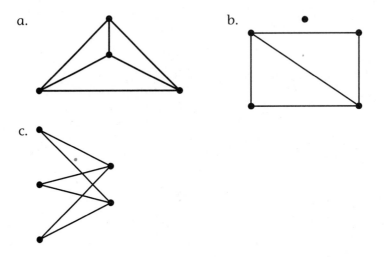

a.

b.

c.

2. a. Draw a graph that has four vertices and a chromatic number of 3.
 b. Draw a graph that has four vertices and a chromatic number of 1.

3. As the number of vertices in a graph increases, a systematic method of labeling (coloring) the vertices becomes necessary. One way to do this is to create a coloring algorithm.
 a. It is possible to begin the coloring process in several different ways, but one way is to color first the vertices with the most conflict. How can the vertices be ranked from those with the most conflict to those with the least?
 b. After having colored the vertex with the most conflict, which other vertices can receive that same color?
 c. Which vertex would then get the second color? Which other vertices could get that same second color?
 d. When would the coloring process be complete?
 e. Refer back to parts a to d of this exercise and create an algorithm that colors a graph.

4. Use the algorithm that you developed in Exercise 3 to color the following graph. What is the chromatic number of the graph? Did your algorithm find the correct chromatic number?

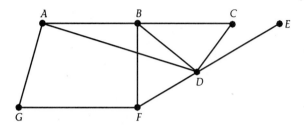

5. a. What is the chromatic number of K_2? K_3? K_4? K_5?
    b. What can you say about the number of colors needed to color a complete graph? Explain your reasoning.
    c. Draw a complete graph with four vertices. Use your algorithm from Exercise 3 to color it. Does your algorithm give the correct chromatic number?

6. A **cycle** is a path that begins and ends at the same vertex and does not use any edge or vertex more than once.
    a. If a cycle has an even number of vertices, what is its chromatic number?
    b. What is the chromatic number of a cycle with an odd number of vertices?
    c. Draw a cycle with six vertices. Use your algorithm to color it. Does your algorithm give the correct chromatic number?

7. If your algorithm failed to give a minimal coloring for one of the graphs you tried, do not be too concerned, because it does not necessarily mean that you have a poor algorithm. Mathematicians continue to search for "good" coloring algorithms, but so far, they have been unable to find one that colors every graph in the fewest number of colors possible. If you've not found a graph that causes your algorithm to fail, try to draw a graph that will do so.

8. Mrs. Suzuki is planning to take her history class to the art museum. Following is a graph showing those students who are not compatible. Assuming that the seating capacity of the cars is not a problem, what is the minimum number of cars necessary to take the students to the museum?

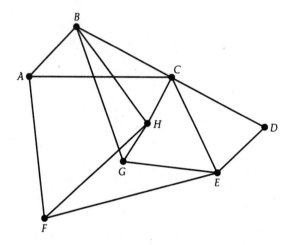

9. Refer to Exercise 1, Lesson 4.3, page 167. What is the minimum number of fish tanks needed to house the fish?

10. Following is a list of chemicals and the chemicals with which each cannot be stored.

Chemicals	Cannot Be Stored with
1	2, 5, 7
2	1, 3, 5
3	2, 4
4	3, 7
5	1, 2, 6, 7
6	5
7	1, 4, 5

How many different storage facilities are necessary in order to keep all seven chemicals?

11. Color the following map using only three colors.

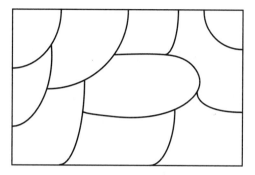

12. Draw graphs to represent the following maps. Color the graphs. What is the minimum number of colors needed to color each map?

a.

b.

Projects
13. Research and report on the mapping industry. Explore questions such as: On the average, how often do maps change? Who are the people/agencies that create maps? How are maps designed and produced? What kind of educational background do cartographers need?

Eulerizing Graphs

Now that you've completed Chapter 4, you have the ability to examine a graph, to determine whether it has an Euler circuit, and to find a circuit if one exists. But, unfortunately, most graphs that represent real-world information do not have Euler circuits. So what does the city street inspector, garbage collector, and utility meter reader do when their street-by-street travels must be made in an optimal manner?

In Lesson 4.4, Exercise 7 (page 180), you solved a problem such as this. The street network graph for the exercise was small, and by careful examination, you were able to find a way to inspect all streets and repeat only a minimal number of them.

This type of problem is often referred to as the Chinese Postman Problem and was first studied by the Chinese mathematician Meigu Guan in 1962. The Chinese Postman Problem differs slightly from the situation that you solved in Lesson 4.4. In the postman problem actual street lengths are examined and attempts are made to minimize the total retraced lengths. In Exercise 7, the assumption was made that all the streets were

of equal length, so minimizing total length was equivalent to minimizing the number of reused edges.

Consider the following representation of a street network.

Since this graph has eight vertices of odd degree, there is no way to begin a circuit at *A*, trace each of the edges, and return to *A* without retracing some of the edges. One way to find a circuit that allows for the necessary retracing is to find the degree for each of the vertices and eliminate all vertices of odd degree by connecting them with additional edges. This process is called **eulerizing the graph**. Once the graph is eulerized, an Euler circuit can be found, and the duplicate edges can then be viewed as streets that must be traveled more than once.

What is the "best" way to add the edges in the eulerization process? When adding edges, it is desirable to add ones that duplicate the fewest number of edges in the original graph. If an edge is added that spans more than one existing edge, then possibly a better eulerization can be found. Since only duplicates of edges in the original graph can be added, be careful not to add "new edges." See the following graphs.

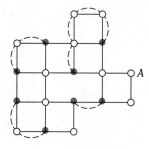

One possible eulerization with nine duplicate edges.

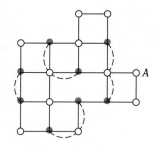

A better eulerization with seven duplicate edges.

As you might have guessed, this problem becomes much more difficult as the size of the graph and the number of vertices with odd degree increase. It can be proved that if a graph has n vertices of odd degree, any circuit in the graph that covers every edge at least once must have at least $n/2$ duplications. Be aware that this theorem gives only the lower bound and not the exact number of duplications. For example, in the preceding graph there are eight odd vertices, so at the very least there must be $8/2 = 4$ duplications. In the preceding figure an eulerization was found that has seven retraced edges. The theorem says that it might be possible to do better. Can you find a better eulerization?

If you are interested in knowing more about eulerizing graphs, additional information can be obtained by doing an Internet search or referring to the bibliography at the end of this chapter.

1. Write a summary of what you think are the important points of this chapter.

2. Draw a graph for the following task table.

Task	Time	Prerequisites
Start	0	—
A	2	None
B	4	A
C	4	A
D	3	B
E	2	C
F	1	C
G	2	D
H	5	D,E,F
I	3	G,H
Finish		

3. Complete the task table for the following graph.

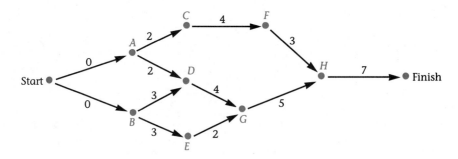

Task	Time	Prerequisites
Start	0	—
A		
B		
C		
D		
E		
F		
G		
H		
Finish		

4. a. List the vertices of the following graph and give their earliest-start time.

b. Determine the minimum project time.

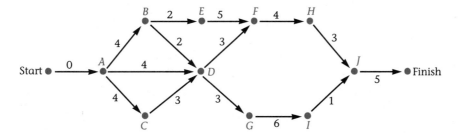

5. Use your graph from Exercise 2.

a. Recopy it and label each vertex with its EST.

b. Determine the critical path and the minimum project time.

6.

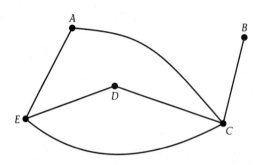

    a. Is this graph connected? Explain why or why not.
    b. Is this graph complete? Explain why or why not.
    c. Name two vertices that are adjacent to vertex E.
    d. Name a path from B to E of length 3.
    e. What is the degree of vertex C?
    f. Determine an adjacency matrix for the graph.

7. Tell whether the following graphs have an Euler circuit, an Euler path, or neither. Explain your answers.

a.

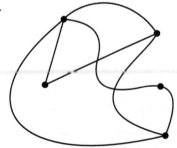

b.

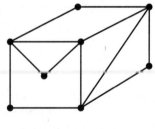

8. Construct a graph for each of the following.
    a. $V = \{A, B, C, D, E\}$
       $E = \{AE, AB, CD, BC, DE\}$

b.

	A	B	C	D
A	0	0	1	1
B	0	0	1	1
C	1	1	0	1
D	1	1	1	0

9. Following is a multigraph that represents the downtown area of a small city. The local post office has decided that the mail drop boxes, which are located at the intersection of each street, must be monitored twice daily.

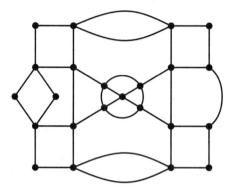

   a. Is it possible to find a circuit that begins and ends at the same intersection and visits each drop box exactly once?

   b. If not, is there a path that begins at one drop box, visits each drop box exactly once, and ends at a different drop box?

   c. If either route exists, copy the figure onto your paper and darken the edges of your proposed route.

10. Use the graph in Exercise 9.

   a. Is it possible for the local street inspector to begin at an intersection and inspect each street exactly once?

   b. Is it possible for the inspector to finish her route at the same intersection from which she began? Explain why or why not.

11. Consider the following set of preference schedules.

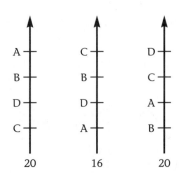

a. Represent this election with a cumulative preference tournament.
b. Is there a Condorcet winner? Explain why or why not.
c. Find several Hamiltonian paths for your graph.
d. Show how to use a Hamiltonian path to construct a pairwise voting scheme (see Exercise 3, Lesson 1.3, page 20) that results in B's winning the election.

12. In scheduling the final exam for summer school at Central High, six different tests must be scheduled. The following table shows the exams that are needed for seven different students.

	Students						
Exam	1	2	3	4	5	6	7
(M) Math	X	—	X	—	X	—	X
(A) Art	—	X	—	X	—	X	—
(S) Science	X	X	—	—	—	—	X
(H) History	—	—	X	—	—	X	—
(F) French	—	—	—	X	X	—	—
(R) Reading	X	X	—	X	X	—	X

a. Draw a graph that illustrates which exams have students in common with other exams.
b. What is the minimum number of time slots needed to schedule the six exams?

13. The Federal Communications Commission (FCC) is in charge of assigning frequencies to radio stations so that broadcasts from one station do not interfere with broadcasts from other stations. Suppose the FCC needs to assign frequencies to eight stations. The following table shows which stations cannot share frequencies.

Station	Cannot Share with
A	B, F, H
B	A, C, F, H
C	B, D, G
D	C, E, G
E	D
F	A, B, H
G	C, D
H	A, B, F

a. Represent this situation with a graph.

b. Find the minimum number of frequencies needed by the FCC.

14. Consider the following digraph.

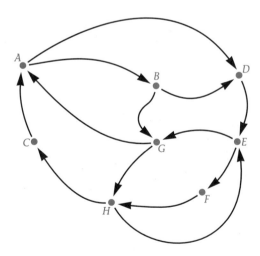

a. Does it have a directed Euler circuit? Explain why or why not. If it does, list one.

b. Does it have a directed Euler path? Explain why or why not. If it does, list one.

15. a. Represent the following map with a graph.
   b. Color your graph.
   c. What is the minimum number of colors needed to color the map?

Bibliography

Biggs, N. L., E. K. Lloyd, and R. J. Wilson. 1976. *Graph Theory 1736–1936*. Oxford: Clarendon.

Busch, D. 1991. *The New Critical Path Method*. Chicago: Probus.

Chavey, Darrah. 1992. *Drawing Pictures with One Line*. (HistoMAP Module #21). Lexington, MA: COMAP.

COMAP. 1997. *For All Practical Purposes: Introduction to Contemporary Mathematics*. 4th ed. New York: W. H. Freeman.

Copes, W., C. Sloyer, R. Stark, and W. Sacco. 1987. *Graph Theory: Euler's Rich Legacy*. Providence, RI: Janson.

Cozzens, Margaret B., and R. Porter. 1987. *Mathematics and Its Applications*. Lexington, MA: Heath.

Cozzens, Margaret B., and R. Porter. 1987. *Problem Solving Using Graphs*. (HiMAP Module #6). Lexington, MA: COMAP.

Dossey, John, A. Otto, L. Spence, and C. Vanden Eynden. 1996. *Discrete Mathematics*. 3rd ed. New York: Harper Collins.

Francis, Richard L. 1989. *The Mathematicians' Coloring Book*. (HiMAP Module #10). Lexington, MA: COMAP.

Kenney, Margaret J., ed. 1991. "Discrete Mathematics across the Curriculum, K-12." *1991 Yearbook of the National Council of Teachers of Mathematics*. Reston, VA: NCTM.

Malkevitch, J., and W. Meyer. 1974. *Graphs, Models, and Finite Mathematics*. Englewood Cliffs, NJ: Prentice-Hall.

Tannenbaum, P., and R. Arnold. 1995. *Excursions in Modern Mathematics*. 2nd ed. Englewood Cliffs, NJ: Prentice-Hall.

More Graphs, Subgraphs, and Trees

Because of modern technology, elaborate communication networks span the country and most of the earth. These networks allow instant transmission of information between almost any two locations. They affect many aspects of our lives, including the way we work, the way we learn, and the way in which we are entertained.

How does one construct a communication network that links several locations together at the lowest possible cost? How is the most efficient route between two locations in a network found? Can the methods used to find the best route between points in a communication network also be used to plan the best route for an automobile or plane trip? The mathematics of graph theory plays an important role in solving these and many other problems that are important in our ever-changing world.

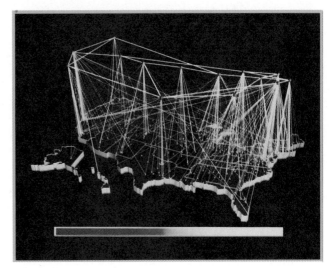

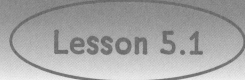
Planarity and Coloring

In Lesson 4.6, problems involving conflict were solved by modeling them with graphs and then coloring the graphs. The four-color theorem states that any map that can be drawn on the surface of a sphere can be colored with four colors or fewer. If this is true, then why does it take more than four colors to color some graphs?

Explore This

Try to redraw the graphs in Figures 5.1 and 5.2 so that their edges intersect only at the vertices. Try to think of the edges of the graph as rubber bands.

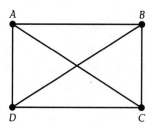

Figure 5.1 K_4 graph.

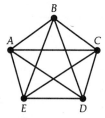

Figure 5.2 K_5 graph.

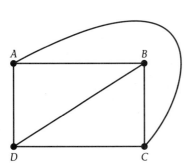

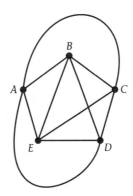

Figure 5.3 Graph in Figure 5.1 redrawn with edges not intersecting.

Figure 5.4 An attempt to redraw Figure 5.2 with edges not intersecting.

It is relatively easy to redraw Figure 5.1 so that the edges do not cross (see Figure 5.3), but no matter how hard you try, at least two edges of Figure 5.2 will always intersect (see Figure 5.4). A graph that can be drawn so that no two edges intersect except at a vertex is called a **planar graph**. Figure 5.1 shows a planar graph, and Figure 5.2 shows a graph that is not planar.

When regions of a map are represented by vertices of a graph and edges are drawn between vertices if boundaries exist between regions, the resulting graph is planar. In other words, when a map in a plane or on a sphere is represented by a graph, the resulting graph is always planar. Hence, the four-color theorem can be stated in a different way:

> Every planar graph has a chromatic number that is less than or equal to four.

The question asked earlier about why some graphs require more than four colors can now be answered. Planarity is the key. If a graph is not planar, we do not know how many colors it will take to color it.

One type of graph that is not planar, a K_5, is shown in Figure 5.2. Another nonplanar graph about which many problems have been written is shown in Figure 5.5. Try to redraw it without the edges crossing. Once again you will discover that this is not possible.

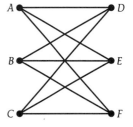

Figure 5.5 $K_{3,3}$ graph.

Bipartite Graphs

The graph in Figure 5.5 has interesting characteristics other than the fact that it is not planar. It is one example of a group of graphs known as **bipartite** graphs. A graph is bipartite if its vertices can be divided into two distinct sets so that each edge of the graph has one vertex in each set. A bipartite graph is said to be complete if it contains all possible edges between the pairs of vertices in the two distinct sets. Complete bipartite graphs can be denoted by $K_{m,n}$, where m and n are the number of vertices in the two distinct sets. Hence, Figure 5.5 is a $K_{3,3}$ graph.

Figure 5.6 is an example of a complete bipartite graph $K_{3,2}$, since its vertices can be separated into two distinct sets $\{A, B, C\}$ and $\{X, Y\}$, every edge has one vertex in each set, and all possible edges from one set to the other are drawn.

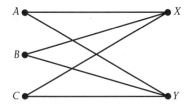

Figure 5.6 $K_{3,2}$ graph.

One way to determine whether a given graph is planar is to try to redraw the graph without edges crossing. For a very large graph, this could prove to be both difficult and time-consuming. In 1930 Kazimierz Kuratowski, a Polish mathematician, provided a partial resolution to this problem of determining the planarity of a graph. He proved that if a graph has a K_5 or $K_{3,3}$ subgraph, it is not planar. A graph G' is said to be a **subgraph** of graph G if all of the vertices and edges of G' are contained in G.

In addition to proving that graphs with K_5 or $K_{3,3}$ subgraphs are not planar, Kuratowski proved that the inverse of his theorem is not true. That is, the lack of a K_5 or $K_{3,3}$ subgraph does not guarantee that the graph is planar (see Exercises 22 and 23 on page 221).

Point of Interest

In practice, approximately 99% of all nonplanar graphs of modest size can be shown to be nonplanar because of a $K_{3,3}$ subgraph rather than a K_5 subgraph.

Example

Determine whether the following graph is planar or nonplanar.

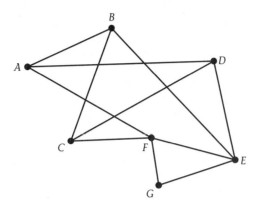

On close inspection of the graph and vertices, A, B, C, D, E, and F, a $K_{3,3}$ subgraph can be found. Therefore, according to Kuratowski's theorem, the graph is nonplanar.

Exercises

In Exercises 1 through 3, decide whether the graph is planar or nonplanar. If the graph is planar, redraw it without edge crossings.

1. 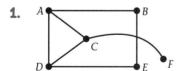   **2.**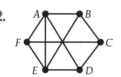

3.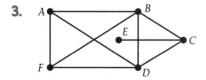

4. The following graph is planar. Draw it without edge crossings.

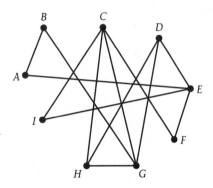

5. By looking at the graph in Exercise 4, how can you tell that it does not contain a K_5 subgraph?

6. Devise a systematic method of searching a graph for a K_5 subgraph. Describe your method in a short paragraph and try it on the following graph. Does the graph contain a K_5 subgraph?

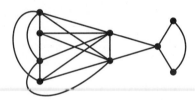

7. The **complement** of a graph G is customarily denoted by $\overline{G}$. The complement $\overline{G}$ has the same vertices as G, but its edges are those not in G. The edges of G and $\overline{G}$ along with vertices from either set would make a complete graph. Draw the complement of the following graph.

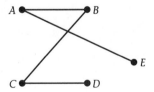

8. Every planar graph with nine vertices has a nonplanar complement. Verify this statement for one case by drawing a planar graph with nine vertices and then drawing its complement. For your case, is the complement nonplanar?

9. The concept of planarity is extremely important to printing circuit boards for the electronics industry. Explain why.

10. Construct the following bipartite graphs.
    a. $K_{2,3}$ b. $K_{2,4}$

11. For each of the following bipartite graphs, list the two distinct sets into which the vertices can be divided.

    a.    b.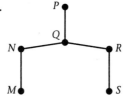

    c.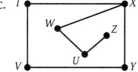

12. State whether the following graphs are bipartite. Explain why or why not.

    a.    b.   c.

13. Devise a method of telling whether a graph is bipartite. Write a short algorithm for your method.

14. How many edges are in a $K_{2,3}$ graph? A $K_{4,3}$? A $K_{m,n}$?

15. When does a bipartite graph $K_{m,n}$ have an Euler circuit?

16. What is the chromatic number of a $K_{m,n}$ graph?

17. At Ms. Johnson's party, six men and five women walk into the dining room. If each man shakes hands with each woman, how many handshakes will occur? Represent this situation with a graph. What kind of a graph is it?

18. Describe a situation that can be represented by a bipartite graph that is not complete.

19. The following puzzle is often referred to as the *Wells and Houses problem* or the *Utilities problem*.

   Three houses and three wells are built on a piece of land in an arid country. Because it seldom rains, the wells often run dry, and so each house must have access to each well. Unfortunately, the occupants of the three houses dislike one another and want to construct paths to the wells so that no two paths cross.

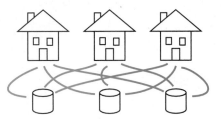

   Draw a graph to illustrate this problem. Is it possible to satisfy the wishes of the feuding families? Explain why or why not.

20.

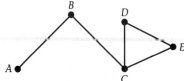

   The preceding graph is a subgraph of which following graph(s). Explain how you know.

a.
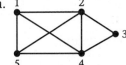

b.

c.

d.
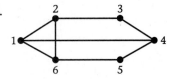

21. Kuratowski proved the following conditional statement.

> If a graph contains a K_5 or a $K_{3,3}$ subgraph, then it is not planar.

He also proved that the inverse of the statement is false:

> If a graph does not contain a K_5 or a $K_{3,3}$ subgraph, then it is planar.

State the converse of the conditional statement. Do you think that it is true or false?

22. Is the following graph planar? Does it contain a K_5 or a $K_{3,3}$ as a subgraph?

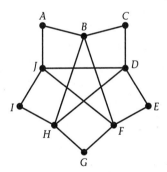

23. For many years mathematicians thought that all nonplanar graphs had either a K_5 or a $K_{3,3}$ as a subgraph. Then nonplanar graphs such as the one in Exercise 22 were found. The graph in Exercise 22 is said to be an **extension** of a K_5 because it was formed by adding a vertex or vertices to the edges of the K_5 graph. Extensions of K_5 and $K_{3,3}$ graphs are nonplanar. This discovery shows that the converse of Kuratowski's theorem is false (Exercise 21).

Redraw the graph in Exercise 22 to show that it is an extension of a K_5.

The Traveling Salesperson Problem

In Lesson 4.5, you explored circuits that visited each vertex of your graph exactly once (Hamiltonian circuits). In this lesson, you will extend your thinking on this original problem and examine a type of problem known as a **traveling salesperson problem** (TSP). These TSPs involve finding a Hamiltonian circuit of minimum value such as time, distance, or cost. Optimization problems of this type are becoming increasingly important in the world of communications, warehousing, airlines networking, road networking, and building wiring.

Explore This

Suppose you are a salesperson who lives in St. Louis. Once a week you have to travel to Minneapolis, Chicago, and New Orleans and then return home to St. Louis. The graph in Figure 5.7 represents the trips that are available to you. The edges of the graph are labeled with the cost of each possible trip. For instance, the cost of making a trip from Chicago to New Orleans is $910. When each edge of a graph is assigned a number (weight), the graph is called a **weighted graph**. The numbers associated with the edges often represent such things as distance, cost, or time.

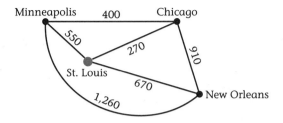

Figure 5.7 Graph of four cities with the costs of traveling between them.

Since you own your own business, it is important that you minimize travel costs. To help save money, find the least expensive route that begins in St. Louis, visits each of the other cities exactly once, and returns to St. Louis.

One way to solve the problem of finding the least expensive route in Figure 5.7 is to list every possible circuit, along with its cost. A tree diagram like the one in Figure 5.8 is helpful in organizing the possibilities. Inspection of all possible routes shows that the optimal solution is the circuit *S, M, C, N, S* or the circuit in reverse order *S, N, C, M, S.*

Solving the TSP for four vertices is not too difficult or time-consuming because only six possibilities need to be considered, but as the number of vertices increases, so does the number of possible circuits. Hence, checking

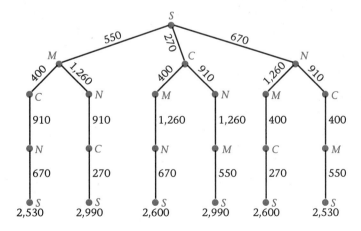

Figure 5.8 Tree diagram of every possible circuit from St. Louis to each of the other cities and back to St. Louis.

all possibilities soon becomes impractical, if not impossible. Even with the help of a computer that can do computations at the rate of 1 billion per second, it would take more than 19 million years for the computer the find the weights of every circuit for a graph with 25 vertices!

Using Computers and Theories of Evolution to Solve Problems

INDIANAPOLIS STAR
AND NEWS
by Eric B. Schoch

Indianapolis— When nature has a problem to solve, it uses natural selection, genetics, mutations and other tools of evolution to give us everything from Mozart to microbes.

The International Conference on Evolutionary Computations brought together about 220 researchers—more than half of them from outside North America—to discuss software that uses the rules of biology to work more intelligently.

Consider the traveling salesman problem. Even for a computer, this trial-and-error system could take a long time.

The alternative is to turn the problem over to computerized evolution: Let the computer create a bunch of possible routes at random, then rank them. Have the computer reproduce the routes, giving the better ones more chances of reproducing—that's survival of the fittest. During the reproduction process, sometimes have the routes swap pieces, the way chromosomes sometimes swap sections when cells divide. Make some random tiny changes in some routes, and call them mutations.

Over time, as generation follows generation, a good answer to the route problem will evolve.

More Efficient Algorithms

Is it possible that a faster, more efficient algorithm exists? You might try, for example, to begin at a vertex, look for the vertex nearest to your starting vertex, move to it, and continue until you complete the circuit.

To explore this method for the graph in Figure 5.7 on page 223, begin at St. Louis, move to the nearest neighboring vertex (Chicago), then to the nearest vertex not yet visited, and return to St. Louis when you have visited all of the other cities (270 + 400 + 1,260 + 670 = 2,600). This procedure is known as the **nearest-neighbor algorithm**. Although the solution was reached quickly, notice that in this case, it is not the best possible one. A method such as this, which produces a quick and reasonably close-to-optimal solution, is known as a **heuristic method**.

The choice of method now becomes a trade-off. The method of inspecting every possible path guarantees the best route but is prohibitively slow. The nearest-neighbor method is quick but does not necessarily produce the optimal solution. There is no known computationally efficient method of solving all traveling salesperson problems. But with the discovery of more and more efficient algorithms, hope has increased that better solutions will be forthcoming. These solutions to TSPs are of great interest because they translate into savings of millions of dollars for certain areas of the economy.

Point of Interest

The current record-setting solution for a TSP was set in May 1998 by Robert Bixby and David Applegate of Rice University and their colleagues. This 13,509-node problem consisted of all cities in the United States with populations ≥ 500 and is one of the problems from a library of TSP problems set up by Gerhard Reinelt, a professor at the University of Heidelberg, to provide a standard set of test problems. According to Applegate, "the story isn't over—there is still work to be done, and there are a few smaller problems which are still unsolved."

Exercises

In Exercises 1 through 4,
a. Construct a tree diagram showing all possible circuits that begin at S, visit each vertex of the graph exactly once, and end at S.
b. Find the total weight of each route.
c. Identify the shortest circuit.
d. Use the nearest-neighbor algorithm to find the shortest circuit.
e. Does the nearest-neighbor algorithm produce the optimal solution?

1.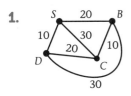

2.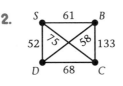

3.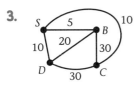

4.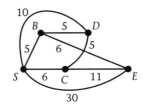

5. In a graph with 10 vertices, 9!, or $9 \cdot 8 \cdot 7 \cdot 6 \cdot 5 \cdot 4 \cdot 3 \cdot 2 \cdot 1$, possible Hamiltonian circuits exist if the beginning vertex is known.
   a. Assume that a computer can perform calculations at the rate of 1 million per second. About how long will it take the computer to check 9! possibilities? What if the graph had 15 vertices (14! possible circuits)?
   b. According to October 29, 1998 news reports, a computer now exists that can do 1 trillion computations per second! How long will it take this new computer to check a graph with 9! possible circuits? With 14! possible circuits? With 20! possible circuits?

6. Give two examples of a situation in which a solution to the traveling salesperson problem would be beneficial.

Point of Interest

Mathematicians in IBM's Tokyo Research Laboratory have developed a drill route optimization system that has a TSP approximation algorithm as its core. The new system reduces the drill route length on a circuit board by an average of 80% and the operation time by 15% on the average.

7. The following figure shows a circuit board and the distances in millimeters between holes that must be drilled by a drilling machine. Since it is advantageous in terms of time to minimize the distance traveled, find the shortest possible circuit for the machine to travel and the total distance for that circuit. (Assume the machine has to begin and end at point S.)

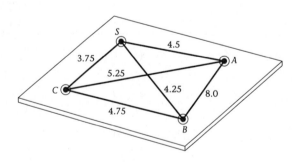

Projects

8. Create a TSP poster that illustrates the use of the nearest-neighbor algorithm. At a minimum, your poster should include the following elements.

   • A map of your neighborhood, town, or state with five or six selected locations highlighted.

   • A table that shows the distances between each pair of selected locations.

   • A weighted graph that models the situation.

   • An explanation of how the algorithm is used to find the minimum distance from one selected starting location to each of the others and returns to the original starting point.

9. Select five colleges or universities that you might be interested in attending. After researching the schools of choice, write a paper that incorporates the following.

   • A short paragraph on each school explaining your interest.

   • A weighted, complete graph that shows the five schools and your home.

   • An optimal route based on the shortest path that starts at your home, visits each of the five schools, and returns to your home.

Car of Future: Virtual Design and Intelligent Navigation

CHRISTIAN SCIENCE
MONITOR,
August 23, 1996,
by Paul A. Eisenstein

The sleek two-seater merges into the dense freeway traffic, nudging into the express lane.

Settling into the flow of Monday morning rush hour, the driver slips her hands off the steering wheel, pulls out the morning paper and settles back to read as her car races along at 100 miles per hour.

A scenario for disaster? Certainly on today's highways. But by the mid-21st century, this may be a perfectly common sight. Or so goes the vision of highway and automotive planners in the United States.

The federal government has authorized a national smart-car program that could be in place within a decade. It would create an intelligent highway system in as many as 75 major US cities. They're likely to look a lot like ADVANCE. Vehicles participating in the program are equipped with on-board navigation computers. Punch in a destination, and the system automatically plots out the best route. If there's a tie-up along the way, the regional center radios an alert, and the car's computer plots an alternate route.

ONE-SEAT WONDER: Honda Motor Company developed this prototype single-seat car for commuting. The hybrid-engine ICVS-1 was designed as part of a government-aided 'intelligent community vehicle system' in Japan.

Finding the Shortest Route

The traveling salesperson problem asks that a Hamiltonian circuit of least total weight be found for a graph. What if you didn't need to visit every vertex in the graph and return back to the starting point, but instead, you needed only to find the shortest path from one vertex in the graph to another? Does an efficient method of solving this type of problem exist? The answer is yes, and the algorithm used in finding the shortest path from a given vertex of a graph to any other vertex in that graph is attributed to E. W. Dijkstra.

Mathematician of Note

E. W. Dijkstra, born in the Netherlands in 1930, is considered one of the original theorists of modern computer science. He first published his shortest path algorithm in a German mathematics journal in 1959, and since that time, he has received many honors and awards. Dijkstra currently serves as professor of mathematics at the University of Texas, Austin.

The following algorithm is a modification of Dijkstra's algorithm.

Shortest Path Algorithm

1. Label the starting vertex S and circle it. Examine all edges that have S as an endpoint. Darken the edge with the shortest length and circle the vertex at the other endpoint of the darkened edge.

2. Examine all uncircled vertices that are adjacent to the circled vertices in the graph.

3. Using only circled vertices and darkened edges between the vertices that are circled, find the lengths of all paths from S to each vertex being examined. Choose the vertex and the edge that yield the shortest path. Circle this vertex and darken this edge. Ties are broken arbitrarily.

4. Repeat steps 2 and 3 until all vertices are circled. The darkened edges of the graph form the shortest routes from S to every other vertex in the graph.

Example

Use the shortest path algorithm to find the shortest path from A to F in the graph.

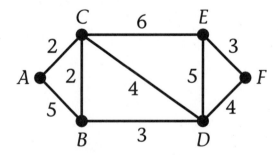

To find the solution to this problem, begin by circling vertex A and labeling it S. Examine all vertices that are adjacent to S.

	Adjacent Vertices	Path from S to Vertex	Length of Path
Adjacent to S	B	SB	5
	C	SC	2

1. Circle C, darken edge SC.

	Adjacent Vertices	Path from S to Vertex	Length of Path
Adjacent to S	B	SB	5
Adjacent to C	B	SCB	4
	E	SCE	8
	D	SCD	6

2. Circle B, darken edge CB.

	Adjacent Vertices	Path from S to Vertex	Length of Path
Adjacent to C	E	SCE	8
	D	SCD	6
Adjacent to B	D	$SCBD$	7

3. Circle D, darken edge CD.

	Adjacent Vertices	Path from S to Vertex	Length of Path
Adjacent to C	E	SCE	8
Adjacent to D	E	$SCDE$	11
	F	$SCDF$	10

4. Circle E, darken edge CE.

	Adjacent Vertices	Path from S to Vertex	Length of Path
Adjacent to E	F	$SCEF$	11
Adjacent to D	F	$SCDF$	10

5. Circle F, darken edge DF.

The shortest route from A to F is A,C,D,F, and the length is 10. The darkened edges also show the shortest routes from A to the other vertices in the graph.

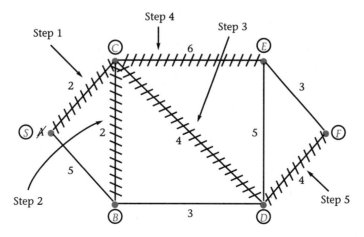

Exercises

1.

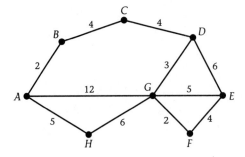

Julian began using the shortest path algorithm to find the shortest route from *A* to *E* for the preceding graph. The work that he was able to complete before he had to stop is shown here.

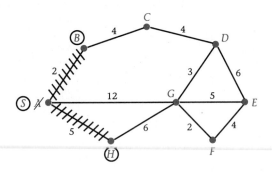

1. *SB* − ②
   SC − 5   Circle *B*, darken *SB*.
   SG − 12
2. *SBC* − 6
   SG − 12   Circle *H*, darken *SH*.
   SH − ⑤
3. *SBC* − ?
   SG − ?   Circle?, darken?
   SHG − ?

Fill in the missing distances, vertex, and edge in step 3. Then complete Julian's problem of using the shortest path algorithm to find the shortest path from *A* to *E*.

2. Use the shortest path algorithm to find the shortest route from *A* to *F*.

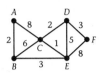

3. When might it not be necessary to repeat the procedure in the algorithm until all of the vertices are circled?

4. Use the shortest path algorithm to determine the shortest distance from *S* to each of the other vertices in the following graph.

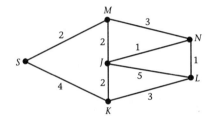

5.

   a. Use the shortest path algorithm to find the shortest route from Albany to Ladue in the preceding graph.
   b. Assume that it is necessary to travel from Albany to Fenton to deliver a package and then to continue from there to Ladue. Find the shortest route for this trip. Explain why the solution to this question might be different than the shortest route from Albany to Ladue.

6. In the shortest path algorithm, each time you examine the uncircled vertices that are adjacent to the circled ones, you have to recalculate the lengths of the paths from the starting vertex. Explain how the efficiency of the algorithm might be improved by modifying it to avoid such recalculation.

7. The shortest path algorithm can be applied to digraphs if slight modifications are made. Make the appropriate changes, and try your revised algorithm on the following digraph to find the shortest route from *A* to *F*.

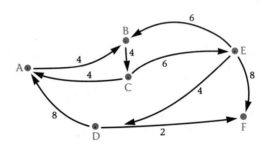

8. Mail Packages, Inc., ships from certain cities in the United States to others. A table of the company's shipping costs follows.

		To						
		Albany	Biloxi	Center	Denver	Evert	Fargo	Gale
	Albany	—	7	—	—	4	—	—
	Biloxi	—	—	—	—	—	—	6
	Center	2	—	—	—	2	—	—
From	Denver	—	—	1	—	—	—	—
	Evert	—	—	—	—	—	—	4
	Fargo	—	—	—	—	3	—	2
	Gale	1	6	—	—	—	1	—

Since a package can't be shipped directly from Denver to Biloxi, construct a digraph to represent the cost table and apply the shortest path algorithm to find the least charge for shipping the package.

Projects

9. Interview several firefighters, ambulance drivers, or paramedics in your community to find out how they determine the shortest route from their facility to an emergency situation. Write a short report on your findings.

Trees and Their Properties

In Lesson 5.2, a special type of graph called a tree was used to organize information and list all possible routes for a traveling salesperson problem.

Tree diagrams have been used since ancient times, but it wasn't until the nineteenth century that their properties were studied in detail. In 1847, Gustav Kirchoff used trees in his study of electrical networks. Ten years later, Arthur Cayley used them in his investigation of certain chemical compounds. Today, trees are one of the most useful structures in discrete mathematics and are invaluable to computer scientists. Many computer programmers depend on trees when creating sorting and searching programs.

Before exploring some of the properties and applications of this type of graph, it is necessary to define a tree. Recall that a **cycle** in a graph is any path that begins and ends at the same vertex and no other vertex is repeated. A **tree** is then defined as a connected graph with no cycles.

Example

Which of the following graphs are trees? Why?

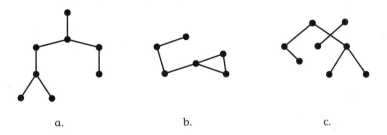

       a.   b.   c.

Figure 5.9 Possible trees.

1. Figure 5.9a is a tree because it is a connected graph with no cycles.

2. Figure 5.9b is not a tree because it has a cycle.

3. Figure 5.9c is not a tree because it is not connected.

Trees have many applications in the real world. They can be used to list and count possibilities, as was done in the traveling salesperson problem and as will be done in Chapter 6. They can also be used to model family genealogical histories (Figure 5.10), to structure decision-making processes (Figure 5.11), and to represent chemical compounds (Figure 5.12).

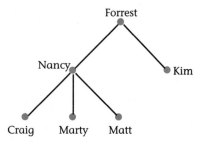

Figure 5.10 Family tree.

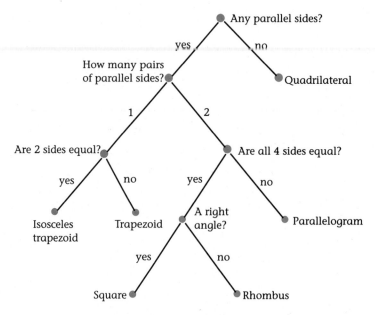

Figure 5.11 Sorting quadrilaterals.

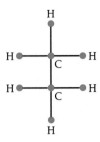

Figure 5.12 C_2H_6 (ethane).

Exercises

1. Examine the following graph for cycles. List as many as you can find.

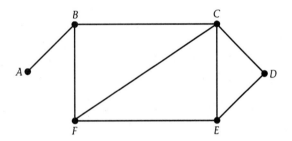

2. Determine whether the following graphs are trees. If the figure is not a tree, explain why.

a.   b.   c.

d.   e.

3. There is only one way to draw a tree with two vertices and only one way to draw a tree with three vertices, but there are two distinct trees that can be formed from four vertices. Draw all of the trees that are possible for five vertices. For six vertices.

Number of Vertices	Tree Diagrams
2	
3	
4	

4. Complete the following table for trees with the indicated numbers of vertices.

Number of Vertices	Number of Edges
1	0
2	1
3	
4	
n	

   a. How many edges does a tree with 19 vertices have?
   b. How many vertices does a tree with 15 edges have?
   c. What is the relationship between the number of vertices of a tree and the number of edges?

5. What happens to a tree if an edge is removed from it?

6. Draw a tree with six vertices that has exactly three vertices of degree 1.

7. Complete the following table for trees with the indicated numbers of vertices.

Number of Vertices	Sum of the Degrees of the Vertices	Recurrence Relation
1	0	$S_1 = 0$
2	2	$S_2 = S_1 + 2$
3	4	$S_3 = $ _____
4	_____	_____
5	_____	_____
6	_____	_____

Write a recurrence relation that expresses the relationship between the sum of the degrees of the vertices of a tree with n vertices and the sum of the degrees of the vertices of a tree with $n - 1$ vertices.

8. Explain why the sum of the degrees of the vertices of a tree with n vertices is equal to twice the number of vertices minus 2. (Hint: Draw several different trees.)

9. Much of the terminology connected with trees is botanical in nature. For instance, a graph that consists of a set of trees is called a **forest**, and a vertex of degree 1 in a tree is called a **leaf**. Draw a forest of three trees. Circle the leaves of your graph.

10. In a hierarchical tree as in Figures 5.10 and 5.11 on page 236, it is natural to speak of the **root** of the tree. A tree is rooted when all of the edges are directed away from the chosen vertex (root). In Figures 5.10 and 5.11, the edges are directed downward. Draw a family tree for your family beginning with one of your grandfathers as the root of the tree. What do the leaves of your tree have in common?

11. Refer to the tree in Figure 5.11 on page 236. This graph is called a **decision tree**, and the leaves represent the final outcomes of the different decisions. Using this tree, what is the name of a quadrilateral with two pairs of parallel sides, four sides equal, and no right angles?

12. For the following graph, find two different subgraphs that are trees.

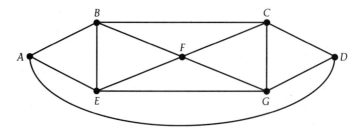

13. Suppose the children in a certain neighborhood want to communicate with one another via their very own communication network. To avoid the expense of connecting each house with every other house, a system needs to be devised that uses as few lines as possible yet allows messages to get to each person. Create such a network for the following houses.

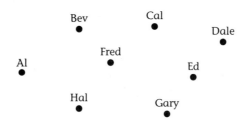

14. Clock solitaire is a card game in which the 52 cards are dealt face down into 13 piles that correspond to the 12 numbers and the center on the face of a clock. In Figure 5.13, clock positions 11 and 12 are identified with the jack and queen, and the thirteenth, or center, pile is identified with the king.

Figure 5.13 Diagram of the 13 piles for the game of clock solitaire.

The game is played by first turning over the top card of the king pile and putting it face up under the pile that corresponds to its value on the clock. Now turn up the top card of the pile under which you just put the card. Continue in this manner. The game is won when you have turned up all 52 cards. If a fourth king is turned up before this happens, play cannot continue, and the game is lost.

In his book *Fundamental Algorithms*, mathematician Donald E. Knuth noted two very interesting things about this game. One is that the probability of winning is 1/13. The other is that by checking the bottom card of the 12 clock piles, you can determine whether you will win the game.

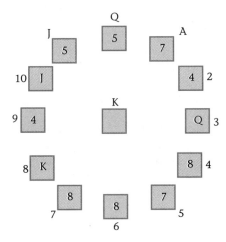

Figure 5.14 One possible configuration of the bottom cards in the game.

To determine whether you can win a game, turn over the bottom card of each pile except for the king pile (see Figure 5.14). Draw an edge from each of the 12 cards to the clock position that corresponds to the card's numeric value. For example, draw an edge from A to 7 (see Figure 5.15). Now redraw the graph with the vertices labeled A, 2, 3, . . . J, Q, and K. You will win the game if the resulting graph is a tree that includes all 13 piles (see Figure 5.16).

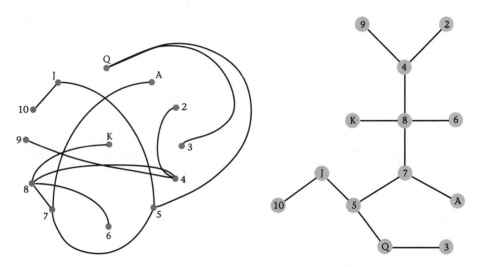

Figure 5.15 Graph showing the edges from the cards to the clock positions.

Figure 5.16 Figure showing that the graph in Figure 5.15 is a tree.

Give the game a try, but before you do, record the bottom 12 cards. Predict whether you will win or lose the game by drawing a graph. Notice that only the bottom 12 cards determine your success. The arrangement of the other 40 cards makes no difference.

Projects

15. Research your family tree and report on your discoveries. Be sure to include a graph (tree) in your report to illustrate your findings.

Minimum Spanning Trees

As in many of the previous lessons, this lesson focuses on optimization. Problems and applications here center on two types of problems: finding ways of connecting the vertices of a graph with the least number of edges and finding ways of connecting them with the least number of edges that have the smallest total weight.

Explore This

In making earthquake preparedness plans, the St. Charles County government needs a design for repairing the county roads in case of an emergency. Figure 5.17 is a map of the towns in the county and the existing major roads between them. Devise a plan that repairs the least number of roads but keeps a route open between each pair of towns.

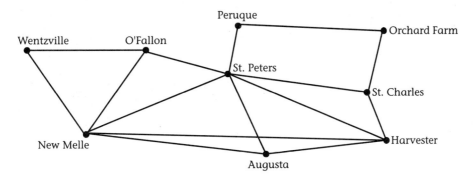

Figure 5.17 The towns in St. Charles County.

Examine your graph. If it connects each of the towns (vertices) and has no cycles, you've found a spanning tree. A **spanning tree** of a connected graph G is a tree that is a subgraph of G and contains every vertex of G. A spanning tree of the graph in Figure 5.17 would represent the minimum number of roads (edges) needed to connect each town in case of an emergency.

Compare your plan with the other plans developed in your class. They should all contain the *same number* of edges but not necessarily the *same* edges. It is possible for a graph to have many different spanning trees. And as you may have guessed, for a graph that is not connected, no spanning tree is possible.

One systematic way to find a spanning tree for a graph is to delete an edge from each cycle in the graph. Unfortunately, this is not an easy procedure for a very large graph. But there are other ways of finding a spanning tree for a graph if one exists. One such method that can be easily adapted to computers is called the *breadth-first search algorithm*.

> ## Breadth-First Search Algorithm for Finding Spanning Trees
>
> 1. Pick a starting vertex, S, and label it with a 0.
>
> 2. Find all vertices that are adjacent to S and label them with a 1.
>
> 3. For each vertex labeled with a 1, find an edge that connects it with the vertex labeled 0. Darken those edges.
>
> 4. Look for unlabeled vertices adjacent to those with the label 1 and label them 2. For each vertex labeled 2, find an edge that connects it with a vertex labeled 1. Darken that edge. If more than one edge exists, choose one arbitrarily.
>
> 5. Continue this process until there are no more unlabeled vertices adjacent to labeled ones. If not all vertices of the graph are labeled, then a spanning tree for the graph does not exist. If all vertices are labeled, the vertices and darkened edges are a spanning tree of the graph.

Example

Use the breadth-first search algorithm to find a spanning tree for the following graph.

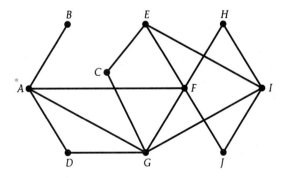

As shown in the following figure, the algorithm begins by picking a starting vertex, calling it *S*, and labeling it with a 0. The labeling and darkening of edges then proceed according to steps 2 to 5 of the algorithm. As you probably noticed, this is not a unique solution. It is just one of the graph's many spanning trees.

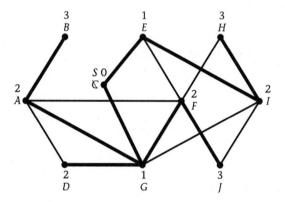

Many applications are best modeled with weighted graphs. When this is the case, it is often not sufficient to find just any spanning tree, but to find one with minimal or maximal weight.

Return to the earthquake preparedness situation and reconsider the problem when distances between towns are added to the graph (see Figure 5.18).

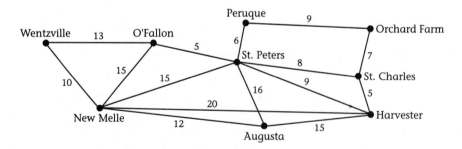

Figure 5.18 Map of St. Charles County with mileage shown.

Refer back to your solution of the original problem and find the total number of miles of road that would need to be repaired if your plan were implemented. Compare your plan with others in your class. Which plan or plans yield the minimum number of miles?

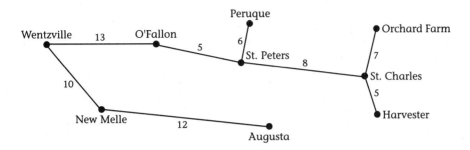

Figure 5.19 Spanning tree of minimum weight for the towns in St. Charles County.

For this particular problem, the minimum possible number of miles of road is 66, and a spanning tree with that total weight is shown in Figure 5.19.

A spanning tree of least, or minimal, weight is called a **minimum spanning tree**. One algorithm for finding a minimum spanning tree for a graph is known as Kruskal's algorithm, named after its designer, Joseph B. Kruskal, a leading mathematician at Bell Laboratories.

Kruskal's Minimum Spanning Tree Algorithm

1. Examine the graph. If it is not connected, there will be no minimum spanning tree.

2. List the edges in order from shortest to longest. Ties are broken arbitrarily.

3. Darken the first edge on the list.

4. Select the next edge on the list. If it does not form a cycle with the darkened edges, darken it.

5. For a graph with n vertices, continue step 4 until $n - 1$ edges of the graph have been darkened. The vertices and the darkened edges are a minimum spanning tree for the graph.

Example

Use Kruskal's algorithm to find a minimum spanning tree for the following graph.

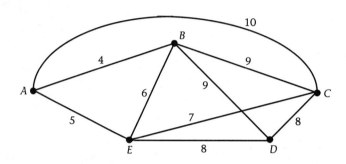

List of Edges from Shortest to Longest	
Edge	Length
AB	4
AE	5
BE	6
EC	7
CD	8
ED	8
BD	9
BC	9
AC	10

There are five vertices in the graph, so four edges must be chosen. List the edges from shortest to longest. First on the list is *AB* (4). Darken it. Then darken *AE* (5). The next shortest edge is *BE*, but if picked, it will form a cycle. So pick *EC* (7). For the last edge there are two edges of length 8. Either *CD* or *ED* can be darkened. The darkened edges of the following graph form one of the minimum spanning trees of the graph. It has a minimal weight of 4 + 5 + 7 + 8 = 24.

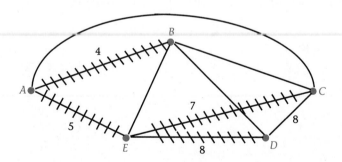

Notice that both Kruskal's and Dijkstra's (Lesson 5.4) algorithms produce spanning trees. But unlike Dijkstra's shortest path algorithm, which gives you a spanning tree of shortest paths, Kruskal's algorithm yields a spanning tree of minimal total weight.

Exercises

In Exercises 1 through 4, find a spanning tree for each graph if one exists.

1. 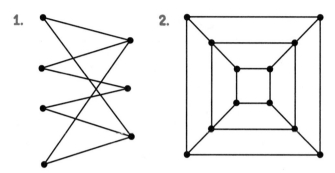 2.

3.

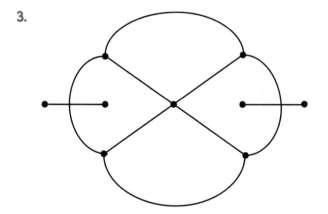

4.

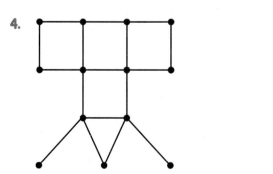

5. Draw a spanning tree for a K_4 graph.

6.

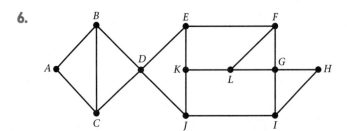

Sid began using the breadth-first search algorithm to try to find a spanning tree for the preceding graph. He began with vertex A, labeled it with a 0, and labeled B and C with 1s. He then darkened edges AB and AC, looked for vertices adjacent to the 1s, and selected vertex D. He labeled it with a 2 and darkened edge BD.

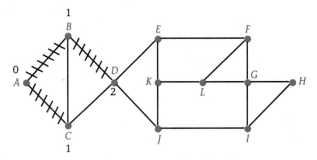

a. Could Sid have darkened CD instead of BD?

Copy Sid's graph and complete the search for Sid by answering the following questions.

b. Which vertices receive 3s for labels? Label these vertices.
c. Which edges subsequently are darkened? Darken these edges.
d. Three vertices should be labeled 4. Which ones? Label these vertices and darken the appropriate edges. Your graph could look like:

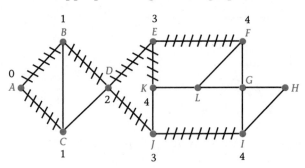

Continue the algorithm until all vertices are labeled. Check your darkened edges to make sure they form a spanning tree.

7.

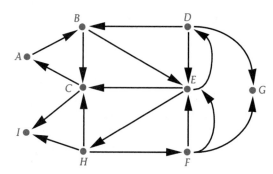

Use the breadth-first search algorithm to find a spanning tree for this graph. Begin at vertex A.

8. Use mathematical induction on the number of edges to prove that every connected graph has a spanning tree.

9. The breadth-first search algorithm can be applied to digraphs if slight changes are made. Modify the algorithm on page 245 so that it can be used with digraphs. Apply your modified breadth-first search algorithm to the following digraph.

Use Kruskal's algorithm to find a minimum spanning tree for the graphs in Exercises 10 through 12. What is the minimal weight in each case?

10.

11.

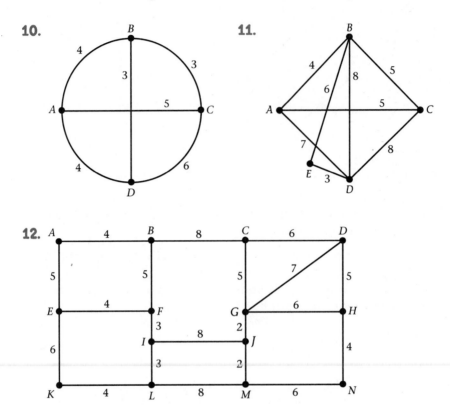

12.

13. The computers in each of the offices at Pattonville High School need to be linked by cable. The following map shows the cost of each link in hundreds of dollars. What is the minimum cost of linking the five offices?

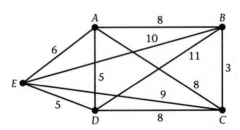

14. Suppose that the cable in Exercise 13 was installed by a disreputable firm that used only the most expensive links. What would be the maximum cost for the four links?

15. How might Kruskal's minimum spanning tree algorithm be modified to make it a maximum spanning tree algorithm?

Another algorithm that can be used to find a minimum spanning tree is attributed to R. C. Prim, a mathematician at the Mathematics Center at Bell Labs.

Prim's Minimum Spanning Tree Algorithm

1. Find the shortest edge of the graph. Darken it and circle its two vertices. Ties are broken arbitrarily.

2. Find the shortest remaining undarkened edge having one circled vertex and one uncircled vertex. Darken this edge and circle its uncircled vertex.

3. Repeat step 2 until all vertices are circled.

16. Use Prim's algorithm to find the minimum spanning tree for Exercise 10 and then for Exercise 11.

17. When the shortest path algorithm from Lesson 5.3 is applied until all vertices of a graph are used, it yields a spanning tree of the graph. Is it always a minimum spanning tree?

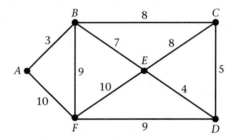

Check your answer to this question by doing the following.

a. Find a minimum spanning tree of the graph above using either Kruskal's or Prim's algorithm. What is the total weight of the minimum spanning tree?

b. Find the shortest route from *A* to each of the other vertices using the shortest path algorithm (page 230). Give the lengths of each of these routes.

c. Is the shortest route tree from *A* to each of the other vertices a minimum spanning tree of this graph? Explain why or why not.

18. Traveling salesperson problems, shortest route problems, and minimum spanning tree problems are often confused because each type of problem can be solved by finding a subgraph that includes all of the vertices of the graph. Compare and contrast what each type of problem asks and when each type of problem is used.

Project

19. In this lesson, you have applied two of the three classical minimum spanning tree algorithms, Kruskal's and Prim's. The third algorithm of this group was designed by O. Borůvka. Investigate Borůvka's algorithm, learn to apply it, and report on how it differs from Kruskal's and Prim's algorithms.

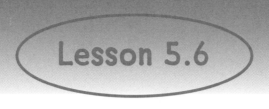
Binary Trees, Expression Trees, and Traversals

Decision trees and family trees are two examples of a special kind of tree known as a **rooted tree**. A rooted tree is a directed tree in which every vertex except the root has an indegree of 1, while the root has an indegree of 0. Figure 5.20 shows an example of a rooted tree in which vertex *R* is the root.

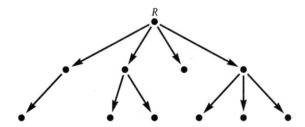

Figure 5.20 Rooted tree.

Since all edges are directed away from the root, it is not necessary to draw the arrowheads on the ends of the edges (see Figure 5.21).

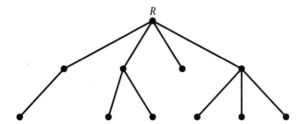

Figure 5.21 Rooted tree without arrowheads.

Rooted trees are used to model situations that are multistaged or hierarchical in structure.

Example

A couple decides to have three children. What are the possible outcomes?

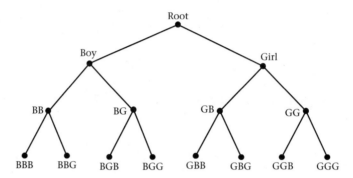

The eight possible outcomes are shown on the vertices of the tree which have outdegrees of 0.

In a rooted tree, a vertex V is said to be at level K if there are K edges on the path from the root to V. The root is at level 0, and the vertices adjacent to the root are at level 1. If a vertex V is at level 4, then any vertex adjacent to V at level 3 is called the **parent** of V, and any adjacent vertex of level 5 is called a **child** of V. A rooted tree in which each vertex has at most two children is called a **binary tree**.

Example

Which trees are binary trees? For those that are binary trees, name the parent of V and the children of V.

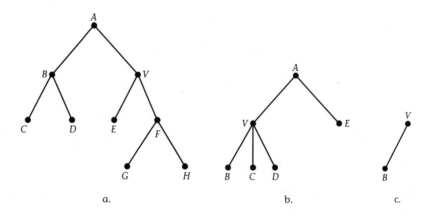

All three of these graphs are trees, but only Figures a and c are binary trees. In a, the parent of *V* is *A* and the children are *E* and *F*. In c, *V* has no parent and *B* is the only child.

In computer science applications, binary trees are used to evaluate arithmetic expressions. When you write the expression (4 + 6) ∗ 8 − 4/2, you understand how to find its value because you are familiar with the order of operations for expressions. Unfortunately, a computer cannot efficiently imitate your methods. However, if an expression is represented as a binary tree, the computer can quickly and efficiently evaluate it.

To represent the expression (4 + 6) ∗ 8 − 4/2 as a binary tree, first find the operation in the expression that is performed last. Make that operation the root of the tree. The right and left sides of this operation become the children of the root (see Figure 5.22).

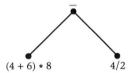

Figure 5.22 First step of representing an expression as a binary tree.

Continue this recursive process of placing operations at each internal vertex (the children) and putting operands on the leaves until no expression that contains operations appears on the leaves (see Figure 5.23).

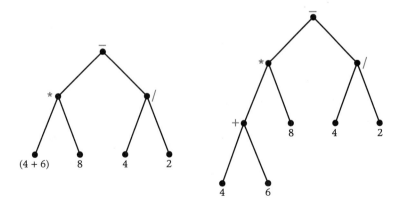

Figure 5.23 Continuing the recursive process.

The final binary tree in Figure 5.23 is called an **expression tree**.

Example

Represent $A/B + C * (D - E)$ as an expression tree.
The solution is:

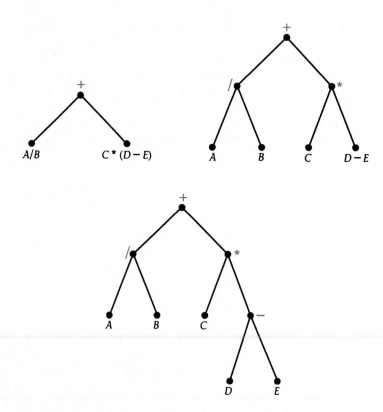

Tree Traversals

Once an expression is represented by a binary tree, the computer must have some systematic way of "looking at" the tree in order to find the value of the original expression. This organized procedure for obtaining information by visiting each vertex of the tree exactly once is called a **traversal** of the graph.

There are many different types of traversals, including one called a **postorder traversal**. This traversal differs from other traversals in that it visits the left child of the tree first, then the right child, and finally the parent or root (see Figure 5.24).

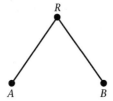

Figure 5.24 Postorder traversal A, B, R.

To find the postorder listing of the vertices of the tree in Figure 5.25, begin by moving to the left subtree of A and doing a postorder traversal on that subtree, which has B as its root. This requires you to branch to the left subtree of B and do a postorder traversal. Since the left subtree of B consists of only the vertex D, visit that vertex by numbering it with a 1. Now go to the right subtree of B and do a postorder traversal. Again, this subtree consists of only one vertex. Visit E and number it with a 2. Since the left and right children of B have been traversed, visit B (the root of that subtree). Number it 3.

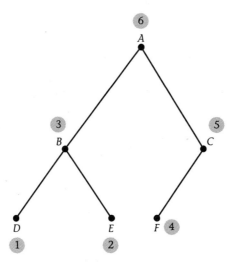

Figure 5.25 Postorder traversal of a binary tree.

You have traversed the left subtree of *A* and must now traverse the right subtree. To do this, you begin with a postorder traversal on the subtree which has *C* as its root. Move to the left subtree of *C*, visit *F*, and number it 4. Since there is no right subtree of *C*, visit the root *C*, and number it 5. Since both the left and right subtrees of root *A* have been visited, *A* can now be visited and numbered with a 6. The postorder traversal is complete and the postorder listing is *DEBFCA*.

Example

Give a postorder listing for the following expression tree.

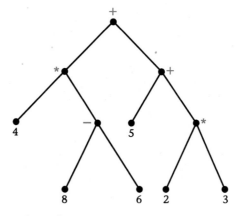

The following is the solution.

Left subtree of *A*   Right subtree of *A*

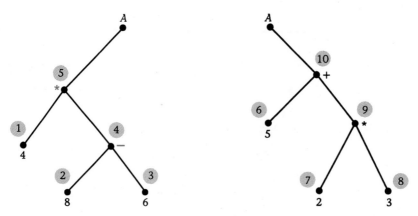

The root is visited last and 11 is assigned to *A*. The postorder listing is $4\ 8\ 6 - *5\ 2\ 3 * + +$.

The notation obtained by doing a postorder traversal is known as **reverse Polish notation (RPN)**. The notation was so named because it was introduced by the Polish mathematician Jan Lukasiewiez. This notation with operations next to each other and no parentheses may look strange to you, but to owners of certain calculators, this notation is familiar and easy to use. RPN works well with calculators and computers because no parentheses are ever needed to indicate the desired order of operations.

So how do you find the value of the expression 4 8 6 − * 5 2 3 * + +? To evaluate RPN, scan the expression from the left until you find two numbers followed by an operation sign, in this case, 8 6 −. This says to you to take the 8 and 6 and subtract. Substitute the result, 2, back in the expression and repeat the process. This continues until you have evaluated the expression.

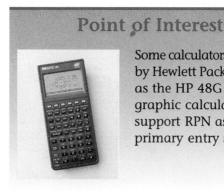

Point of Interest

Some calculators designed by Hewlett Packard, such as the HP 48G series of graphic calculators, support RPN as their primary entry system.

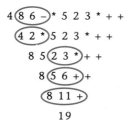

19

People who become accustomed to using this type of notation find it very quick and convenient to use because there is never a question about the order in which to perform the operations.

Exercises

1. Tony wants to buy a car. He has the options of two different brands of radios and four different exterior colors. Draw a tree diagram to show all possible outcomes of choosing a radio and a color for the car.

2. A coin is tossed three times. Draw a tree diagram to show the possible outcomes.

In Exercises 3 through 6, examine each tree. If the tree is a binary tree, (a) give the level of vertex *V*, (b) name the parent of *V*, and (c) name the children of *V*.

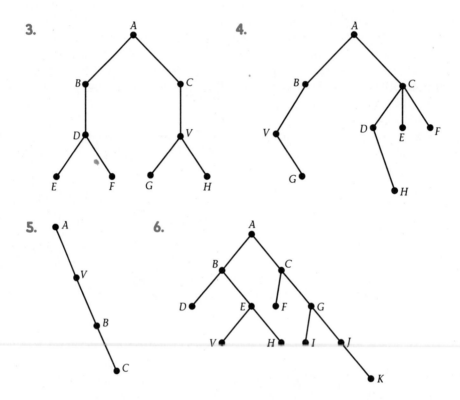

3.

4.

5.

6.

7. Jeff's brother Tom has a first-grade spelling book that contains five chapters. Each odd-numbered chapter has two lessons and each even-numbered chapter has three lessons. The second lesson of each chapter has two questions whereas all others have one. Draw a rooted tree that models Tom's book. How many questions are in the book?

In Exercises 8 through 11, represent the expression as a binary expression tree.

8. $(2 - 5) * (4 + 7)$ **9.** $(2 + 3) * 4$

10. $2 + 3 * 4 - 6/2$ **11.** $A * B + (C - D/E)$

In Exercises 12 through 14, find the postorder listings for each binary tree.

12.

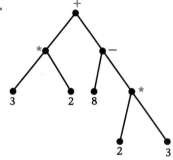

13.

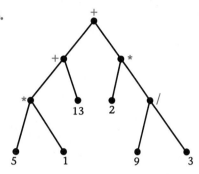

14.

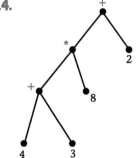

15. Evaluate the following reverse Polish notations.
   a. $6\ 2 - 7 * 3\ 2 + +$ b. $6\ 5\ 4\ 3\ 2 - + / +$
   c. $1\ 2 + 4\ 3 - + 6\ 2 / 2 + +$ d. $4\ 3 + 8\ 2 - + 4 + 3 -$

16. Give the reverse Polish notation for each of the following expressions.
   a. $2 + 3 * 6 - (4 + 1)$ b. $(5 - 3) * 2 + (7 - 6/2)$

17. Construct an expression tree that would have reverse Polish notation

$$A\ B * C\ D + E - +$$

18. Construct a binary tree that has *ABC* as its postorder listing. Is your answer unique? If not, construct an additional tree(s).

19. A traversal that visits first the parent or root of the tree, then the left child, and finally the right child is called a **preorder traversal**.

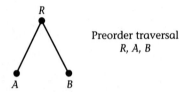

Preorder traversal
R, A, B

The preorder listing for the following binary tree is *ABDECFG*.

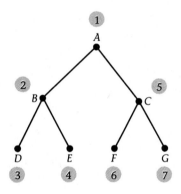

Find the preorder listing for the following binary tree.

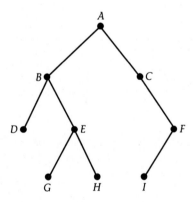

20. Find the preorder listings for the binary trees in Exercises 12 through 14.

21. The notation obtained from a preorder traversal is called **Polish notation**. To evaluate the expression, scan it from the left until you come to an operation followed by two numbers. Perform that operation, place the result back in the expression and continue. For example, * 2 3 is evaluated as 6. Complete the following evaluation.

$$+ + 4 \left(\!* \; 2 3\!\right) + 5 \, / \, 6 \, 3$$
$$+ \left(\!+ \; 4 6\!\right) + 5 \, / \, 6 \, 3$$
$$+ \; 10 + 5 \, / \, 6 \, 3$$

22. Evaluate the following Polish notations.
   a. $+ * 3 \, 2 - 8 * 2 \, 3$ b. $+ \, / \, 6 \, 3 + 4 \, 3$

Project

23. A postorder traversal of an expression tree yields reverse Polish notation, in which the operations follow the operands. A preorder traversal yields Polish notation, in which the operations precede the operands. Create a traversal rule that yields a notation in which the operations are between the operands. In a report, describe your procedure and show how it works on an expression tree. Also find a rule that evaluates your listings. Try your rule on several different expressions. Does it work for all of them? If not, explain why you think it's flawed.

Steiner Trees

Dr. Terry has three computers in her office which need to be networked. If the following graph shows the shortest distance between each computer, what is the minimum amount of cable required?

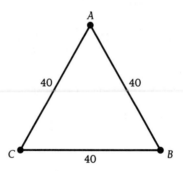

Finding a minimum spanning tree to link the three computers yields a total length of 80 feet of cable, but is there a better way?

The answer to this question is yes, if you do not have to follow the edges in the graph, and creating a junction someplace other than at one of the computers is not a problem. For example, if the cables for these computers can be placed and joined anywhere in the room, you can use less than the 80 feet required for the minimum spanning tree solution (see Figures a and b).

a.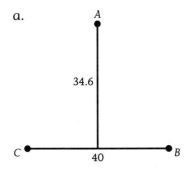

Total length of cables: 74.6 ft.

b.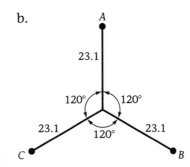

Total length of cables: 69.3 ft.

The tree using the newly created junction point, such as the one shown in Figure b, is known as a **Steiner tree,** and the junction point where the three edges meet at 120° angles is called a **Steiner point.**

A Steiner point inside a triangle can be found using the following procedure, known as the Torricelli procedure.

> Assume that the largest angle of triangle *ABC* is less than 120° and that *BC* is the longest side of the triangle. On the longest side of the triangle *BC* construct an equilateral triangle *BCP*. Circumscribe a circle around triangle *BCP*. Join *P* to *A*. The point where the line segment *PA* intersects the circle is the desired Steiner point.

Creating Steiner trees brings up many questions that can be explored using a drawing utility such as the Geometer's Sketchpad or Cabri:

- If the computers are positioned on the vertices of a triangle that is not equilateral, will there be a Steiner tree?

- In some three-point cases, the minimum spanning tree solution is the best networking solution. When will this happen?

- If there were four computers to network instead of three, how many Steiner points would there be?

- Will there ever be more than one Steiner tree? If so, will they always be equal in length?

Of course, the questions continue to expand as the number of vertices increases. As you may have already discovered, the optimal way of connecting the vertices of a graph (or computers, in this case) is either to find the minimum spanning tree if no interior junctions are allowed or to construct an optimal Steiner tree.

As the number of vertices in a graph increase, the number of Steiner trees increases very rapidly. For example, the number of Steiner trees in a graph with ten vertices totals in the millions. Therefore, finding the Steiner tree with the least weight becomes very complicated. Since their discovery, mathematicians have made significant findings about Steiner trees and have even created efficient algorithms that approximate optimal solutions. But unfortunately, as with the traveling salesperson problem, there is no efficient algorithm that always finds the desired minimum distance.

1. Write a summary of what you think are the important points of this chapter.

2. Show that the graph described by the following adjacency matrix is planar.

$$
\begin{array}{c c}
 & \begin{array}{c c c c c} A & B & C & D & E \end{array} \\
\begin{array}{c} A \\ B \\ C \\ D \\ E \end{array} &
\left[\begin{array}{c c c c c}
0 & 1 & 1 & 1 & 0 \\
1 & 0 & 0 & 1 & 1 \\
1 & 0 & 0 & 1 & 1 \\
1 & 1 & 1 & 0 & 1 \\
0 & 1 & 1 & 1 & 0
\end{array}\right]
\end{array}
$$

3. What is the chromatic number for a tree with five vertices? Any odd number of vertices? Any even number of vertices? Any tree with two or more vertices?

4.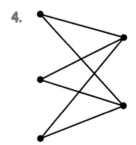

   a. Explain why this graph is a bipartite graph.
   b. Is this graph a complete bipartite graph? Explain why or why not.
   c. Is this graph planar? If so, find a planar drawing for the graph.
   d. What is the chromatic number for this graph?

5. Mr. Gonzalez, the principal at Central High School, leaves his office once an hour to visit the math, science, and social studies classrooms, and then returns to his office. The distances between rooms are shown on the following graph.

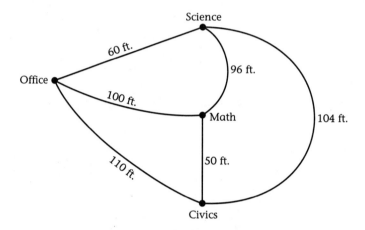

a. Find the shortest route possible for Mr. Gonzalez.
b. What is the total distance of the shortest route in part a?
c. What kind of circuit does Mr. Gonzalez make?

6. Find a spanning tree for the following graph if one exists.

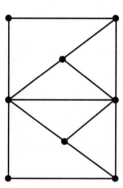

7. How many different spanning trees are there for a cycle with 3 vertices? With 4 vertices? With 5 vertices? With *n* vertices?

8. When given the position where it is currently located and its destination, a certain robot car is programmed to find the shortest path for the trip. The routes that the car can travel are shown on the following graph.

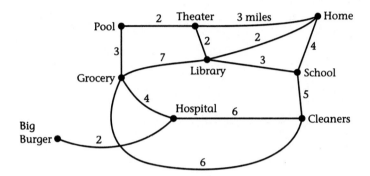

a. Use inspection to find the shortest path from Home to Big Burger.
b. What is the minimum distance from Home to Big Burger?
c. Use the shortest path algorithm (page 230) to find the shortest paths from Home to each of the other locations on the graph.

9. Assume that all locations represented by the graph in Exercise 8 need to be connected by cable. Find the minimum amount of cable needed to link the nine locations.

10. Are the following graphs trees? Explain why or why not.

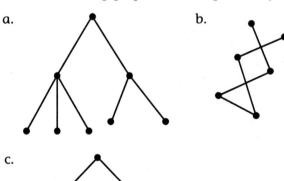

a.

b.

c.

11. Use the breadth-first search algorithm from page 245 to find a spanning tree for the following graph. Begin the algorithm at the vertex labeled S.

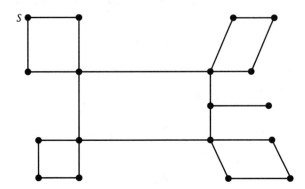

12. Draw a tree with eight vertices that has exactly four vertices of degree 1.

13. The vertices of the following graph represent buildings on a small college campus. Administrators at the campus want to connect the buildings with fiber-optic cable and are interested in finding the least expensive way of doing so. The costs of connecting buildings (in thousands of dollars) are shown as weighted edges of the graph.

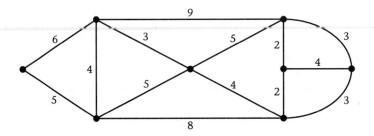

a. Use one of the spanning tree algorithms to find a minimum spanning tree for the graph.
b. What is the total cost of connecting the buildings?

14. Create a problem that can be solved by using one of the minimum spanning tree algorithms and find the solution to your problem. Then give your problem to a classmate and ask him or her to solve it. If the answer differs from yours, determine the correct solution.

15. Represent $(4 - 3) * 8 + 5$ as a binary expression tree.

16. Evaluate the reverse Polish notation

$$7\ 1 + 3 * 2\ 4 + -$$

17. Create a mathematical expression of your own, represent it with an expression tree, and find the postorder listing for the tree. Give your listing to another student in your class and have him or her evaluate the reverse Polish notation. Check the value of the notation with the value of your original mathematical expression.

Bibliography

Chartrand, Gary. 1977. *Introductory Graph Theory*. New York: Dover.

Chavey, Darrah. 1992. *Drawing Pictures with One Line*. (HistoMAP Module #21). Lexington, MA: COMAP.

COMAP. 1997. *For All Practical Purposes: Introduction to Contemporary Mathematics*. 4th ed. New York: W. H. Freeman.

Copes, W., C. Sloyer, R. Stark, and W. Sacco. 1987. *Graph Theory: Euler's Rich Legacy*. Providence, RI: Janson.

Cozzens, Margaret B., and R. Porter. 1987. *Mathematics and Its Applications*. Lexington, MA: COMAP.

Cozzens, Margaret B., and R. Porter. 1987. *Problem Solving Using Graphs*. Lexington, MA: COMAP.

Crisler, Nancy, and Walter Meyer. 1993. *Shortest Paths*. Lexington, MA: COMAP.

Dossey, John, A. Otto, L. Spence, and C. Vanden Eynden. 1996. *Discrete Mathematics*. 3rd ed. New York: Harper Collins.

Francis, Richard L. 1989. *The Mathematician's Coloring Book*. (HiMAP Module #10). Lexington, MA: COMAP.

Ore, O. 1990. *Graphs and Their Uses*. Washington, DC: Mathematical Association of America.

Tannenbaum, Peter, and Robert Arnold. 1992. *Excursions in Modern Mathematics*. Englewood Cliffs, NJ: Prentice-Hall.

Counting and Probability

Lotteries have become quite popular in the United States: consumer spending on lotteries grew at an average annual rate of 11% from 1982 to 1996, reaching a high of about $16 billion in 1996. Each week millions of tickets are sold to people who pick several numbers in hopes that the ones they chose will match those generated by a random process. The lucky few who match several or all of the numbers win anywhere from a few dollars to several million dollars.

In how many ways can a lottery participant choose several numbers from those on a lottery ticket? What is the probability of winning the jackpot in a lottery? How can methods that mathematicians use to determine the probability of winning a gambling game also be used to determine the probability that a medical test's results are correct? How has an understanding of probability improved the reliability of U.S. space shuttle launches?

A Counting Activity

Probability calculations are important in many endeavors. A meteorologist, for example, must calculate the probability of rain, a lottery commission must calculate the probability a player will win, and a medical researcher must calculate the probability that the results of a test are accurate.

Many probability calculations require knowing the number of ways in which an event can happen, such as the number of ways a lottery player can fill out a lottery card. Often the numbers involved are quite large, and careful methods must be used to be sure counting is done properly. However, the best way to begin your work is by considering some situations involving relatively small numbers.

Explore This

The Central High School student council is discussing three fund-raising proposals. Pierre suggests that the council operate a game at the annual

WIZARD OF ID   BY BRANT PARKER & JOHNNY HART

Reprinted by permission of Creators Syndicate, Inc.

school fair. His idea is to write each of the letters of the school's team name, Lions, on a Ping-Pong ball and have participants draw two of the balls from an opaque container. If the letters spell (in the order drawn) a legal word, the participant will win a prize. His proposal has been criticized by some council members who feel it would be too easy to win.

Hilary proposes printing cards with the numbers 1 through 9 displayed in a square matrix and having participants mark two of the numbers (see Figure 6.1). A winning pair would be generated at random, and a prize given to any participant who matches both winning numbers. Her scheme has left council members uncertain about how many winners might be expected in a school of 1,000 students.

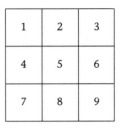

Figure 6.1 A card for Hilary's game.

Chuck also wants to operate a game at the school fair. His game would involve a board with the numbers 1 through 6 displayed (see Figure 6.2). A participant would place $1 on any of the numbers, roll two dice, and win a dollar for each time the chosen number appeared. Several council members feel that the organization would lose money on this game.

1	2	3	4	5	6

Figure 6.2 The board for Chuck's game.

Following are three sets of questions related to the games suggested by Pierre, Hilary, and Chuck. As time permits, discuss one or more of the three sets with a few other people.

Here is one way to divide the sets of questions among small groups in your class. At the direction of your instructor, divide your class into groups of three people. Write the numbers 1, 2, or 3 on each of several slips of paper. Have each group draw one of the slips from a bag or box. Each group should consider the set of questions whose number corresponds to the number drawn.

After all groups have finished their discussions, a spokesperson for each group should present the results of the group's discussion to the class. The groups that discussed set 1 should report first, and so forth.

1. Analyze Pierre's proposal. How many different two-letter "words" are there? How many of them are real words? If each of the school's 1,000 students enters exactly once and pays a $1 entry fee, how many winners might there be? How much should each winner receive if the council hopes to raise $500?

State Counts Serious Revenue Loss with 8-0-0 as Winning Number

Hartford, CT (AP), November 13, 1996

Lottery players in Connecticut ended up breaking the bank over the weekend. On Saturday, thousands of people played the daily lottery and put their money down on number 8-0-0. When that number hit, lottery officials said they had to pay out just over $1 million in winnings, about three times what was wagered.

The state took in $344,277 on the three-digit game and lost $678,170.

The state normally pulls in an average of $340,000 for the daily number and makes a profit of between $170,000 and $270,000.

Of those who bought tickets for Saturday's drawing, 3,026 people bet the "800" number "straight," and another 3,182 chose the number in a three-way box, which means they would win with 800, 080, or 008.

Gamblers who spent 50 cents on a straight bet won $250 while people with three-way bets won $83.50 per bet.

2. Analyze Hilary's proposal. In how many ways could a student fill in the entry form? If each of the school's 1,000 students enters exactly once and pays a $1 entry fee, how many winners might there be? How much should each winner receive if the council hopes to raise $500?

3. Analyze Chuck's proposal. In how many ways could the two dice fall? How often would the council pay the participant $1? $2? How often would the council make $1? Do you think that the council could raise $500 if that is the goal? (If you want to try the game, check with your teacher to see if dice are available in your classroom.)

Exercises

1. One of the goals of this chapter is to develop a few techniques that can be used to determine the number of ways in which an event can happen. The simplest such technique is making a complete list of all possible ways. This is a reasonable method as long as the number of things in the list is not too large. Make a list of all possible "words" that can be made by using two letters of the word *Lions*.

2. Suppose the Ping-Pong balls in Pierre's game are drawn one at a time and the first is kept out of the container while the second is drawn. How many different letters could appear on the first Ping-Pong ball? The second? What is the connection between these numbers and the number of "words" in the list you made in Exercise 1? If the school's

team name were Tigers, how many "words" of two letters would there be?

3. Make a list of all possible ways of choosing two numbers from the nine available on one of Hilary's cards. How many are there?

4. If you are filling in one of Hilary's cards, in how many ways could you select your first number? After you've picked your first number, in how many ways could you pick your second number? How are these two numbers related to the number of pairs you listed in Exercise 3?

5. Since Chuck's game involves two dice, it is important to be able to distinguish them. Therefore, imagine the dice are two different colors, say red and green. One way the dice could fall is the red die a 3 and the green die a 4. This can be written in shorthand as (3, 4). This outcome is different from the red die a 4 and the green die a 3, which can be written as (4, 3). Make a list of all possible ways the red die and the green die could fall together. How many pairs are in your list?

6. The red die could land in six different ways, as could the green. How are these two sixes related to the number of things in the list you made in Exercise 5?

> A second counting technique is known in mathematics as the *fundamental multiplication principle*, which says that if events *A* and *B* can occur in *a* and *b* ways separately, then there are *a* × *b* ways that the events can occur together.

To use the principle, make a blank for each event and write the number of ways each event can occur in a blank. Then multiply these numbers. For example, to determine the number of ways that a die and a coin can fall together, make two blanks: _____ _____, then write the number of possibilities for the die and the coin in the blanks: __6__ __2__ , and multiply to get 12.

If a full list of the 12 is needed, a systematic way to ensure that all items are listed is to make a tree diagram like that shown here.

Die	Coin	Outcome
1	H	1, H
	T	1, T
2	H	2, H
	T	2, T
3	H	3, H
	T	3, T
4	H	4, H
	T	4, T
5	H	5, H
	T	5, T
6	H	6, H
	T	6, T

7. Explain how the multiplication principle can be applied to determine the number of different "words" of two letters that can be made from the letters of *Lions*.

8. A utility company in North Dakota once sponsored a contest to promote energy conservation. The contest was to find all legal words that could be made from the letters of *insulate*.
   a. Use the multiplication principle to determine the number of "words" of two letters that are possible.
   b. The multiplication principle can be extended to three or more events. Show how to apply the principle to determine the number of "words" that can be made from three letters of *insulate*.

9. It is possible to modify the multiplication principle to determine the number of ways of selecting two numbers on one of Hilary's cards. Explain how this can be done.

10. Why was it necessary to modify the multiplication principle in Exercise 9?

11. Lotteries often require the participant to select several numbers from a collection of numbers printed on a card. If a state lottery has the numbers 1 through 25 printed in a square matrix, in how many ways can a participant select two of them? Explain.

12. Explain how the multiplication principle can be used to determine the number of ways in which two dice can fall.

13. Make a tree diagram to show all the outcomes when a red die and a green die are tossed together.

14. Make a tree diagram to show all the possibilities when filling out one of Hilary's cards.

15. You are playing Chuck's game and decide to bet on the number 5.
   a. Use the tree diagram you made in Exercise 13 to count the number of ways in which you could win or lose. In how many ways could you win $1? $2? In how many ways could you lose $1?
   b. In the long run, if you played this game many times, do you think you would win or lose money? Explain.

16. In a common carnival dice game, three dice are rolled. Use the multiplication principle to determine the number of ways in which three dice can fall.

17. Counting techniques are useful in many areas other than the analysis of games. An example is genetics. As you may know from your study of biology, a female inherits an X chromosome from her mother and another X chromosome from her father. A male inherits an X chromosome from his mother and a Y chromosome from his father. Use the counting techniques you learned in this lesson to explain the different ways in which chromosomes can be passed from parents to an offspring.

Computer/Calculator Explorations

18. Many calculators have built-in random-number generators that can be modified to simulate random situations. Adapt the random-number generator of a calculator to simulate the games proposed by Pierre, Hilary, and Chuck. Present your work to the members of your class.

Projects

19. Research and report on the impact of lotteries in the United States. What are the benefits and problems associated with lotteries?

Counting Techniques, Part 1

Lesson 6.1 examined several situations in which the answer to the question, In how many ways can this be done? is important. Two important techniques that can be used to answer this question are the multiplication principle and making a complete list. The latter can often be done with the aid of a tree diagram or some other systematic procedure such as an algorithm.

This lesson continues the consideration of counting techniques that began in the last lesson, beginning with the addition principle.

The Addition Principle

Recall that the multiplication principle says that if events A and B can occur in a and b ways, respectively, then events A and B can occur together in $a \times b$ ways. The **addition principle** says that if A and B can occur in a and b ways, respectively, then either event A or event B can occur in $a + b$ ways.

For example, if the student council at Central High consists of 17 members, of which 9 are girls and 8 are boys, and if one girl *and* one boy are to be selected to hold two different offices on the council, then there are 9 $\times$ 8 = 72 ways of filling the two offices. If a single student, who may be either a boy *or* a girl, is to be selected to hold a single office, then there are 9 + 8 = 17 ways of making the selection.

> The word *and* in the description of an event often indicates that the multiplication principle should be used, and the word *or* often indicates that the addition principle should be used.

The events "selecting a boy" and "selecting a girl" are called **mutually exclusive** or **disjoint** because a person cannot be both a boy and a girl. On the other hand, events such as "selecting a member of your school's football team" and "selecting a member of your school's basketball team" are not mutually exclusive if there is a person who is a member of both teams. When events are not mutually exclusive, the addition principle requires a modification that you will consider in this lesson's exercises.

Using the Multiplication and Addition Principles Together

The multiplication and addition principles are often used together, as the following example shows. The Central High council members are considering a contest in which words of any length are made from the team name *Lions*.

A word of one letter may be composed in only five ways: *l, i, o, n,* and *s.* A word of two letters requires a first letter *and* a second letter, which must be different from the first letter. Thus, there are 5 × 4 = 20 ways of composing a word of two letters. A word of three letters requires a first letter and a second letter and a third letter, so there are 5 × 4 × 3 = 60 ways of composing a word of three letters. Similarly, there are 5 × 4 × 3 × 2 = 120 ways of composing a word of four letters and 5 × 4 × 3 × 2 × 1 = 120 ways of composing a word of five letters.

A word may be composed by using one letter *or* by using two letters *or* by using three letters *or* by using four letters *or* by using five letters. Therefore, the total number of words is 5 + 20 + 60 + 120 + 120 = 325. (Note that the events composing a word by using one letter, composing a word by using two letters, composing a word by using three letters, and composing a word by using four letters are mutually exclusive.)

Factorials, Permutations, and Probability

The calculation of the number of words of five letters than can be made from the letters of *Lions* requires multiplying all integers from 1 through

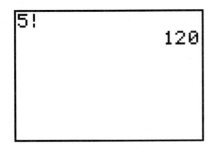

A factorial calculation on a graphing calculator.

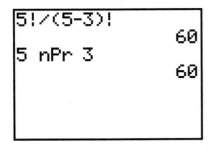

Two ways to calculate a permutation on a graphing calculator.

5. This product is an extension of the multiplication principle and is known as the **factorial** of 5 or, more simply, 5 factorial. A factorial is symbolized by an exclamation mark: 5!. Most calculators have a factorial key or function. If you have never done a factorial on your calculator, try doing so now.

The term **permutation** is often used to describe a counting procedure in which order matters. For example, the game proposed by Pierre is one in which order matters: the words *is* and *si* are not the same. (Situations in which order does not matter are considered in Lesson 6.3.)

Permutations can be computed by using your calculator's factorial key. For example, to find the number of "words" of three letters that can be formed from the letters of *lions*, divide 5! by 2!. Note that this calculation produces the correct result because

$$5 \times 4 \times 3 = \frac{5 \times 4 \times 3 \times 2 \times 1}{2 \times 1}.$$

There are two commonly used symbolic expressions for a permutation of three things selected from a group of five: $P(5, 3)$ or $_5P_3$. Either is calculated by evaluating the expression $5!/(5 - 3)!$. Some calculators have a special permutation function.

In general, $P(n, m)$ is calculated by evaluating the expression $\dfrac{n!}{(n - m)!}$.

Lesson 6.1 began with several questions about the frequency with which certain events could occur. An event's probability is simply the ratio of the number of ways the event can occur to the total number of possibilities in that situation. For example, there are 325 "words" that can be formed from the letters of *Lions*, but only 20 of them have two letters. Thus, the probability of forming a two-letter "word" from the letters of *Lions* is 20/325. Probabilities can be expressed as fractions, decimals, or percentages. As a decimal, the probability of forming a two-letter "word" from *Lions* is .0615, so you would expect a two-letter word about 6 times out of 100.

Because the numerator of a probability is never smaller than 0 and never larger than the denominator, probabilities always range between 0 and 1, inclusive. An event that cannot happen has a probability of 0, and an event that is certain to happen has a probability of 1.

You now have several counting techniques at your disposal:

1. Making a list of all possibilities, for which tree diagrams are often helpful.

2. The multiplication principle and the related factorial and permutation formulas.

3. The addition principle.

Skill at using these techniques develops with practice. The following exercises help develop that skill and also demonstrate some refinements of the three techniques.

Exercises

1. Which is equivalent to $P(10, 4)$, $10!/4!$ or $10!/6!$? Find the value of $P(10, 4)$.

2. At right are the final *USA Today* 1997–1998 season rankings of high school girls soccer teams. If you are a sportswriter voting for the top teams and you can rank only your top 5, in how many ways can you form your ranking from the 25 teams shown?

High School Girls Soccer Rankings

USA TODAY, February 25, 1998

1. Henderson (West Chester, PA) (26-0-0)
2. Winchester (MA) (24-0-0)
3. St. Francis De Sales (Columbus OH) (23-0-0)
4. La Cueva (Albequerque NM) (22-0-0)
5. Massapequa (NY) (18-2-1)
6. Baltimore Catholic (19-1-1)
7. Ramapo (Franklin Lakes NJ) (26-0-0)
8. Jesuit (Portland OR) (20-0-0)
9. Carmel (IN) (23-0-1)
10. Needham (MA) (20-1-3)
11. South Side (Rockville Centre NY) (17-2-1)
12. Governor Thomas Johnson (Frederick MD) (18-1-1)
13. Trumbull (CT) (18-0-1)
14. Santa Rosa (CA) (20-1-0)
15. Mahtomedi (MN) (20-1-2)
16. Roxsbury (Succasunna NJ) (24-1-0)
17. Northport (NY) (16-2-1)
18. Brentwood (TN) (19-1-1)
19. Northfield (VT) (17-0-0)
20. Strath Haven (Wallingford PA) (23-2-1)
21. Brighton (Salt Lake City) (15-1-0)
22. Medina (OH) (19-2-2)
23. Bishop Dennis J. O'Connell (Arlington VA) (18-2-1)
24. Walt Whitman (South Huntington NY) (15-3-1)
25. Simsbury (CT) (16-1-3)

3. A multispeed bicycle has a chain that can be moved to change the bicycle's speed. The rider uses the bicycle's front and rear shift mechanisms to move the chain from one front or rear sprocket to another.

   a. If a bicycle has three front sprockets and five rear sprockets, how many speeds does it have?
   b. Is it correct to say that a particular speed requires a particular front sprocket *and* a particular rear sprocket, or is it correct to say that a particular speed requires a particular front sprocket *or* a particular rear sprocket?

4. (See Exercise 8 in Lesson 6.1, page 280.)
   a. How many different "words" of any length can be made from the letters of *insulate*? (Hint: You can make a word of one letter *or* a word of two letters *or* a word of three letters . . . *or* a word of eight letters.)
   b. A group of students is considering entering the contest by programming a computer to print all possible "words" that can be made from the letters of *insulate* and then checking the list against an unabridged dictionary. If the computer prints the words in four columns of 50 words each on a page of paper, how many pages would there be?

5. Some states have vehicle license numbers that consist of three letters followed by three digits. Often the letters *I, O,* and *Z* are not used because they can be confused with the numerals 1, 0, and 2, respectively.
   a. If these restrictions apply and if characters may be repeated, how many different license plates are possible?
   b. What is the probability that a vehicle selected at random will have a license number that begins with *CAT*?

6. a. In how many ways can the coach of a baseball team arrange the batting order of nine starting players?
   b. A sportscaster once suggested that a baseball team try every possible batting order for its nine starters in order to determine which one worked best. If a team decides to do so and plays one game each day of the week with a different batting order in each game, how long will it take to complete the experiment?

7. Three math students and three science students are taking final exams. They must be seated at six desks so that no two math students are next to each other and no two science students are next to each other.
   a. In how many ways can the students be seated if the desks are in a single row? (Hint: Draw six blanks and use the multiplication principle.)
   b. What is the probability that a math student will occupy the first seat in the row?
   c. What is probability that math students will occupy the first seat and the last seat?
   d. What is the probability that a math student will occupy either the first seat or the last seat?

8. The multiplication principle states that the number of permutations of the letters of the word *math* is 4!. The permutation formula says that $P(4, 4)$ is $\dfrac{4!}{(4-4)!}$. The denominator of this expression is 0!, which is meaningless. However, in order that 4! and $\dfrac{4!}{(4-4)!}$ yield the same result, what value must 0! have? Explain.

9. The U. S. Postal Service began using five-digit zip codes in 1963. Every post office was given its own zip code, which ranged from 00601 in Adjuntas, Puerto Rico, to 99950 in Ketchikan, Alaska.

   a. If the only five-digit zip code that could not be used was 00000, how many zip codes were possible in 1963?

   b. Some five-digit zip codes are prone to errors because they are still legal five-digit zip codes when read upside down. When this happens, a letter goes to the wrong post office and must be returned. How many zip codes are legal when read upside down? (Hint: Draw five blanks, think carefully about which digits could go in each blank, and apply the multiplication principle.)

   c. How many of the zip codes you identified in part b are not prone to errors because they read the same when turned upside down?

10. a. In how many ways can a person draw 2 cards from a standard 52-card deck if the first card is returned to the deck before the second card is drawn?

    b. In how many ways can 2 cards be drawn if the first card is not put back?

11. The addition principle cannot be used as stated in this lesson if the two events are not mutually exclusive. For example, if there are 20 people on your school's basketball team and 45 people on your school's football team, then there are 65 ways of choosing a person from either group only if there are no people on both teams.

    a. If there are ten people who play both football and basketball, in how many ways can a person be selected from either team?

    b. Write the appropriate number of people in each of the three regions of the following diagram.

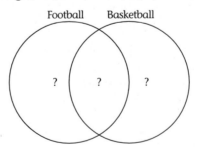

    c. Describe how the addition principle is applied when two events are mutually exclusive and how it is applied when two events are not mutually exclusive.

d. Event *A* and event *B* can occur in *a* and *b* ways, respectively, and events *A* and *B* have *c* items in common. Write an algebraic expression for the number of ways in which event *A* or event *B* can occur.

e. Central High's soccer team has 38 members, and its basketball team has 13 members. If there are a total of 42 students involved, how many are on both teams. Explain.

12. Before the 1992 major league baseball season began, Joe Torre, who was then the manager of the St. Louis Cardinals, said he'd picked his starting lineup. He also said he had determined his first three batters but not the order in which they would bat. In how many ways can Joe arrange his batting order if the pitcher must bat last? (Hint: Draw nine blanks and apply the multiplication principle.)

13. a. In how many different ways can a teacher arrange 30 students in a classroom with 30 desks?

b. The radius of the earth is approximately 6,370 kilometers, and a standard medical drop is $\frac{1}{10}$ cubic centimeter. Use the formula for the volume of a sphere, $V = \frac{4}{3}\pi r^3$, to find the volume of the earth in drops of water. Compare this with the number of seating arrangements in part a.

Joe Torre managed the St. Louis Cardinals before becoming manager of the New York Yankees.

14. There are three highways from Claremont to Upland and two highways from Upland to Pasadena.

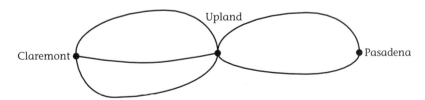

a. In how many ways can a driver select a route from Claremont to Pasadena?

b. Is it correct to say that a trip from Claremont to Pasadena requires a road from Claremont to Upland *and* a road from Upland to Pasadena, or is it correct to say that the trip requires a road from Claremont to Upland *or* a road from Upland to Pasadena?

c. In how many ways can a driver plan a round-trip from Claremont to Upland and back?

The logo for Fordham University radio station WFUV.

15. Radio station call letters in the United States consist of three or four letters, of which the first must be either a *K* or a *W*. Assuming that letters may be repeated, determine the number of radio stations that can be assigned call letters.

16. Six different prizes are given by drawing names from the 72 Central High orchestra members attending the orchestra's annual picnic. In how many ways can the prizes be given if no one can receive more than one prize?

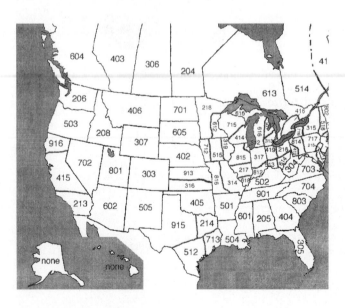

The original area codes.

17. Three-digit telephone area codes were introduced in 1947. At that time, the first digit could not be a 0 or a 1, the second digit could be only a 0 or a 1, and the third could be any digit except 0.

a. How many area codes were possible?

b. Because of shortage of area codes, beginning in 1995 any digit became a legal second digit. How many area codes were possible after 1995?

18. The UPC bar codes consist of two groups of five digits each. One group, as assigned by the Uniform Code Council in Dayton, Ohio, represents the manufacturer, and the other group represents the products of that manufacturer.

0 16000 68790 5

a. How many different manufacturers can be encoded?

b. How many products can each manufacturer encode?

19. Some newspapers publish weekly word puzzles that offer an accumulating cash prize. These puzzles are modified crosswords in which clues are given and usually only two choices are offered.

a. Suppose a prize word puzzle contains 20 questions, each having two possible answers. How many different entries are possible? (Hint: Imagine 20 blanks and use the multiplication principle.)

b. Suppose that you embark on the ambitious project of submitting every possible entry. Further suppose that it takes you 5 minutes to do an entry and that the piece of paper on which you submit an entry is 0.003 inch thick. How long would it take you to prepare the entries, and how thick a stack of paper would you deliver to the newspaper?

For Many, Area Codes Are Not Created Equal

CHRISTIAN SCIENCE MONITOR,
February 6, 1998

Like book ends to the Digital Revolution, residents of New York and San Francisco, and all those in between, will have to get used to different telephone area codes this year.

As of this past weekend, the nice roll-off-your tongue sound of 415 for the San Francisco Bay area has a new companion: 650. And sometime this year, getting ahold of someone in Manhattan won't be as easy as remembering 212.

Driving the need for new area codes is the explosion, particularly over the past two years, in communication needs.

Americans bought 15 million pagers in 1996, to accompany the 43 million cell phones already in use. Fax machines have become nearly as common as an office wastebasket. Some 13 percent of the adult population now cruises the Internet. And retail stores started making life easier with machines that required only a swipe of a credit card to pay the bill. What they all have in common is the need for a phone number.

California, for example, had accumulated 13 area codes through its first 50 years of long-distance dialing. During the next two, it will add another 10.

20. A carnival game called Chuk-a-Luk is similar to the one proposed by Chuck in the Lesson 6.1, except that three dice are used.

a. In how many ways can three dice fall? Explain.

b. Determine the number of ways you could win $1, win $2, win $3, or lose $1 in the game of Chuk-a-Luk if you win $1 each time your number shows. (Hint: You win $2 if the first *and* second dice show your number *and* the third die doesn't, *or* if the first *and* third dice show your number *and* the second die doesn't, *or* if the second *and* third dice show your number *and* the first die doesn't. Draw several sets of three blanks and then use the multiplication and addition principles.)

c. In the long run, do you think a player would win or lose money in the game of Chuk-a-Luk? Explain.

21. The news article on page 278 describes two types of bets that can be made by selecting three digits from 0 through 9. Discuss the difference in the two types of bets and determine the probability of winning for each type.

22. Factorials can be described recursively. Let $f(n)$ represent $n!$. Write a recurrence relation that expresses the relationship between $f(n)$ and $f(n-1)$.

Projects

23. Research the history of the study of probability. How did it begin? What roles did Jerome Cardan, Blaise Pascal, and Pierre Fermat play? What problems interested them?

24. Investigate the number of permutations of several objects, of which some look alike. For example, the letters of *math* can be arranged in $4! = 24$ ways; how many different permutations can there be of the letters of *look?* What if there are several sets of identical letters, such as in *Mississippi?* Write a summary that includes a general principle for handling such situations and several examples.

25. Investigate the number of permutations of several objects arranged in circular fashion. For example, Ann, Sean, Juanita, and Herb can be seated along one side of a rectangular table in $4! = 24$ ways. In how many different ways can they be seated around a circular table? Write a summary that includes a general principle for handling such situations and several examples. Explain your interpretation of the meaning of the word *different.*

26. Investigate the use of the addition principle with three events that are not mutually exclusive. Suppose, for example, that the football, basketball, and track teams of Central High have 41, 15, and 34 members, respectively. If 6 people play both football and basketball, 7 are on both the basketball and track teams, 15 are on both the football and track teams, and 4 play all three sports, how many people are involved in one sport or another? Develop a general principle for handling situations of this type and draw a diagram to represent it. Can the principle be extended to four or more events? How?

Counting Techniques, Part 2

The counting techniques developed in Lessons 6.1 and 6.2 are only one method short of forming a fairly complete tool kit for analyzing a variety of probability problems. In this lesson, you consider a technique for counting in situations in which the order of occurrence is unimportant.

Combinations

The game proposed by Hilary in Lesson 6.1 is a simple lottery in which participants must select two numbers from nine printed on a card. The order in which the participant selects the numbers is unimportant. That is, if the winning numbers are 2 and 6, it does not matter whether the owner of a winning ticket selected 2 or 6 first.

A counting situation in which the order does not matter is called a **combination**. Combinations are counted by modifying the technique used to count permutations. For example, had the order of selection mattered in Hilary's game, then the number of ways of filling out a ticket would be counted as a permutation: $P(9, 2) = \dfrac{9!}{(9-2)!} = 72$. Because this permutation counts the selection of a pair such as 2 and 6 as different from the pair 6 and 2, every possible pair is counted twice. Thus, the number of combinations of two things selected from a group of 9 is $72/2 = 36$. This combination is expressed symbolically as $C(9, 2)$ or $_9C_2$.

If Hilary's game required the selection of three numbers and the order mattered, the number of ways of filling out a card would be $P(9, 3) = 504$. If the order does not matter, then 504 is too large. For example, if the winning numbers are 2, 5, and 8, then 504 counts any arrangement of 2, 5, and 8 as different. The number of ways of arranging 2, 5, and 8 is $3 \times$

$2 \times 1 = 6$. Therefore, 504 is six times too large, and $C(9, 3) = \dfrac{P(9, 3)}{3!} = 504/6 = 84$.

In general, $C(n, m)$ is calculated by evaluating the expression $\dfrac{P(n, m)}{m!}$. But, $P(n, m) = \dfrac{n!}{(n - m)!}$, so

$$C(n, m) = \dfrac{n!}{(n - m)!m!}.$$

Point of Interest

```
9!/(6!3!)
                84
9 nCr 3
                84
```

Combinations can be calculated by using a calculator's factorial function or, on some calculators, by using a combination function.

Since there are 36 ways of filling in one of Hilary's lottery tickets, the probability that any one ticket will win is 1/36, or about .028. If 1,000 tickets are sold, Hilary can expect about $1,000 \times .028 = 28$ winners. If the game requires the selection of three numbers, the probability a single ticket will win is 1/84, or about .012. If 1,000 tickets are sold, about $1,000 \times .012 = 12$ winners can be expected.

Using Combinations with Other Counting Techniques

Combinations are often used along with other counting techniques. For example, the 17-member student council at Central High consists of 9 girls and 8 boys. A committee of 4 council members is being selected. If the positions on the committee are not different in any way, then the order of selection is unimportant, and the number of ways the committee can be selected is $C(17, 4) = \dfrac{17!}{13!4!} = 2,380$.

If the committee must have two girls and two boys, there are $C(9, 2) =$ $\frac{9!}{7!2!} = 36$ ways of selecting the 2 girls and $C(8, 2) = \frac{8!}{6!2!} = 28$ ways of selecting the two boys. Because the committee must consist of 2 girls *and* 2 boys, apply the multiplication principle to conclude that there are $36 \times 28 = 1{,}008$ ways of forming the committee. If the 4 committee members are selected at random, the probability that the committee will consist of 2 girls and 2 boys is $\frac{1{,}008}{2{,}380}$, or about .424.

Now suppose that the committee must consist of either all boys or all girls. There are $C(9, 4) = \frac{9!}{5!4!} = 126$ ways of selecting 4 girls and $C(8, 4) = \frac{8!}{4!4!} = 70$ ways of selecting 4 boys. Because the committee must consist of either 4 girls *or* 4 boys and because all-boy and all-girl committees are mutually exclusive, apply the addition principle to conclude that there are $126 + 70 = 196$ ways of forming the committee. Again, if the 4 committee members are selected at random, the probability the committee will consist of either all boys or all girls is $\frac{196}{2{,}380}$, or about .082.

Exercises

1. Which is larger: $C(10, 2)$ or $C(10, 8)$?

2. Find the sum of all possible combinations of four things. That is, find $C(4, 0) + C(4, 1) + C(4, 2) + C(4, 3) + C(4, 4)$. Do the same for all possible combinations of three things and all possible combinations of five things. On the basis of your results, make a guess about the sum of all possible combinations of six things. Describe any pattern you noticed.

3. In this lesson the number of all-boy four-person committees on the Central High student council was calculated as $C(8, 4) = 70$, the number of all-girl four-person committees was calculated as $C(9, 4) = 126$, and the number of four-person committees that are half boys and half girls was calculated as $C(8, 2) \times C(9, 2) = 1{,}008$.
   a. How many four-person committees consist of three girls and one boy?
   b. How many committees consist of one girl and three boys?

   c. Find the sum of the numbers of committees that consist of four boys, no boys, two boys, three boys, and one boy. Compare this sum with the total number of four-person committees calculated by $C(17, 4)$ in the lesson (see page 295).

4. Darrell Dewey has just left his Central High social studies class and bumped into his friend Carla Cheetham. Darrell informs Carla that Ms. Howe is giving a ten-question true/false quiz today. When Carla asks about the quiz, Darrell says he thought that the quiz was easy and four of the answers were false.

   a. When Carla takes the quiz, in how many ways can she select four questions to mark false?

   b. In how many ways can Carla select six questions to mark true?

   c. In how many ways can Carla fill in the quiz if she ignores Darrel's hint?

5. A standard deck of cards contains 13 different cards from each of four suits: spades and clubs, which are black in color, and diamonds and hearts, which are red in color.

   a. In how many ways can 2 cards be dealt from a standard 52-card deck?

   b. In how many ways can 2 red cards be dealt from a standard 52-card deck?

   c. What is the probability that 2 cards dealt from a standard 52-card deck are both red?

6. Maria has a part-time summer job selling ice cream from a small vehicle she drives through residential areas of her community. She carries six different flavors and sells a two-scoop cone for $1.60.

   a. How many two-scoop cones are possible if both scoops are the same flavor?

   b. How many two-scoop cones are possible if each scoop is a different flavor?

   c. All together, how many two-scoop cones are possible?

7. Hedy Foans, who writes a music column in the Central High Scribbler, has decided to poll students on their favorite songs. She has prepared a list of ten current favorites, from which students will be asked to rank their top three. In how many ways can a student pick a first, second, and third choice from Hedy's ten?

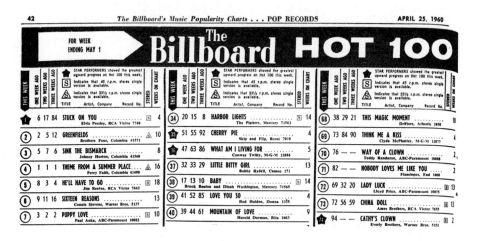

8. Ms. Howe has a planter in one of her classroom windows that is divided into five sections. She has purchased two geraniums and three marigolds to plant in the five spaces.
  a. In how many ways can Ms. Howe select the two sections in which to plant the geraniums?
  b. In how many ways can Ms. Howe select the three sections in which to plant the marigolds?

9. In February 1992, an Australian company sent representatives to Virginia in an attempt to purchase every possible ticket in the state's lottery. The representatives spread their purchases among eight retail chains that had a total of 125 outlets; one representative bought a total of 2.4 million tickets at a single retail chain headquarters. When time ran out, the group had purchased 5 million tickets, or about 70% of all possible combinations.

One of the tickets purchased by the group matched the winning combination of 8, 11, 13, 15, 19, 20. After a controversy over the legality of the purchase, the lottery decided to award the $27 million jackpot to the Australian group, which represented about 2,500 investors who paid an average of $3,000 each. Each investor stood to receive an average of $10,800, at the rate of $540 a year over the 20-year payment period.

a. A Virginia lottery ticket contains the numbers 1 through 44, from which a participant must select six. In how many ways can the selection be made?

b. If it takes 5 seconds to fill out a Virginia lottery ticket, how long would it take one person working 40 hours a week to fill out all possible tickets?

c. If each Virginia lottery form has space for five entries and if each form has a thickness of 0.003 inch, how thick would a stack of forms of all possible entries be?

d. A Florida lottery ticket contains the numbers 1 through 49, from which a participant selects six. In how many ways can this be done?

e. An individual once bought 80,000 tickets in the Florida lottery. What was this person's probability of sharing the $94 million jackpot that had accrued at that time?

f. How does the probability of winning the jackpot in the Virginia lottery compare with the probability of winning the jackpot in the Florida lottery?

Cornering a Lottery: Foolproof, or Foolish?

WASHINGTON POST,
February 28, 1992, by
Albert B. Crenshaw

You might think that betting on every possible number combination in a state lottery would guarantee a big payoff.

But it doesn't.

A look at what happened in Virginia last week, when an Australian syndicate apparently attempted to corner the lottery, indicates that to be successful, players must not only bet every number, but they must also be lucky enough to be the only winners.

The Australians' system is far from a sure-fire money-maker, and as soon as more than one group tries it, the system is almost certain to be a money-loser. . .

The difference between the size of the investment—around $7 million to cover all the numbers in Virginia—and the $27 million jackpot is not as great as it seems. The grand-prize winnings in most lotteries are paid out over a long time, making them far less valuable than if they were paid all at once.

In fact, after adjusting for inflation and potential investment earnings, $27 million paid out over 20 years is the equivalent of only about $9 million to $12 million if it were paid out today.

Reprinted by permission of United Feature Syndicate.

10. Most lotteries include several prizes besides the jackpot. For example, the Virginia lottery gives second prizes to tickets that match 5 of the 6 winning numbers, third prizes to those that match 4 of the 6, and fourth prizes to those that match 3 of the 6.
  a. How many different ways are there to receive a second prize? (Hint: The ticket must match 5 of the 6 winning numbers *and* 1 of the 38 nonwinning numbers.)
  b. How many different ways are there to receive a third prize?
  c. How many different ways are there to receive a fourth prize?

11. Chapter 1 discussed various ways of voting and of determining a winner in elections. Suppose there are seven choices on a ballot.
  a. In how many ways can a voter rank the seven choices?
  b. Recall that when approval voting is used, the choices are not ranked. In how many ways can you select three choices of which to approve?

12. Dee Noat, the director of Central High's music department, is holding tryouts for the school's jazz band. There are 7 students competing for the three saxophone positions, 8 for the two piano spots, 5 for the two percussion spots, and 12 for the three places as guitarists. In how many ways can Dee select her band?

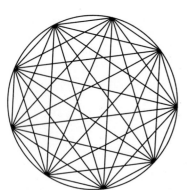

13. The figure at left was drawn by marking nine equally spaced points on the circumference of a circle and connecting every pair of points.
  a. How many chords are there?
  b. Number the points from 1 through 9, and explain why drawing the chords is analogous to filling out every possible ticket in Hilary's lottery.

c. Recall that a complete graph is one in which every pair of vertices is connected with an edge. How many edges are there in a complete graph with ten vertices?

14. Emily's Pizza Emporium can prepare a pizza with any one or more of eight available ingredients. In how many different ways can a pizza be ordered at Emily's? (Hint: A pizza can be ordered with one ingredient, or two ingredients, or three ingredients, ...)

15. College Inn Pizza claims that it offers 105 different two-topping pizzas. How many different toppings do you think College Inn Pizza uses? Explain.

16. Carl Burns, coach of the Central High Lions basketball team, has 12 players on his squad. Of these, 3 are centers, 4 are forwards, and 5 are guards.
   a. Is it correct to say that a team requires a center *and* 2 forwards *and* 2 guards, or is it correct to say that a team requires a center *or* 2 forwards *or* 2 guards?
   b. In how many ways can Coach Burns select his starting team?

17. A telephone exchange consists of all phone numbers with the same three-digit prefix.
   a. How many different phone numbers are possible in a given exchange?
   b. If a community has 95,000 telephone subscribers, what is the minimum number of exchanges needed?
   c. How many phone numbers are possible in a given three-digit area code? (Assume all possible exchanges are permitted.)

18. Allison Gerber, a math teacher at Central High, gives prizes to students in her class who have improved their average grade. At the end of each term she places the names of all qualifying students in a container and draws three.
a. If there are 19 qualifying students and the prizes are three Central High Lions T-shirts, in how many ways can the prizes be awarded?
b. If there are 19 qualifying students and the prizes are a new calculator, a Lions T-shirt, and a book on discrete mathematics, in how many ways can the prizes be awarded?

19. Dominoes come in different-sized sets, of which a double-six set is the most common. In a double-six set, each half of a domino may have any number of spots from 0 through 6. The two halves of a given domino in the set pair a number of spots with itself or with another number of spots.
a. How many dominoes with the same number of spots on each half are there in a double-six set?
b. If every possible pairing is included in the set, how many dominoes with a different number of spots on each half are there in a double-six set?
c. What is the total number of dominoes in a double-six set?
d. Write a description of the way a domino in a double-six set is formed. Explain how the words *and* and *or* in your description reflect the calculations you made to obtain your answer to part c.
e. If you select a domino at random from a double-six set, what is the probability that it will have the same number of spots on each half?
f. How many dominoes are there in a double-twelve set?

20. How many different sums of money can be made from a $1 bill, a $5 bill, a $10 bill, and a $20 bill? (Hint: You can use one bill at a time or two bills at a time or three bills at a time or four bills at a time.)

21. Many card games involve 5-card hands. (See the description of a standard deck of cards in Exercise 5.)
a. How many different 5-card hands can be dealt from a standard 52-card deck?
b. In how many ways can a selection of 3 aces be made from the 4 aces that are found in a standard deck?
c. In how many ways can 3 cards of the same kind (aces, twos, threes, and so forth) be dealt from a standard deck? (Hint: You can deal 2 aces or 3 twos or 3 threes or . . . .)

d. Repeat part c for 2 cards of the same kind.
e. In how many ways can a hand consisting of 3 of one kind and 2 of another (a full house) be dealt from a standard deck?

22. To win the jackpot in the California lottery, a participant must match 6 numbers from 51 that are available. If you buy ten tickets per week in the California lottery, about how often could you expect to win the jackpot? Explain.

23. Some bike locks allow the user to set a four-digit code that opens the lock. These locks are convenient because the user can select a familiar number and is thereby less likely to forget the code.

a. How many different codes are possible?
b. Locks like these are often called combination locks. Do you think this is a good name? Explain.

Projects

24. Research one or more of the lotteries in your part of the country. How large are the jackpots? How many tickets are usually sold? What portion of the proceeds goes to the players? What happens to the rest of the money? Are there any rules to prevent the kind of purchase made by the Australian group in the Virginia lottery? What kinds of strategies are known to be used by players?

25. Investigate probabilities of common card hands. Show how to calculate as many as possible.

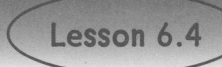

Probability, Part 1

The counting techniques discussed in the first three lessons of this chapter can be used to find probabilities of simple events. However, many applications of probability involve compound events that are formed by combining two or more simple events. This lesson is concerned with rules that govern operations on two or more probabilities.

Recall that the probability of an event is the ratio of the number of ways the event can occur to the total number of possibilities. For example, the probability that a die will fall with an even number showing is 3/6 because three of the six possibilities are even. For convenience, the statement "the probability that a die will fall even" is abbreviated p(a die will fall even).

The Addition Principle for Probabilities

The addition and multiplication principles you previously used to find the number of ways in which events can occur have counterparts in probability.

> The addition principle for probabilities states that
> $p(A \text{ or } B) = p(A) + p(B) - p(A \text{ and } B)$, for any two
> events A and B; $p(A \text{ or } B) = p(A) + p(B)$, for mutually
> exclusive (disjoint) events A and B.

The subtraction of $p(A \text{ and } B)$ is unnecessary when events are mutually exclusive because $p(A \text{ and } B) = 0$. When applying the addition principle, do not assume events are mutually exclusive unless you are certain.

Consider Exercise 11 from Lesson 6.2 (pages 288 and 289). In that exercise, 45 people were on a football team, 20 were on the basketball team, and 10 people were on both teams. To determine the total number of people involved, you added 45 and 20, then subtracted 10 to get 55.

If a person is chosen at random from this group, the probability of selecting a football player is $\frac{45}{55}$, the probability of selecting a basketball player is $\frac{20}{55}$; and the probability of selecting someone who is both a football and basketball player is $\frac{10}{55}$. To determine the probability of selecting either a football or basketball player, perform a calculation similar to the one in the previous paragraph: $\frac{40}{55} + \frac{20}{55} - \frac{10}{55} = \frac{55}{55}$, or 1 (see Figure 6.3). Note that without the subtraction of $\frac{10}{55}$, the answer would exceed 1, which is impossible.

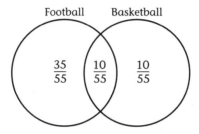

Figure 6.3 The addition principle for probabilities.

The Multiplication Principle and Conditional Probability

The multiplication principle is similar to the addition principle in that it is applied differently depending on the kind of events involved. The addition principle can be shortened if the events are mutually exclusive; the multiplication principle can be shortened if the events are independent.

As an example, consider the following data on the student population at Central High, which has exactly 1,000 students.

	Male	Female	Total
Seniors	156	144	300 (30%)
Juniors	168	172	340 (34%)
Sophomores	196	164	360 (36%)
Total	520 (52%)	480 (48%)	

Many probabilities can be found from these data. For example, the probability of selecting a junior is .34, and the probability of selecting a girl from the junior class is $\frac{172}{340} = .5059$.

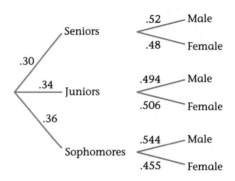

Figure 6.4 Organizing probabilities with a tree diagram.

A tree diagram with appropriate probabilities written along the branches is a convenient way to organize these probabilities (see Figure 6.4).

The probability of selecting a girl from the junior class is called a **conditional probability** because the event describes the selection of a girl under the condition that the person selected be a junior. (Verbal descriptions of conditional events often use the word *from*. Other commonly used words are *if, when,* and *given that*.) The probability of A from B is sometimes written symbolically as $p(A/B)$.

Note that conditional probabilities change if the order of events is reversed. For example, the probability of selecting a girl from the juniors $\left(\frac{172}{340} = .5059\right)$ is different from the probability of selecting a junior from the girls $\left(\frac{172}{480} = .3583\right)$.

Since 172 of the 1,000 students are girls, the probability of selecting a student who is both a junior and a girl is .172. Refer to the tree diagram in Figure 6.4 and note that the probabilities written along the junior branch and the female branch that follows it are .34 and .506, respectively. The product of .34 and .506 is approximately .172. Thus, the probability of selecting a student who is a junior and a girl equals the product of the probability of selecting a junior and the probability of selecting a girl from the juniors.

> The **multiplication principle** for probabilities states that for two events A and B, $p(A \text{ and } B) = p(A) \times p(B \text{ from } A)$. The products that result from applying the multiplication principle can be written to the right of the tree diagram, as shown in Figure 6.5.

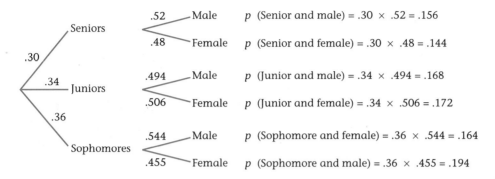

Figure 6.5 The multiplication principle for probabilities.

Refer back to the table and note that 48% of the students are girls. Refer to the tree diagram in Figure 6.4; you'll note something unusual about the senior class: the percentage of seniors who are girls is exactly the same as the percentage of girls in the entire school. For this reason, the events selecting a girl and selecting a senior are called independent.

When events are independent, the probability that one occurs is not affected by the occurrence of the other: If you want to know the probability of selecting a girl, it makes no difference if the selection is made from the entire school or from the senior class. However, the probability of selecting a girl from the entire school (.48) is not the same as the probability of selecting a girl from the junior class $\left(\frac{172}{340} = .5059\right)$. Therefore, the events selecting a girl and selecting a junior are not independent.

> Two events A and B are **independent** if $p(B$ from $A) = p(B)$ or if $p(A$ from $B) = p(A)$. Thus, for independent events, the multiplication principles states that $p(A$ and $B) = p(A) \times p(B)$.

When applying the multiplication principle, be careful not to assume that events are independent unless you are certain. In some cases, independence is quite obvious. For example, a toss of a coin has nothing to do with the outcomes of previous tosses: the probability that a particular toss is heads is $\frac{1}{2}$ regardless of the previous toss. This independence means that it is correct to calculate the probability of obtaining two heads in a row by multiplying $\frac{1}{2}$ by $\frac{1}{2}$ and obtaining $\frac{1}{4}$.

Just as independence is obvious in some cases, so is the lack thereof. For example, a man who has a beard is more likely to have a mustache than are men in general. Therefore, having a beard and having a mustache are not independent. It would be incorrect to calculate the probability that a man has both a beard and a mustache by multiplying the probability that a man has a beard by the probability that a man has a mustache.

In other cases, it can be difficult to determine whether two events are independent without inspecting data or probabilities. For example, are people who own poodles more or less likely to own pink cars than are people in general? If pink car ownership is either more or less common among poodle owners than among the general public, then the events owning a poodle and owning a pink car are not independent.

There are two ways to determine whether events A and B are independent:

1. Compare $p(A)$ with $p(A$ from $B)$. If they are the same, A and B are independent. (You can also compare $p(B)$ with $p(B$ from $A)$, but it is not necessary to make both comparisons.)

2. Multiply the probability of A and the probability of B. If the result is the same as the probability of both A's and B's occurring, then A and B are independent.

Example: Checking for Independence

Consider the following data on ownership of pink cars and poodles in a community. Are owning a pink car and owning a poodle independent in this community?

	Own Poodles	Don't Own Poodles
Own pink cars	250	450
Don't own pink cars	1,250	18,350
Totals	1,500	18,800

To apply the first method of checking for independence, note that the probability of selecting someone who owns a pink car is $\frac{700}{20,300} = .0345$, and the probability of selecting someone who owns a pink car from the poodle owners is $\frac{250}{1,500} = .167$. Because the two probabilities are not equal, owning a pink car and owning a poodle are not independent.

To apply the second method of checking for independence, note the following probabilities: p(owning a poodle) $= \frac{1,500}{20,300} = .0739$, p(owning a pink car) $= \frac{700}{20,300} =$

.0345, and p(owning a poodle and owning a pink car) $= \frac{250}{20,300} = .0123$. Calculate the product of the first two probabilities and compare it with the third: $.0739 \times .0345 = .00255 \neq .0123$.

Attention to detail is important when either the addition or multiplication principles are applied. In this lesson's exercises you use both principles in a variety of settings and consider the consequences of improper application.

Exercises

1. Use the table of Central High student population data on page 305.
   a. What is the probability of selecting a male?
   b. What is the probability of selecting a male from the sophomore class?
   c. Use your answers to parts a and b to determine whether the events selecting a male and selecting a sophomore are independent.
   d. What is the probability of selecting a sophomore?
   e. What is the probability of selecting someone who is both a male and a sophomore? Compare this probability with the product of the probability of selecting a male and the probability of selecting a sophomore.
   f. Multiply the probability of selecting a male from the sophomore class by the probability of selecting a sophomore. Compare this with the probability of selecting someone who is both a male and a sophomore.
   g. Is the probability of selecting a male from the sophomore class the same as the probability of selecting a sophomore from the males?

2. Consider events that describe the performance of your school's football team.
   a. Do you think the events "the team wins" and "the team wins at home" are independent? Explain.
   b. Describe the conditions that would have to exist if the events "the team wins" and "the team wins in bad weather" are independent.

3. Use the table of Central High student population data on page 305.
   a. What is the probability of selecting someone who is either a male or a sophomore?
   b. Add the probability of selecting a sophomore and the probability of selecting a male. Compare this with the probability of selecting someone who is either a sophomore or a male.
   c. Are the events selecting a male and selecting a sophomore mutually exclusive? Explain.

4. Read the *Florida Today* editorial on page 311 about opposition to the Cassini launch.
   a. How was the probability $\frac{1}{350}$ obtained?
   b. What assumptions were made in the calculation of this probability?

5. The following tree diagram represents the Central High data of this lesson, with the events selecting a male and selecting a female drawn first.

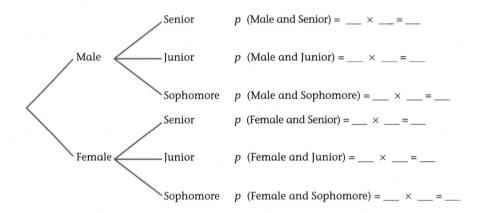

p (Male and Senior) = ___ × ___ = ___

p (Male and Junior) = ___ × ___ = ___

p (Male and Sophomore) = ___ × ___ = ___

p (Female and Senior) = ___ × ___ = ___

p (Female and Junior) = ___ × ___ = ___

p (Female and Sophomore) = ___ × ___ = ___

   a. Write the correct probabilities along each branch and complete the calculations at the right.
   b. Find the sum of the probabilities you calculated on the right.

6. A card is drawn at random from a standard 52-card deck.
   a. What is the probability of drawing an ace?
   b. What is the probability of drawing an ace from the diamonds only?
   c. Are the events drawing an ace and drawing a diamond independent? Explain.
   d. What is the probability of drawing a diamond?
   e. What is the probability of drawing a card that is both a diamond and an ace? Compare this probability with the product of the probability of drawing an ace and the probability of drawing a diamond.
   f. What is the probability of drawing a card that is either an ace or a diamond?
   g. Are the events drawing an ace and drawing a diamond mutually exclusive?

h. Analyze the events drawing a king and drawing a face card (jack, queen, or king). Are they independent? Are they mutually exclusive? Explain your answers.

7. Two cards are drawn separately from a standard deck.

    a. What is the probability that the first card is red?

    b. What is the probability that the second card is red if the first card is red and is not put back in the deck before the second card is drawn?

    c. What is the probability that the first card is red and the second card is red if the first card is not put back in the deck before the second card is drawn?

    d. Compare your answer for part c with that for part c of Exercise 5 from Lesson 6.3 (page 297).

    e. The following tree diagram represents the colors of cards when two cards are drawn in succession from a standard deck without the first card's being put back. Write the correct probabilities along each branch and complete the calculations at the right.

Cassini Foes Exaggerate Risk of Saturn Mission

FLORIDA TODAY,
October 9, 1997
By Louis Friedman

Washington—A small but vociferous group of antinuclear activists is fighting against Monday's launch of the international Cassini mission, destined for Saturn, because they fear a potential release of plutonium from the on-board power supply. While their concern is understandable, an examination of the issue shows the safety and environmental risks to be very small and the knowledge to be gained very large.

The Cassini spacecraft is designed with a power system that has been employed on 23 planetary missions over the past three decades. It uses plutonium to generate heat, which is converted to electricity to operate the probe. To protect against an accident, the plutonium is encased in special containers that can withstand high impact and temperatures.

To be sure, plutonium is radioactive and toxic. There is a measurable but small danger from Cassini's plutonium. The probability of an accident during initial launch in which there could be a release of plutonium is 1 in 1,500. The chance that there could be a release in the final launch phase is about 1 in 450. When the probe swings by Earth in a gravity assist two years after launch, the likelihood there could be a release from an accidental re-entry is less than 1 in 1 million.

Put together, the total probability of plutonium release is estimated at about 1 in 350.

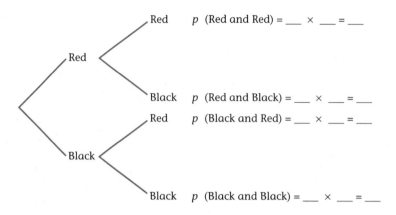

Red p (Red and Red) = ___ × ___ = ___

Black p (Red and Black) = ___ × ___ = ___

Red p (Black and Red) = ___ × ___ = ___

Black p (Black and Black) = ___ × ___ = ___

    f. What is the probability that exactly one of the two cards is red? This occurs if the first card is red and the second card is black or if the first card is black and the second card is red. Find this probability by adding two of the probabilities you calculated and wrote to the right of the tree diagram.

8. Again, two cards are drawn separately from a standard deck.
    a. What is the probability that the first card is red?
    b. What is the probability that the second card is red if the first card is red and is put back before the second card is drawn?
    c. What is the probability that the first card is red and the second card is red if the first card is put back in the deck?
    d. The following tree diagram represents the colors of cards when two cards are drawn in succession from a standard deck with the first card replaced before the second is drawn. Write the correct probabilities along each branch and complete the calculations at the right.

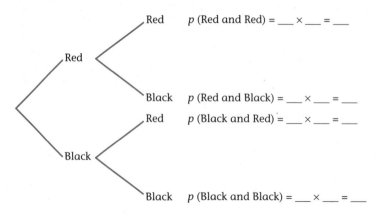

Red p (Red and Red) = ___ × ___ = ___

Black p (Red and Black) = ___ × ___ = ___

Red p (Black and Red) = ___ × ___ = ___

Black p (Black and Black) = ___ × ___ = ___

e. What is the probability that exactly one of the two cards is red?

f. Is the second draw independent of the first in the situation of Exercise 7 or in the situation of this exercise?

9. The probability that Coach Burns's Central High basketball team will win its first game of the season is .9, and the probability the team will win its second game of the season is .6.

a. What is the probability the team will win both its first and second games?

b. What assumption did you make about the outcomes of the first and second games in answering the previous question? Do you think that this is a safe assumption? Explain.

10. The probability of rain today is .3. Also, 40% of all rainy days are followed by rainy days and 20% of all days without rain are followed by rainy days. The following tree diagram represents the weather for today and tomorrow.

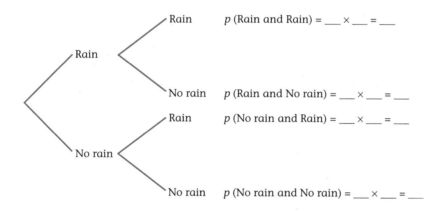

a. Write the correct probabilities along each branch and complete the calculations at the right.

b. What is the probability that it will rain on both days?

c. What is the probability that it will rain on one of the two days?

d. What is the probability that it will not rain on either day?

e. Is tomorrow's weather independent of today's? Explain.

11. Some Americans favor mandatory HIV screening for workers in certain professions such as health care. However, medical tests are seldom perfect. When medical tests that are not perfect are used on people who lack symptoms, the results must be carefully interpreted because false

positives are common. (A false positive is a positive test result for a person who does not have the disease being tested for.) For example, consider a test that is 98% accurate. That is, it fails to report the existence of the disease in only 2% of those who have it, and it incorrectly reports the existence of the disease in 2% of those who do not have it. About 2 people out of 1,000 have the disease.

a. Write the appropriate probabilities along each branch of the tree diagram and complete the calculations at the right.

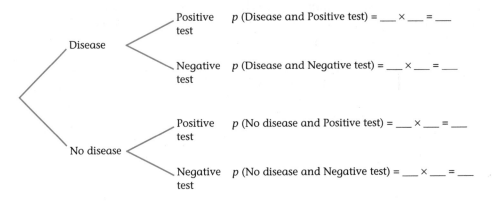

Positive test p (Disease and Positive test) = ___ × ___ = ___

Negative test p (Disease and Negative test) = ___ × ___ = ___

Positive test p (No disease and Positive test) = ___ × ___ = ___

Negative test p (No disease and Negative test) = ___ × ___ = ___

b. If 100,000 people are screened for the disease, about how many can be expected to test positive for the disease?

c. Of the 100,000 people, how many people who test positive actually have the disease?

d. What is the probability that a person who tests positive for the disease actually has it?

e. Why is the tree diagram in part a ordered the way it is? That is, why would it not make sense to show test results along the first branch and the presence of disease or lack thereof second?

12. Consider the dice game proposed by Chuck in Lesson 6.1 and suppose that you have bet on the number 5.

a. Are the outcomes of the two dice independent of each other? Explain.

b. What is the probability of a 5's appearing on the first die and on the second?

c. The following tree diagram represents the outcomes of the two dice. Write the correct probabilities along each branch and complete the calculations at the right.

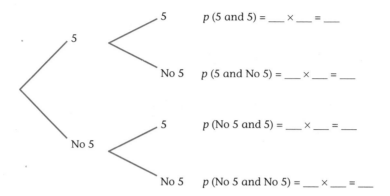

$p\,(5 \text{ and } 5) = \underline{} \times \underline{} = \underline{}$

$p\,(5 \text{ and No } 5) = \underline{} \times \underline{} = \underline{}$

$p\,(\text{No } 5 \text{ and } 5) = \underline{} \times \underline{} = \underline{}$

$p\,(\text{No } 5 \text{ and No } 5) = \underline{} \times \underline{} = \underline{}$

d. A single 5 can appear if either the first die shows a 5 or the second die shows a 5. What is the probability that exactly one 5 will appear?

e. What is the probability that no 5s will appear? How is this probability related to the probabilities you found in parts b and d?

f. Do you think you would win or lose money in the long run if you played this game? Explain.

13. The multiplication principle for independent events can be used for more than two such events. The carnival game Chuk-a-Luk, for example, is similar to the game proposed by Chuck in Lesson 6.1, except that three dice are rolled. If you bet on 5, you can calculate the probability that all three dice will show a 5 by cubing the probability that a single die will show a 5. What is the probability that all three dice will show a 5?

14. The probability that one of Ms. Howe's plants will bloom is .9. What is the probability that all five will bloom?

15. European roulette wheels contain the numbers 0 through 36. The ball spun around the perimeter of the wheel has an equal chance of landing on any of these numbers. The British actor Sean Connery once bet on the number 17 three times in a row, and the ball landed on 17 all three times. What is the probability that the ball will land on the same number three times in a row?

Unlike the European roulette wheel, an American wheel has a 00.

Two Columbia O-ring Seals Found Singed

Cape Canaveral,
March 5, 1996,
(Associated Press)

Seals in Columbia's booster rockets were singed during liftoff nearly two weeks ago, the latest in a series of problems with space shuttle O-rings.

The damage, caused by hot rocket gas, posed no threat to Columbia or its seven astronauts, NASA said Monday. Nonetheless, the space agency was treating the problem with ultracaution because of the tragic results of O-ring failure on Challenger 10 years ago.

The problem has been seen nine times before, most recently last fall. But it's the first time that gas snaked through the adhesive in more than one place in the so-called case-to-nozzle joint.

Last summer, O-ring seals in different nozzle joints were singed by rocket gas during back-to-back launches. NASA performed extensive repairs before allowing the shuttle to fly again. That problem has not resurfaced.

A leak of hot gas through O-rings in yet another booster joint caused the Challenger to explode in 1986, killing all seven crew members.

This photo was taken just seconds after the space shuttle *Challenger* exploded.

16. On January 28, 1986, the space shuttle *Challenger* exploded over Florida, killing astronauts Greg Jarvis, Christa McAuliffe, Ron McNair, Ellison Onizuka, Judy Resnik, Dick Scobee, and Mike Smith. Many authorities feel this accident could have been prevented if closer attention had been paid to the laws of probability. The rocket that carried the shuttle aloft was separated into sections that were sealed by large rubber O-rings. A presidential commission found that the accident was caused by the leakage of burning gases from the O-rings. Studies found that the probability of a single O-ring's working properly was about .977.

a. If the *Challenger's* six O-rings were truly independent of one another, what is the probability that all six would function properly?

b. After the commission issued its report, some people likened the probabilities in the *Challenger* accident to those in the game of Russian roulette. Compare the probability that all six O-rings will function properly with the probability that a six-chamber revolver will fire if only one chamber contains a bullet and a chamber is selected at random.

c. Each joint was sealed by a system of two O-rings that were supposed to be independent of each other but were not. If the probability that one O-ring fails is $1 - .977 = .023$, what is the probability that both O-rings in a system of two will fail if the two are independent of each other?

d. Subtract your previous answer from 1 to determine the probability that a single joint's seal will work and find the probability that all six will function properly. (Keep in mind that your answer is based on a faulty assumption of independence of the pair of O-rings in a single joint.)

17. It has been estimated that about 1 automobile trip in 100,000 ends in an injury accident.

a. What is the probability that a given automobile trip will not end in an injury accident?

b. If you make approximately 3 automobile trips a day, about how many would you make over a 30-year period?

c. If the outcome of a particular automobile trip is independent of the previous one, what is the probability none of the trips you make over a 30-year period will end in an injury accident?

18. Of the inhabitants of Wilderland, 40% are Hobbits and 60% are humans. Furthermore, 20% of all Hobbits wear shoes and 90% of all humans wear shoes.

a. Make a tree diagram to show the breakdown of residents into Hobbits who either do or do not wear shoes and humans who either do or do not wear shoes. Write the appropriate probabilities along the branches and write the appropriate events and their calculated probabilities to the right of the diagram.

b. Suppose 10,000 residents are selected at random. About how many would you expect to be Hobbits who wear shoes?

c. About how many of the 10,000 would you expect to be shoeless Hobbits?

d. About how many of the 10,000 would you expect to be humans who wear shoes?

e. About how many of the 10,000 would you expect to be shoeless humans?

f. What percentage of the inhabitants who wear shoes are Hobbits?

g. If an inhabitant is selected at random, are any of the events selecting a Hobbit, selecting a human, selecting someone who wears shoes, or selecting someone who doesn't wear shoes independent? Are any of them mutually exclusive? Explain.

19. A witness to a crime sees a man with red hair fleeing the scene and escaping in a blue car. Suppose that one man in ten has red hair and that one man in eight owns a blue car.

a. What is the probability that a man selected at random has red hair and owns a blue car?

b. What assumption did you make when you answered part a?

c. The following table represents the men in the community in question. Inspect the data and present an argument either in favor of or against the assumption.

	Blue Car	No Blue Car
Red hair	1,990	14,010
Nonred hair	18,015	125,985

20. Many games involve rolling a pair of dice.

a. In how many ways can a pair of dice fall?

b. What is the probability of rolling a pair of 6s?

c. What is the probability of rolling a pair of 6s twice in succession?

d. What is the probability of not rolling a pair of 6s?

e. What is the probability of not rolling a pair of 6s twice in succession?

f. A New York gambler nicknamed "Fat the Butch" once bet that he could roll at least one pair of 6s in 21 chances. Find the probability of not rolling a pair of 6s in 21 successive rolls of a pair of dice. (By the way, he lost about $50,000 in the course of several bets.)

21. The circles in the following diagram represent events A and B; the variables x, y, and z represent the number of things in each region. Use

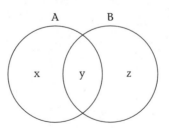

the diagram to explain why the multiplication principle for probabilities: $p(A \text{ and } B) = p(A) \times p(B \text{ from } A)$, is true.

Projects

22. Gather data on your school's student population broken down into the categories of the Central High population in this lesson. Determine whether the data exhibit any events that approximate independence.

23. Research the probabilities of about a dozen real-world events, such as being killed in an automobile accident, winning a lottery in your area, and being killed in a plane accident. Rank them according to their probability of occurring. Do you think the attention given to events in our culture is proportionate to their rate of occurrence? Explain.

24. Real-world data can exhibit events that come close to passing a test for independence. How close is close enough? If you have taken or are taking a statistics course, prepare a report to share with other members of your class that discusses how statisticians decide whether a difference is statistically significant.

25. Research and report on misuses of probability in the courtroom. For example, in a 1968 California case, People v. Collins (68 Cal 2d319), probabilities were erroneously multiplied. What precedent was set by this case? What are other probability-related precedents that have been established?

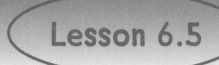
Probability, Part 2

This chapter began with an examination of several games proposed for a school fund-raiser. The counting and probability techniques you learned in the preceding sections are important tools in analyzing these games and many real-world situations that involve probability. However, the questions of whether the school can expect to make money on these games and how much it can expect to make (or lose) have not been answered completely. This lesson considers two important ideas—probability distributions and expected value—that will enable you to complete your analysis of the games proposed by Pierre, Hilary, and Chuck.

Binomial Probability Distributions

As a first example, consider a brief quiz of three questions in which the answers are either true or false. Because there are only two possible outcomes for any one question, the process of answering a single question is called **binomial**. When the process is repeated several times, however, the multiplication principle states that there are more outcomes than the two that are possible in a single trial. For example, with three trials there are $2 \times 2 \times 2 = 8$ outcomes. A tree diagram can be used as an aid to listing all eight outcomes (see Figure 6.6).

Because each of the outcomes is as likely as any other, the probability associated with each of the eight is $\frac{1}{8}$. The probabilities can also be calculated by writing a probability of $\frac{1}{2}$ on each branch of the tree diagram and multiplying (see Figure 6.7).

	Outcome	Number of trues
T	TTT	3
F	TTF	2
T	TFT	2
F	TFF	1
T	FTT	2
F	FTF	1
T	FFT	1
F	FFF	0

Figure 6.6 A three-question true/false quiz.

The results can be collected into a table called a **probability distribution** table. The way in which the table is constructed depends on whether you are interested in the number of true answers or the number of false answers. If you are interested in the number of true answers, then the table

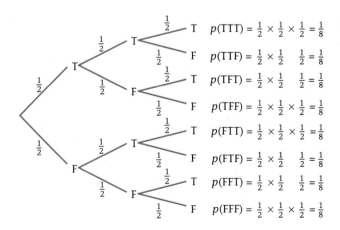

$p(\text{TTT}) = \frac{1}{2} \times \frac{1}{2} \times \frac{1}{2} = \frac{1}{8}$

$p(\text{TTF}) = \frac{1}{2} \times \frac{1}{2} \quad \frac{1}{2} = \frac{1}{8}$

$p(\text{TFT}) = \frac{1}{2} \times \frac{1}{2} \quad \frac{1}{2} = \frac{1}{8}$

$p(\text{TFF}) = \frac{1}{2} \times \frac{1}{2} \times \frac{1}{2} = \frac{1}{8}$

$p(\text{FTT}) = \frac{1}{2} \times \frac{1}{2} \times \frac{1}{2} = \frac{1}{8}$

$p(\text{FTF}) = \frac{1}{2} \times \frac{1}{2} \quad \frac{1}{2} = \frac{1}{8}$

$p(\text{FFT}) = \frac{1}{2} \times \frac{1}{2} \quad \frac{1}{2} = \frac{1}{8}$

$p(\text{FFF}) = \frac{1}{2} \times \frac{1}{2} \times \frac{1}{2} = \frac{1}{8}$

Figure 6.7 Probabilities in a three-question true/false quiz.

looks like this:

Number of True	0	1	2	3
Probability	$\frac{1}{8}$	$\frac{3}{8}$	$\frac{3}{8}$	$\frac{1}{8}$

A Binomial Probability Shortcut

The tree diagram approach isn't useful if the number of outcomes is large, which is often the case. The counting techniques and the multiplication principle you learned in previous lessons of this chapter can be used to develop an alternative to the tree diagram approach.

For example, consider the probability that exactly two of the three answers are true. The blanks shown below represent the three questions. There are three ways you can select the two to mark true.

Possibility 1	Possibility 2	Possibility 3
_____ ←	_____ ←	_____
_____ ←	_____	_____ ←
_____	_____ ←	_____ ←

Because the order of selection of the two questions to mark true does not matter, the number of ways of selecting two questions to mark true can be counted as $C(3, 2) = \dfrac{3!}{2!1!} = 3$. The probability associated with each true answer is $\frac{1}{2}$, and the probability associated with each false answer is also $\frac{1}{2}$, so the probability of two true answers followed by one false answer is $\left(\frac{1}{2}\right)^2 \left(\frac{1}{2}\right)$. This probability is multiplied by 3 because there are three ways two trues can occur.

As a second example, consider a quiz of ten questions and the probability that, say, exactly four of them are true. If there are four true, there must be six false, so the probabilities that must be multiplied are four $\frac{1}{2}$s for the true answers and six $\frac{1}{2}$s for the false, which gives $\left(\frac{1}{2}\right)^4 \left(\frac{1}{2}\right)^6$. The number of ways that the four true questions can be selected is $C(10, 4)$. The correct probability, therefore, is $C(10, 4) \left(\frac{1}{2}\right)^4 \left(\frac{1}{2}\right)^6 = 210 \left(\frac{1}{64}\right) \left(\frac{1}{64}\right)$, or about .205.

Note that the denominators of the combination formula match the powers of the probabilities.

$$C(10, 4)\left(\tfrac{1}{2}\right)^4\left(\tfrac{1}{2}\right)^6 = \frac{10!}{4!\,6!}\left(\tfrac{1}{2}\right)^4\left(\tfrac{1}{2}\right)^6$$

Point of Interest

```
binompdf(10,.5,4
)
        .205078125
```

```
10 nCr 4*.5^4*.5
^6
        .205078125
```

The first screen shows a binomial probability calculation on a calculator with a binominal probability function. The number of trials is the first number, the probability of a single true is the second, and the number of trues is the third. The second screen shows a binomial probability calculation using a calculator's combination function.

When using this method of calculating binomial probabilities, the probability of a single question's being true is multiplied by itself several times. It is essential, therefore, that the individual outcomes be independent of one another. If, for example, the answer to one question depends on an answer to another, then this technique should not be used.

The probability associated with a given outcome is often different from $\frac{1}{2}$. Consider, for example, the dice game proposed by Chuck in Lesson 6.1. Because this game involves only two dice, it is not difficult to analyze it with tree diagrams. However, to serve as an example, the following analysis uses the counting technique just discussed.

If you bet on the number 5 in Chuck's game, then you will win \$2 if two 5s appear, win \$1 if a single 5 appears, or lose \$1 if no 5s appear. Consider the possibility that one 5 will appear. There are $C(2, 1)$ ways to select the die that shows a 5, and the probability of one 5 is $\left(\frac{1}{6}\right)\left(\frac{5}{6}\right)$ because one of the dice must be a 5 and the other must be anything but a 5. The correct probability is therefore $C(2, 1) \left(\frac{1}{6}\right)^1 \left(\frac{5}{6}\right)^1$, or about .278.

Similarly, the probabilities of no 5s and two 5s can be calculated as $C(2, 0) \left(\frac{1}{6}\right)^0 \left(\frac{5}{6}\right)^2$ and $C(2, 2) \left(\frac{1}{6}\right)^2 \left(\frac{5}{6}\right)^0$, respectively. The probabilities are summarized in the following distribution table.

Amount Won	−1	1	2
Probability	.694	.278	.028

In general, if p is the probability associated with a single binomial outcome, the probability of n successes in m attempts is $C(m, n)(p)^n (1 - p)^{m-n}$, provided that individual trials are independent.

Expectation

The question that remains to be answered is how a player could expect to do in Chuck's game in the long run. The player's **expectation** (also known as the expected value of the player's probability distribution) is used to answer this question.

The calculation of expectation for a player in Chuck's game weights each amount that Chuck might win (or lose) according to its probability: $-1(.694) + 1(.278) + 2(.028) = -.36$. The expectation can be interpreted as the average amount the player can expect to win per play of the game. If, for example, the game is played 100 times, the player can expect to be about $100(.36) = \$36$ behind.

If the Central High council decides to use this game as a fund-raiser, the council's viewpoint will be the opposite of that of the player: It loses $2 when two 5s appear, loses $1 when one 5 appears, and makes $1 when no 5s appear. Therefore, the council's expectation is +$.36. The council can expect to make about $36 for each 100 times the game is played.

A Binomial Probability/Expectation Example

A quality control engineer at the manufacturing plant of an electronics company randomly selects five compact disc players from the assembly line and tests them for defects. If a problem on the assembly line causes the factory to produce 10% defective, what is the probability that the problem will be detected by the engineer's test?

The following table shows the probability distribution for the number of defective players that the engineer found. The probabilities have been rounded to three decimal places.

Number Defective	Probability Calculation	Probability
0	$C(5, 0)(.1)^0(.9)^5$	.590
1	$C(5, 1)(.1)^1(.9)^4$	.328
2	$C(5, 2)(.1)^2(.9)^3$	.073
3	$C(5, 3)(.1)^3(.9)^2$	.008
4	$C(5, 4)(.1)^4(.9)^1$	.000
5	$C(5, 5)(.1)^5(.9)^0$	.000

There is about a 59% chance that the engineer will fail to find a defective player if the defective level is 10%. The probability that the engineer will find at least one defective player is about 41%, which can be found by subtracting 59% from 100% or by adding the last five probabilities in the table.

The engineer's expectation can be calculated as 0(.590) + 1(.328) + 2(.073) + 3(.008) + 4(.000) + 5(.000), or about .498, which can be interpreted as the average number of defective CD players the engineer detects. To put it another way, if the engineer goes through this routine once each day when the defective level is at 10%, the engineer can expect, on average, to detect about one-half a defective CD player a day, or one every two days.

Point of Interest

```
binomcdf(10,.4,3
)
         .3822806016
```

Some calculators have a function that finds the sum of probabilities in a binomial distribution. This screen shows such a calculation in a situation with a .4 probability of success. The calculated probability is the sum of the probabilities of 0, 1, 2, or 3 successes in 10 trials.

The following exercises treat the use of binomial probability and expectation in a variety of settings.

Exercises

1. Hale Ault, a student at Central High, is known for occasionally neglecting his studies. When he finds a question on an exam that he cannot answer, he uses one of several random processes as an aid. Examine each of Hale's schemes and discuss, first, whether each outcome has the same probability of occurring as does each of the others and, second, whether several successive applications of the scheme are independent of one another.

   a. On a true/false question, flip a coin and answer true if the result is heads and false if it is tails.

   b. On a three-choice multiple-choice question, flip two coins. Mark the first answer if both coins are heads, the second if both are tails, and the third if the result is one head and one tail.

   c. On a four-choice multiple-choice question, associate each of your fingers on one hand with one of the choices, slap your fingers against the desk, and select the one that stings the most.

   d. On a four-choice multiple-choice test that allows the use of scientific calculators, use the calculator's random number generator to display a random-number between 0 and 1. Mark the first answer if the number is below 0.25, the second if it is between 0.25 and 0.5, the third if it is between 0.5 and 0.75, and the fourth if it is between 0.75 and 1.

2. Ms. Howe is giving a five-question true/false quiz.

   a. In how many ways can a student select three of the questions to mark true?

   b. Show how to calculate the probability that exactly three of the questions on the quiz are true.

   c. Complete the probability distribution for the number of true answers:

Number of True	0	1	2	3	4	5
Probability						

   d. Ms. Howe has a bias toward true answers, and so her questions have true answers about 70% of the time. Recalculate the probability distribution:

Number of True	0	1	2	3	4	5
Probability						

3. Hale Ault is taking a ten-question true/false quiz on which the answers have equal chances of being true or false. Hale is doing the quiz by guessing and needs at least six correct in order to pass.
   a. Find the probability of exactly six correct answers.
   b. Find the probability of exactly seven correct answers.
   c. Find the probability of exactly eight correct answers.
   d. Find the probability of exactly nine correct answers.
   e. Find the probability of exactly ten correct answers.
   f. Hale will pass if he gets six right or if he gets seven right or if he gets eight right or if he gets nine right or if he gets ten right. What is the probability that Hale will pass the quiz?

4. Recall that the game of Chuk-a-Luk is played with three dice and that you win $1 for each time your number shows but lose $1 if your number doesn't show.
   a. Suppose you bet on the number 5. Show how to calculate the probability that 5 shows exactly once.
   b. Complete the probability distribution for the number of times that 5 shows:

Number of 5s	0	1	2	3	4	5
Probability						

   c. Complete the following calculation of your expected winnings: (Note that the distribution of winnings is different from the distribution of the number of 5s because zero 5s results in a loss of $1.)

   $$-1(\underline{\qquad}) + 1(\underline{\qquad}) + 2(\underline{\qquad}) + 3(\underline{\qquad}) = \underline{\qquad}.$$

   d. If you play the game 100 times, about how much money should you expect to win or lose? Explain.

5. A list of people eligible for jury duty contains about 40% women. A judge is responsible for selecting six jurors from this list.
   a. If the judge's selection is made at random, what is the probability that three of the six jurors will be women?
   b. Prepare a probability distribution table for the number of women among the six jurors.

Convictions in Jay Bias Homicide Reversed
Prosecutor Violated Equal Rights by Blocking Women From Jury

WASHINGTON POST,
April 28, 1993,
by Jon Jeter

The Maryland Court of Appeals reversed the convictions yesterday of two men in the killing of James S. "Jay" Bias, the brother of the late college basketball star, Len Bias, because the prosecutor blocked more than a dozen women from serving on the jury solely because of their sex.

The state's highest court ruled that a Prince George's County prosecutor who tried the murder case against Gerald Eiland and Jerry S. Tyler violated the state's equal rights law by barring the women from serving on the jury. The assistant state's attorney,

Mark Foley, used 16 of his 20 peremptory strikes—the legal device by which attorneys from both sides may excuse potential jurors without explanation—to bar women from the jury.

The decision extends to women the protection from discrimination provided by a 1986 Supreme Court decision, which reaffirmed that black people cannot be barred from a jury solely because of their race. And it could end a long-standing practice by some prosecutors who eliminate women from juries based on the belief that they are more compassionate than men and less likely to convict, particularly when the accused is a young man.

c. Suppose that the judge's selection includes only one woman. Do you think this is sufficient reason to suspect the judge of discrimination? Explain.

6. Sickle cell anemia is a genetic disease that strikes an estimated 1 in 400 African-American children in the United States. The disease causes red blood cells to have a crescent shape rather than the normal round shape, which inhibits their ability to carry oxygen. Victims suffer from severe pain and are susceptible to pneumonia and organ failure. Children of parents who are both carriers of the sickle cell gene are frequently stricken.

a. Normal parents have two normal A genes, and carrier parents have one normal A gene and one sickle S gene. A victim of the disease has two S genes. A child inherits one gene independently from each parent. Complete the probability calculations in the tree diagram representing parents who are both carriers.

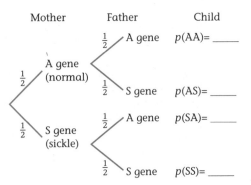

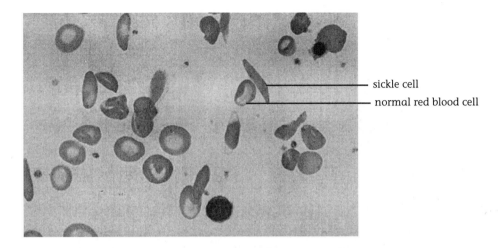

sickle cell
normal red blood cell

b. What percentage of children of two carrier parents will have sickle cell anemia? What percentage are carriers? What percentage are normal?

c. A couple who are both carriers have five children. Complete the following probability distribution for the number of children who have the disease.

Number of Children	0	1	2	3	4	5
Probability						

d. Calculate the expectation. Interpret the expectation in this case.

7. A quality control engineer at a widget factory randomly selects three widgets each day for a thorough inspection. Suppose the assembly process begins producing 20% defective.
   a. Prepare a probability distribution for the number of defective widgets the engineer will find in the sample of three.
   b. Do you think the engineer's quality control scheme is a good one? If not, suggest a way to improve it.

8. A lottery ticket costs $1 and requires you to select 6 numbers from the 44 available.
   a. If the $27 million jackpot is the only prize and you do not have to share the jackpot, complete the following probability distribution for your expected winnings (see Exercise 9 in Lesson 6.3, pages 298 and 299).

Amount Won	$27 million	−$1
Probability		

b. Calculate the expected value for this distribution.

c. Recall that an Australian group purchased 5 million tickets in the Virginia lottery. Assume that the jackpot is the only prize, revise the distribution, and recalculate the expectation for the Australian group's winnings.

9. A country has a series of three radar defense systems that are independent of one another. The probability that an enemy plane escapes any one system is .2.

a. Prepare a probability distribution for the number of radar systems that a plane will escape.

b. What is the probability that an enemy plane will penetrate the country's radar defenses?

10. Recall the lottery that Hilary proposed in Lesson 6.1. It required selecting two numbers from the nine available. Hilary has proposed that the price of a ticket be $1, that the prize for matching both numbers be $20, and that a prize of $1 be given to anyone who matches one of the two winning numbers.

a. Prepare a probability distribution for the amount a player can expect to win.

b. Calculate the player's expectation.

c. Would the Central High council make money on this game? Explain. If the council would lose money, suggest a revision of Hilary's plan for awarding prizes so that the council could make money.

11. A fair coin is tossed several times.

a. Find the probability of obtaining exactly 5 heads in 10 tosses. (Do not do the entire probability distribution.)

b. Find the probability of obtaining exactly 10 heads in 20 tosses. Compare this with the previous answer.

c. Prepare a partial distribution of the probabilities of obtaining 4, 5, or 6 heads in 10 tosses.

d. Prepare a partial distribution of the probabilities of obtaining 8, 9, 10, 11, or 12 heads in 20 tosses.

e. Are you more likely to obtain between 40 and 60% heads in 10 tosses or in 20? Explain.

12. Extra Sensory Perception (ESP) is the ability to communicate with another person without speaking. One common test for ESP requires that while one person concentrates on a card selected at random from a special deck, the other person records the image that is received or felt.

   a. If the deck consists of five of each of the cards shown here, what is the probability of guessing any one card correctly?
   b. Consider a test in which while one person selects a card and concentrates on it, the other person records his or her impression. The card is placed back in the deck, the deck is shuffled, and the experiment is repeated a total of five times. Prepare a probability distribution for the number of cards the receiver can guess correctly.
   c. Suppose the receiver gets more than three correct. What is the probability of this happening by chance?

13. Recall the word game proposed by Pierre in Lesson 6.1. Suppose that the only two-letter words made from the letters of *Lions* that are considered legal are *in, is, on, no,* and *so.*
   a. What is the probability that a player will draw a legal word from the letters recorded on the Ping-Pong balls?
   b. Pierre proposes that the charge for playing the game be $.50 and the prize for selecting a legal word be $1. Prepare a probability distribution for the winnings of someone who plays the game.
   c. Calculate the player's expectation.
   d. Would the council make money on the game? If your answer is no, suggest a revision of Pierre's plan so that the council could make money.

14. Sara Swisher, the Central High Lions' star basketball player, has a field goal percentage of 65%.
   a. Sara attempts seven field goals in the first quarter of tonight's game. Prepare a probability distribution for the number of field goals that Sara makes.

b. What assumption have you made? Do you think this is a realistic assumption?

c. Calculate the expectation. What does it mean in this case?

15. Expected value is used to help people make decisions. For example, a person might decide to make an investment if the expectation is positive but not to do so if the expectation is negative.

a. The price of one share of a stock is $35. You estimate that the probability the price will fall to $30 is .3, the probability the price will fall to $25 is .1, the probability the price will increase to $38 is .4, and the probability the price will increase to $42 is .2. Should you buy the stock? Explain.

b. A lottery ticket costs one dollar. To win the jackpot, a participant must match 6 numbers from 42 available. Assuming that the participant does not share the jackpot with anyone else and that the jackpot is the only prize, how large must the jackpot be in order for the player's expectation to be positive?

c. Is a positive expected value a good reason to play a lottery? Explain.

16. *Pascal's triangle* is an array of numbers that you may have seen in an algebra class. To construct the triangle, begin with a row of two 1s: 1 1. Each new row starts and ends with a 1, and the other numbers are found by adding the numbers above and on either side of them in this way:

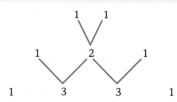

a. Continue the triangle for three additional rows.

b. Calculate all possible combinations of five things: $C(5, 0)$, $C(5, 1)$, $C(5, 2)$, $C(5, 3)$, $C(5, 4)$, and $C(5, 5)$.

c. Where do your answers in part b occur in the triangle?

d. The following tree diagram shows all possible ways of answering true/false quizzes with up to four questions. Fill in the distributions of the number of trues for quizzes of one, two, three, and four questions by tracing the paths of the diagram. Compare these numbers to those in Pascal's triangle.

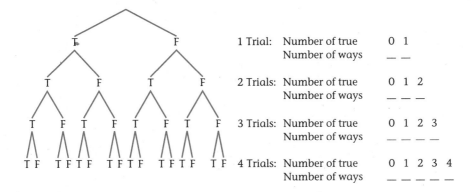

Computer/Calculator Explorations

17. Write a program for your computer or programmable calculator that does probability distributions of the type discussed in this lesson. The program should accept as input the probability of a single success and the number of trials. It should calculate and display each number of successes and the related probability.

Projects

18. Pascal's triangle contains many patterns other than those in Exercise 16. Investigate and report on some of them.

19. Research and report on the use of expected value as a decision-making tool. How, for example, is it used in business?

20. This lesson discusses a combinatorial technique for calculating binomial probabilities. Although useful, this technique has limitations, particularly if the number of trials is very large. Mathematicians sometimes use the Poisson distribution to approximate binomial probabilities. Research and report on the use of the Poisson distribution to approximate binomial probabilities.

21. On September 9, 1990, the Sunday newspaper supplement *Parade* carried a column by Marilyn vos Savant in which a reader posed a problem about the television game show *Let's Make a Deal.* The problem asked whether a contestant should switch doors after the contestant's selection of one of three available doors prompted the show's host to open a door containing a worthless prize. Marilyn's response that the contestant should switch brought a flood of mail, most of which disagreed. Research the controversy and prepare a report on the arguments on both sides. Select the answer with which you agree and defend it.

Monte Carlo Methods

Direct calculation of probabilities of real-world events is sometimes difficult, even for professional mathematicians. Such was the experience of Stanislaw Ulam when he worked in a laboratory in Los Alamos, New Mexico, in the 1940s. Ulam used a computer to simulate the occurrence of events for which he could not calculate probabilities directly. He chose the name "Monte Carlo" to describe his approach.

Mathematician of Note

Stanislaw Ulam came to the United States from Poland in 1935 at the invitation of fellow mathematician John von Neumann. He worked first at the Institute for Advanced Study at Princeton University, then at Los Alamos, where he was instrumental in the development of the hydrogen bomb.

Today's computers are faster and more readily available than those Ulam used at Los Alamos. Even handheld calculators can do many of the tasks only computers could perform a half-century ago.

Consider, for example, the answers to true/false questions. The random selection of a single answer can be simulated by flipping a coin, but the selection can also be simulated by generating a random number. Many calculators have a random-number function that generates random decimals between 0 and 1. The instruction int(2 ∗ rand) doubles this range, then drops the decimal; thus generates random 0s and 1s. (Some calculators have a second random number function that generates random integers

```
0 → T: 0 → F
For (N,1,100)
int(2 ∗ rand) → R
If R = 0
1 + F → F
If R = 1
1 + T → T
End
Disp F,T
```

over a specified range.) A simple program that simulates the random selection of 100 answers to true/false questions is shown on page 335. It runs on most Texas Instruments graphing calculators.

The accuracy of answers obtained with Monte Carlo methods depends on the number of trials that are run: a larger number of trials is more likely to produce an answer close to the correct one than is a smaller number of trials. Thus, you can expect the percentage of trues given by this simulation to be closer to 50 if 1,000 trials are used instead of 100. Computers, it should be noted, can perform a large number of trials much more quickly than calculators.

The following is an example of a difficult problem that can be solved by Monte Carlo methods.

Engineers sometimes need to determine the temperature of a point on a rectangular metal plate from the temperatures at the four edges. Direct calculation requires calculus.

Figure 6.8 represents a 10 centimeter × 6 centimeter plate. Coordinates have been assigned to facilitate the simulation. The point in question is at (6, 4). Edge temperatures are written along the edges.

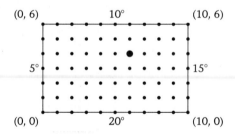

Figure 6.8 A 10 centimeter × 6 centimeter plate.

The Monte Carlo solution to this problem envisions a random walk starting at (6, 4). A random selection determines a movement of either one unit to the right, one unit to the left, one unit upward, or one unit downward. The selection is repeated until the walk terminates at one of the edges. After many trials, the temperature at each edge is weighted according to the portion of trials that terminated at the edge. The weighted average of the four edge temperatures gives an estimate of the temperature at the selected point.

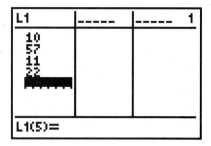

```
RCT H?10
RCT V?6
PT X?6
PT Y?4
TRIALS?100
```

```
L1      -----  -----  1
  10
  57
  11
  22

L1(5)=
```

The above screens show the input (left) and results (right) of the TI-83 program RCTRNDWK, which accompanies this book. The user is prompted for the rectangle's horizontal and vertical dimensions, the coordinates of the point, and the number of trials. The numbers of times the walk ended at the left, top, right, and bottom edges are stored, in that order, in the calculator's list L1. Based on these 100 trials, an estimate of the temperature at the point (6, 4) is 5(.10) + 10(.57) + 15(.11) + 20(.22), or about 12°. A more reliable estimate can be made by running 1,000 trials.

1. Write a summary of what you think are the important points of this chapter.

2. The following table represents ownership of cats among professionals and non-professionals in a community.

	Own a Cat	Don't Own a Cat
Professionals	2,300	11,400
Nonprofessionals	2,600	27,600
Totals	7,900	39,000

   a. If a person is selected at random from this group, what is the probability of selecting someone who owns a cat?
   b. What is the probability of selecting a cat owner from the professionals?
   c. What is the probability of selecting a professional?
   d. What is the probability of selecting a person who owns a cat and is a professional?
   e. What is the probability of selecting someone who owns a cat or is a professional?
   f. Are the events selecting a cat owner and selecting a professional independent? Explain.
   g. Are the events selecting a cat owner and selecting a professional mutually exclusive? Explain.

3. Lesson 1.5 examined voting situations in which some voters received more votes than others. Recall that a coalition is a collection of voters. In how many ways can you form a coalition of one, two, three, four, or five voters from a group of five voters?

4. Teams A and B are playing in the world series, and each has a 50% chance of winning any game.
  a. What is the probability that team A will win the first four games?
  b. What is the probability that team B will win the first four games?
  c. The series will end in four games if either team A or team B wins the first four games. What is the probability that the series will end in four games?

5. a. In how many ways can six books be arranged on a shelf?
  b. If two of the books are math books, how many arrangements have the math books in the first two positions? (Hint: Draw six blanks and use the multiplication principle.)

Reprinted by permission of United Feature Syndicate.

  c. What is the probability that the math books will be in the first two positions?
  d. In how many arrangements are the math books next to each other?
  e. What is the probability that the math books will be next to each other?

6. You are playing a game in which you flip two coins. If both show heads, you win $2; if both show tails, you win $1; but if the coins do not match, you lose $1.
  a. Construct a probability distribution for the amount you could win on a single play of the game.
  b. Calculate your expectation.
  c. If you played the game 100 times, about how much would you expect to win or lose?

7. Being listed first on an election ballot is known to improve a candidate's chances. In order to minimize this effect, ballots could be printed in such a way that each candidate is first on some ballots but not on all. There are three candidates for mayor and five candidates for city council in a local election. How many different orderings are possible if the mayoral candidates must be listed before the council candidates? If 20,000 people vote, about how many would see each ballot?

8. Are you more likely to win the jackpot in a lottery that requires the selection of 5 numbers from 39 or in one that requires the selection of 6 numbers from 36? If 3,000,000 tickets are sold in each lottery, about how many winning tickets would you expect in each?

9. Two cards are drawn from a standard deck, and the first card is not put back before the second card is drawn.
   a. What is the probability that the first card will be a heart?
   b. What is the probability that the second card will be a heart if the first card is a heart?
   c. What is the probability the first card will be a heart and the second card will be a heart?
   d. The following tree diagram represents the two draws. Write the correct probabilities along the branches and calculate the probabilities at the right.

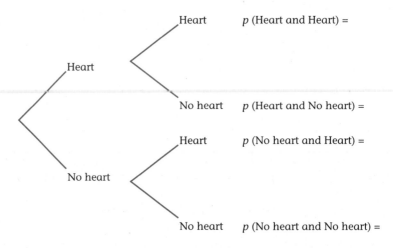

Heart p (Heart and Heart) =

Heart

No heart p (Heart and No heart) =

Heart p (No heart and Heart) =

No heart

No heart p (No heart and No heart) =

   e. Is the second draw independent of the first? Explain.

10. There are five boys and six girls in a group, and a committee of three is being selected.
   a. In how many ways can the committee be formed?
   b. How many committees will consist of one boy and two girls?

c. What is the probability that the committee will have exactly one boy?

d. How many committees will consist of one boy or one girl?

e. What is the probability that the committee will have one boy or one girl?

11. Suppose 90% of all drivers are good, 5% of all good drivers get tickets, and 70% of all bad drivers get tickets.

a. Write the appropriate probabilities along each branch of the following tree diagram and calculate the probabilities shown at the right.

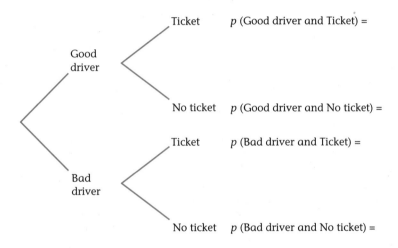

Good driver

Ticket p (Good driver and Ticket) =

No ticket p (Good driver and No ticket) =

Bad driver

Ticket p (Bad driver and Ticket) =

No ticket p (Bad driver and No ticket) =

b. If a community has 50,000 drivers, about how many would be expected to get tickets?

c. How many of the people who get tickets are bad drivers?

d. What is the probability that a person who gets a ticket is a bad driver?

12. Suppose that you and three friends each choose a number between 1 and 10.

a. What is the probability that all three of your friends will pick the same number you pick?

b. What is the probability that all three of your friends will pick a number different from yours?

c. What is the probability that all three of your friends will pick a number that is different from the number picked by any of the others?

13. Two different prizes are being awarded in a group of ten people.
  a. In how many ways can this be done if the same person can win both prizes?
  b. In how many ways can this be done if each person can win no more than one prize?
  c. In how many ways can this be done if each person can win no more than one prize and both prizes are the same?

14. A fair die is rolled five times.
  a. What is the probability that a 6 will appear exactly twice?
  b. What is the probability that a 6 will appear two times or fewer?
  c. What is the probability that a 6 will never appear?

15. Egbert fixes an omelet for breakfast every morning. Depending on what he has in his refrigerator, he adds one or more of the following ingredients to his eggs: mushrooms, green peppers, cheddar cheese, and ham. How many different omelets can Egbert make?

16. The diagram shows the probabilities associated with events *A* and *B*.

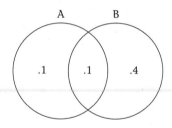

  a. What is $p(A)$?
  b. What is $p(A$ and $B)$?
  c. What is $p(A$ or $B)$?
  d. What is $p(A$ from $B)$?
  e. Are *A* and *B* mutually exclusive? Explain.
  f. Are *A* and *B* independent? Explain.

17. The sensitivity (the ability to detect a disease in a person who has it) of a medical test is 90%. The specificity (the ability to detect the absence of a disease in a person who doesn't have it) is 95%. About 400 people in 100,000 have the disease.
  a. If the disease is used to screen a large group of people who lack any signs of the disease, what percentage will test positive?

b. What is the probability that someone who tests positive does not have the disease (a false positive)?

c. In a group of 20 people who are screened, estimate the number who will have false positive tests.

18. A die is rolled. Consider the events the number rolled is divisible by 2, the number rolled is divisible by 3 and the number rolled is divisible by 5. Which pairs of events are mutually exclusive? Which pairs are independent?

Bibliography

Bernstein, Peter L. 1996. *Against the Gods: The Remarkable Story of Risk.* New York: Wiley.

Campbell, Stephen K. 1974. *Flaws and Fallacies in Statistical Thinking.* Englewood Cliffs, NJ: Prentice-Hall.

Casti, John L. 1996. *Would-Be Worlds: How Simulation is Changing the Frontiers of Science.* New York: Wiley.

David, F. N. 1998. *Games, Gods, and Gambling.* Mineola, NY: Dover.

Dewdney, A. K. 1996. *200% of Nothing.* New York: Wiley.

Freedman, David, et. al. 1998. *Statistics.* 2nd ed. New York: Norton.

Green, Thomas A., and Charles L. Hamburg. 1986. *Pascal's Triangle.* Palo Alto, CA: Dale Seymour.

Huff, Darrell. 1959. *How to Take a Chance.* New York: Norton.

Orkin, Mike. 1991. *Can You Win? The Real Odds for Casino Gambling, Sports Betting, and Lotteries.* New York: W. H. Freeman.

Packel, Edward W. 1981. *The Mathematics of Games and Gambling.* Washington, DC: Mathematical Association of America.

Paulos, John Allen. 1988. *Innumeracy: Mathematical Illiteracy and Its Consequences.* New York: Hill and Wang.

Peterson, Ivars. 1997. *The Jungles of Randomness.* New York: Wiley.

Weaver, Warren. 1963. *Lady Luck: The Theory of Probability.* Garden City, NY: Doubleday.

Ulam, Stanislaw M. 1976. *Adventures of a Mathematician* New York: Scribner's.

vos Savant, Marilyn. 1996. *The Power of Logical Thinking.* New York: St. Martin's.

Matrices Revisited

T he daily business activity that supplies us with the products and services we need generates large quantities of data that often must be organized into matrices to be understood. Proper organization of these data, as reflected in this board at the Chicago Commodity Exchange, is necessary not only for understanding but also for effective planning.

For example, how can a company that provides batteries for another company's compact disc players be sure that it will have enough batteries on hand to fill orders? How does a fast-food chain determine prices that will allow it to do as well as possible against a competitor? How can a meteorologist use data about recent weather activity to predict the weather for tomorrow or a week from now? How can a park service use birth and survival rates of deer in managing herd populations? Matrices demonstrate remarkable versatility in helping to solve these and other real-world problems.

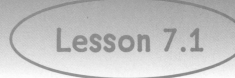

The Leontief Input-Output Model, Part 1

The Leontief input-output model discussed in this lesson is used to analyze the flow of goods among sectors in an economy. This model, developed by Nobel laureate Wassily Leontief of Harvard University in the 1960s, can be applied to extremely complex economies with hundreds of production sectors, such as a country (or even the world), or to a situation as small as a single company that produces only one product.

You begin your exploration of this model by looking at a simple case. Suppose, for example, that the Best Battery Company of Lincoln, Nebraska, manufactures a particular type of battery that is used to power various kinds of electric motors. Not all the batteries produced by the company, however, are available for sale outside the company. For every 100 batteries produced, 3 (3%) are used by various departments within the company. Thus, if the company produces 500 batteries during a week's time, 15 will be used within the company and 485 will be available for external sales. In general, for

a total production of P batteries by this company, $0.03P$ batteries will be used internally and $P - 0.03P$ will be available for external sales to customers.

In other words, the number of batteries available for sale outside the company equals the total production of batteries less 3% of that total production. If D represents the number of batteries available for external sales **(demand)** and P represents the total production of batteries, then the following linear equation can be used to model this situation.

$$P - 0.03P = D. \tag{1}$$

Suppose the company receives an order for 5,000 batteries. What must the total production be to satisfy this external demand for batteries? To find the total production necessary, substitute 5,000 for D in the previous equation and solve for P:

$$P - 0.03P = 5,000 \tag{2}$$

$$P(1 - 0.03) = 5,000 \tag{3}$$

$$P = 5,155 \text{ batteries.} \tag{4}$$

Hence, the company must produce a total of 5,155 batteries to satisfy an external demand for 5,000 batteries.

> Notice that the total production of batteries equals the number of batteries used within the company during production plus the number of batteries necessary to fill the external demand.

Now take a look at a simple two-sector economy. Suppose the Best Battery Company buys an electric motor company and begins producing motors as well as batteries. The company's primary reason for this merger is that electric motors are used to produce batteries. Batteries are also used to manufacture motors. The expanded company has two divisions, the battery division and the motor division. It also has changed its name to the Best Battery and Motor Company.

The production requirements of the newly expanded company's two divisions are:

Battery Division

1. For the battery division to produce 100 batteries, it must use 3 (3%) of its own batteries.

2. For every 100 batteries produced, 1 motor (1% of the number of batteries produced) is required from the motor division.

Raytheon Co. to Buy Business Jet Unit of British Aerospace for $391 Million

WALL STREET JOURNAL,
June 2, 1993
by David Stipp and
Brian Coleman

Raytheon Co., as expected, agreed to buy British Aerospace PLC's business jet unit for about $391 million.

The planned purchase would help Raytheon move beyond its core defense business and enable British Aerospace to reduce debt and better focus on its principal businesses, including automobiles and defense.

The purchase would nearly double Raytheon's corporate-jet market share and increase its overall aircraft sales to about $1.7 billion from $1.3 billion annually. Raytheon owns Beech Aircraft, a Wichita, Kan., maker of turboprop and jet aircraft, which has about 12% of the market for light to medium jets. The industry leader for business jets, Textron Inc.'s Cessna Aircraft unit, has an estimated 60% market share.

Max Bleck, Raytheon's president, said that although the small-plane market "has been in general malaise," Raytheon is positioning itself for an industry turnaround by buying when acquisition prices are low. Mr. Bleck reiterated Raytheon's goal of boosting to 50% from about 30% its portion of operating profit that is derived from non-defense business. He said Raytheon remains interested in possible acquisitions in the engineering, construction and appliance areas.

Motor Division

1. For the motor division to produce 100 motors, it must use 4 (4%) of its own motors.

2. For every 100 motors produced, 8 batteries are required from the battery division. (Notice that the number of batteries required is 8% of the total number of motors produced.)

The production needs within this two-sector economy can be represented visually by using a weighted digraph as shown in Figure 7.1. The digraph shows that if the company needs to produce a total of b batteries and m motors, then:

1. The battery division will require $0.03b$ batteries from its own division and $0.01b$ motors from the motor division.

2. The motor division will require $0.04m$ motors from its division and $0.08m$ batteries from the battery division.

A third way to present the information regarding the company's production needs is with a matrix. This matrix, which shows the required input from each sector of the economy, is called a **consumption matrix** for the economy.

$$\begin{array}{c} \\ \text{From} \end{array} \begin{array}{cc} & \text{To} \\ & \begin{array}{cc} \text{Battery} & \text{Motor} \end{array} \\ \begin{array}{c} \text{Battery} \\ \text{Motor} \end{array} & \begin{bmatrix} 0.03 & 0.08 \\ 0.01 & 0.04 \end{bmatrix} \end{array}$$

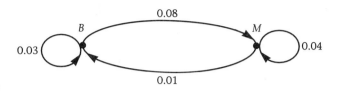

Figure 7.1 Weighted digraph showing the input required by each division.

As an example, to produce 200 batteries, the battery division will need $(0.03)(200) = 6$ batteries from the battery division and $(0.01)(200) = 2$ motors from the motor division.

Explore This

At the direction of your teacher, divide your class into groups of three to four people. Each group is to represent a production management team for the newly merged battery and motor company just described.

The company has given the battery division a daily total production quota of 1,000 batteries. The daily quota for the motor division is 200 motors. The production management team's problem is to determine how many batteries and motors each of the company's divisions will need during production and how many batteries and motors will be available for sales.

To solve the problem, the management team must complete the following tasks:

1. Find the number of batteries and motors needed by the battery division to meet its quota. Repeat for the motor division.

2. Find the number of batteries available for external sales if both divisions meet their daily quotas. Repeat for the number of motors available for sales outside the company.

3. Make up a formula for computing the number of batteries that the company will use internally to meet its production quotas. Repeat for the number of motors used internally. These formulas will be used to program a computer to do future calculations to save the management team some work. Let b represent the total number of batteries produced, and m represent the total number of motors produced by the company.

4. Explore this problem from another point of view. Suppose the company needs to fill an order of 400 batteries and 100 motors. Estimate the total production needed from each division to fill these orders. Check your estimate and revise it if necessary. Hint: Recall that the amount of a product available to fill an external order (outside demand) equals the total production minus the amount of that product consumed internally by the company.

5. If time permits, find a system of two equations in two unknowns that could be used to find the total production for each division required in task 4.

After all groups have finished tasks 1 through 4, a spokesperson for each production management team should present the results of the team's discussion to the class.

Exercises

1. A utility company produces electric energy. Suppose that 5% of the total production of electricity is used up within the company to operate equipment needed to produce the electricity. Complete the following production table for this one-sector economy.

Total Production Units	Units Used Internally	Units for External Sales
500	$.05(500) = $ —	$500 - .05(500) = $ —
900	—	—
—	100	—
—	250	—
—	—	2,375
—	—	7,125
P	—	—

2. Suppose that for every dollar's worth of computer chips produced by a high-tech company, 2 cents' worth is used by the company in the manufacturing process.
   a. What percentage of the company's total production of computer chips is used up within the company?
   b. What would the weighted digraph look like for this situation?

    c. Write an equation that represents the dollars' worth of computer chips available for external demands (*D*) in terms of the total production (*P*) of chips by the company.

    d. What must the total production of computer chips be in order for the company to meet an external demand for $20,000 worth of computer chips?

3. The high-tech company described in Exercise 2 adds another division that produces computers. Each division within the expanded company uses some of the other division's product:

    Computer Chip Division: Every dollar's worth of computer chips produced requires an input of 2 cents' worth of computer chips and 1 cent's worth of computers.

    Computer Division: Every dollar's worth of computers produced requires an input of 20 cents' worth of computer chips and 3 cents' worth of computers.

    a. Draw a weighted digraph that summarizes the production needs for this two-sector economy.

    b. Construct a consumption matrix for this economy.

    c. Suppose the computer chip division produces $1,000 worth of chips. How much input does it need from itself and from the computer division?

    d. Suppose the computer division produces $5,000 worth of computers. How much input does it need from itself and from the computer chip division?

    e. What must the total production of computer chips be for the company to meet an external demand for $25,000 worth of computer chips?

    f. What must the total production of computers be for the company to meet an external demand for $50,000 worth of computers?

4. A company has two departments: service and production. The needs of each department within this company (in cents per dollar's worth of output) are shown in the following consumption matrix.

	Service	Production
Service	0.05	0.20
Production	0.04	0.01

a. Draw a weighted digraph to show the flow of goods and services within this company.

b. Complete the following: For every dollar's worth of output, the service department requires _____ cents' worth of input from its own department and _____ cents' worth of input from the production department. For every dollar's worth of output, the production department requires _____ cents' worth of input from its own department and _____ cents' worth of input from the service department.

c. Suppose that the total output for the service department is $20 million over a certain period of time. How much of this amount is used within the service department? How much input is required from the production department?

d. The total output from the production department is $40 million over the same time period. How much of this total output is used within the production department? How much input is required from the service department?

e. Combine the information in parts c and d to find how much of the total output from the service and production departments will be available for sales demands outside the company.

5. The weighted digraph below represents the flow of goods and services in a two-sector economy involving transportation and agriculture. The numbers on the edges represent cents' worth of product (or service) used per dollar's worth of output.

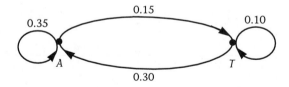

a. Construct a consumption matrix for this situation.

b. Complete the following: For every dollar's worth of output the agriculture sector requires _____ cents' worth of input from its own sector and _____ cents' worth of input from the transportation sector. For every dollar's worth of output the transportation sector requires _____ cents' worth of input from its own sector and _____ cents' worth of input from the agriculture sector.

c. If the total output for the agriculture sector is $50 million dollars and the total output for the transportation sector is $100 million dollars over a certain period of time, find:

     i. The total amount of agricultural goods used internally by this two-sector economy.

    ii. The amount of agricultural goods available for external sales.

    iii. The total amount of transportation services used internally by this two-sector economy.

    iv. The amount of transportation services available for external sales.

  d. Write an equation to represent the internal consumption of agricultural products for this economy. Let P_A represent the total production for agriculture and P_T represent the total output for transportation.

  e. Write an equation to represent the internal consumption of transportation services. Use P_A and P_T as in part d.

  f. Using the information from parts d and e, write equations representing the amount of agricultural products and transportation available to fill external consumer demands. Let D_A represent the total available for external demand for agriculture and D_T represent the total available for external demand for transportation. Recall: The amount of a product available to fill external demands is equal to the total production less the amount that is used internally.

  g. Suppose that this economy has an external demand for $10 million dollars' worth of agricultural products and $15 million dollars in transportation services. Write a system of equations that could be solved to find the total production of agriculture products and transportation services necessary to satisfy these demands.

6. Complete task 5 for the production management team of the Best Battery and Motor Company (see page 350).

Project

7. Research and report to the class on the life and work of Wassily Leontief.

The Leontief Input-Output Model, Part 2

As you can see from the exercises in Lesson 7.1, the Leontief model becomes quite complicated very quickly, even when dealing with only one- and two-sector economies. You are probably thinking that there has to be an easier way. You're right. There is an easier way.

Go back and review your Best Battery and Motor Company exploration in Lesson 7.1 (page 349). The production needs of each of the two sectors are summarized in the consumption matrix (C) for this economy.

$$C = \begin{matrix} \\ \text{Battery} \\ \text{Motor} \end{matrix} \begin{matrix} \text{Battery} \quad \text{Motor} \\ \begin{bmatrix} 0.03 & 0.08 \\ 0.01 & 0.04 \end{bmatrix} \end{matrix}$$

In Exercise 6 (page 353) you were asked to find a system of two equations in two unknowns that could be solved to find the total production of batteries (b) and motors (m) necessary to meet external sales demands of 400 batteries and 100 motors. To find two such equations, you used the fact that the total production for each product equaled the amount of the product used up by the two divisions during production plus the external demand for that product (sales outside the company).

For example, the total number of batteries produced (b) equals the number of batteries used within the battery division (0.03b) plus the number of batteries sent to the motor division (0.08m) plus the number of batteries required for sales outside the company (400). This gives the equation $b = 0.03b + 0.08m + 400$.

Likewise, the total number of motors produced (m) equals the number of motors sent to the battery division ($0.01b$) plus the number of motors used within the motor division ($0.04m$) plus the number of motors required for sales outside the company (100). This gives the equation $m = 0.01b + 0.04m + 100$.

Thus, the system of two equations in two unknowns (unsimplified) looks like

$$b = 0.03b + 0.08m + 400$$

$$m = 0.01b + 0.04m + 100.$$

You know how to solve this system algebraically by combining terms and using linear combinations or some other technique. However, there is an easier way to do it that uses matrices and takes advantage of technology (calculators or computers) to do the work.

Explore, first, how this system of equations could be represented using matrices. A consumption matrix C has already been defined. If you let

$$P = \begin{bmatrix} b \\ m \end{bmatrix}$$

represent a total production matrix (P) and

$$D = \begin{bmatrix} 400 \\ 100 \end{bmatrix}$$

represent a demand matrix (D), then the system of equations can be described with a matrix equation:

$$\begin{bmatrix} b \\ m \end{bmatrix} = \begin{bmatrix} 0.03 & 0.08 \\ 0.01 & 0.04 \end{bmatrix} \begin{bmatrix} b \\ m \end{bmatrix} + \begin{bmatrix} 400 \\ 100 \end{bmatrix}.$$

Perform the matrix multiplication shown here to convince yourself that the matrix equation does, indeed, represent the two algebraic equations previously shown.

This matrix equation can be written more simply as

$$P = CP + D,$$

showing that the total production = internal consumption + external demand.

Notice that this matrix equation resembles the simple linear equation solved in Lesson 7.1 (page 347). Indeed, solving this matrix equation uses the same operations that are used in solving a linear equation. Verify that this is true by tracing the steps for solving a linear equation using ordinary algebra and solving a matrix equation using matrix algebra in the following table.

One-Sector Economy Linear Equation Ordinary Algebra	Two-Sector Economy Matrix Equation Matrix Algebra	Comments
$p = 0.03p + d$	$P = CP + D$	
$p - 0.03p = d$	$P - CP = D$	
$1p - 0.03p = d$	$IP - CP = D$	The identity matrix I times $P = P$.
$(1 - 0.03)p = d$	$(I - C)P = D$	
$\dfrac{1}{1 - 0.03}(1 - 0.03)p$	$(I - C)^{-1}(I - C)P$ $= (I - C)^{-1}D$	*
$= \dfrac{1}{1 - 0.03}d$		
$1p = \dfrac{1}{1 - 0.03}d$	$IP = (I - C)^{-1}D$	Recall that $A^{-1}A = I$. (See Exercise 7 in Lesson 3.3, page 127.)
$p = \dfrac{1}{1 - 0.03}d$	$P = (I - C)^{-1}D$	

* Up to this point the ordinary algebra operations and the matrix operations have been identical. In solving the linear equation, it would be natural to divide both sides of the equation by $(1 - 0.03)$. But, there is no division operation in matrix algebra. The thing to do, then, is to multiply both sides of the linear equation by the multiplicative inverse of $(1 - 0.03)$. Multiplying both sides of the matrix equation by the multiplicative inverse of $(I - C)$ is a valid matrix operation.

In summary, if the consumption matrix (C) and the external demand matrix (D) are known, then the total production matrix (P) can be found using the matrix equation

$$P = (I - C)^{-1}D.$$

For the battery and motor problem, the solution is

$$\begin{bmatrix} b \\ m \end{bmatrix} = \left(\begin{bmatrix} 1 & 0 \\ 0 & 1 \end{bmatrix} - \begin{bmatrix} 0.03 & 0.08 \\ 0.01 & 0.04 \end{bmatrix} \right)^{-1} \begin{bmatrix} 400 \\ 100 \end{bmatrix}.$$

Now, using a calculator or computer to do the computations and rounding to the nearest whole number, you will find that

$$\begin{bmatrix} b \\ m \end{bmatrix} = \begin{bmatrix} 421 \\ 109 \end{bmatrix}.$$

The results show that to fill an order for 400 batteries and 100 motors, the company must produce 421 batteries and 109 motors.

Solving Systems of Linear Equations Using Matrices

The matrix techniques used for solving systems of equations in this lesson can be used to solve any system of n independent equations in n unknowns. Look, for example, at the following system of two equations in two unknowns.

$$2x_1 + 3x_2 = 23$$

$$5x_1 - 2x_2 = 10.$$

This system can be written as a single matrix equation:

$$\begin{bmatrix} 2 & 3 \\ 5 & -2 \end{bmatrix} \begin{bmatrix} x_1 \\ x_2 \end{bmatrix} = \begin{bmatrix} 23 \\ 10 \end{bmatrix}.$$

If we let

$$A = \begin{bmatrix} 2 & 3 \\ 5 & -2 \end{bmatrix}, \qquad X = \begin{bmatrix} x_1 \\ x_2 \end{bmatrix}, \qquad \text{and} \qquad B = \begin{bmatrix} 23 \\ 10 \end{bmatrix},$$

the matrix equation can be written in the form $AX = B$, which is similar to a simple linear equation such as $ax = b$.

One way to solve this linear equation is to multiply both sides of the equation by the multiplicative inverse of a, $\frac{1}{a}$. The same strategy can be used to solve the matrix equation as shown in the following table.

	Linear Equations Ordinary Algebra	Matrix Equations Matrix Algebra
Step 1	$ax = b$	$AX = B$
Step 2	$\frac{1}{a} ax = \frac{1}{a} b$	$A^{-1}AX = A^{-1}B$
Step 3	$1x = \frac{1}{a} b$	$IX = A^{-1}B$
Step 4	$x = \frac{1}{a} b$	$X = A^{-1}B$

Applying this method to the previous system of equations, we have

$$\begin{bmatrix} x_1 \\ x_2 \end{bmatrix} = \begin{bmatrix} 2 & 3 \\ 5 & -2 \end{bmatrix}^{-1} \begin{bmatrix} 23 \\ 10 \end{bmatrix}.$$

Use a calculator or computer to do the calculations and verify that the solution for the given system of linear equations is $x_1 = 4$ and $x_2 = 5$.

Exercises

Use either a calculator or computer software to perform matrix operations in the following exercises.

1. The total production for the high-tech company described in Exercises 2 and 3 in Lesson 7.1 (pages 350 and 351) over a period of time is $40,000 worth of computer chips and $50,000 worth of computers.
   a. Write a consumption matrix, C, for this company. Label the rows and columns of your matrix.
   b. Write a production matrix, P, and label the rows and columns.
   c. Compute the matrix product CP to find the amount of each product that the company uses internally.
   d. Use the information from parts b and c and the matrix equation $D = P - CP$ to compute the amount of computer chips and computers available for sales outside the company (external demand).
   e. The company has a order for $20,000 worth of computer chips and $70,000 worth of computers. Find the total production of computer chips and computers necessary to fill this order. Use the matrix equation $P = (I - C)^{-1}D$.

2. a. Use matrices to compute the results for parts c, d, and e of Exercise 4 in Lesson 7.1 (pages 351 and 352).
   b. The company must meet external demands of $25 million in service and $50 million in products over a period of time. What must the total production in service and products to meet this demand?

3. Use matrices to compute the results for parts c and g of Exercise 5 in Lesson 7.1 (pages 352 and 353).

4. The techniques developed in this lesson using a two-sector economy can easily be extended to solve problems that involve economies of more than two sectors. For example, look at an economy that has three sectors—transportation, energy, and manufacturing. Each of these

sectors uses some of its own products or services as well as some from each of the other sectors, as follows:

Transportation Sector: Every dollar's worth of transportation provided requires an input of 10 cents' worth of transportation services, 15 cents' worth of energy, and 25 cents' worth of manufactured goods.

Energy Sector: Every dollar's worth of energy produced requires an input of 25 cents' worth of transportation services, 10 cents' worth of energy, and 20 cents' worth of manufactured goods.

Manufacturing Sector: Every dollar's worth of manufactured goods produced requires an input of 20 cents' worth of transportation services, 20 cents' worth of energy, and 15 cents' worth of manufactured goods.

a. Draw a weighted digraph for this three-sector economy.
b. Construct a consumption matrix (C) for this economy. Label the rows and columns of your matrix.
c. The total production over a period of time for this economy is $150 million in transportation, $200 million in energy, and $160 million in manufactured goods. Write a production matrix (P) for this economy. Label the rows and columns of your matrix.
d. Compute the matrix product CP to find the amount of each product that is used internally by the economy. Write your answer as a matrix and label the rows and columns.
e. Use the information from parts c and d and the matrix equation $D = P - CP$ to find the amount of goods available for external demand (sales outside the three sectors described here).
f. The estimated consumer demand for transportation, energy, and manufactured goods and services in millions of dollars are 100, 95, and 110, respectively. Find the total production necessary to fulfill these demands. Use the matrix equation $P = (I - C)^{-1}D$.

5. An economy consisting of three sectors (services, manufacturing, and agriculture) has the consumption matrix

$$
C = \begin{matrix} & \text{Services} & \text{Manufacturing} & \text{Agriculture} \\ \begin{matrix} \text{Services} \\ \text{Manufacturing} \\ \text{Agriculture} \end{matrix} & \begin{bmatrix} 0.1 & 0.3 & 0.2 \\ 0.2 & 0.3 & 0.1 \\ 0.2 & 0.1 & 0.2 \end{bmatrix} \end{matrix}
$$

a. Draw a weighted digraph for this economy.
b. On which sector of the economy is manufacturing the most dependent? The least dependent?
c. If the services sector has an output of $40 million dollars, what is the input in dollars from manufacturing? From agriculture?
d. A production matrix, P, in millions of dollars for this economy is as follows. Use the matrix product CP to find the internal consumption of services and products within this economy. Find the external demand matrix D.

$$P = \begin{array}{l} \text{Services} \\ \text{Manufacturing} \\ \text{Agriculture} \end{array} \begin{bmatrix} 20 \\ 25 \\ 15 \end{bmatrix}.$$

e. An external demand matrix, D, in millions of dollars follows. How much must be produced by each sector to meet this demand?

$$D = \begin{array}{l} \text{Services} \\ \text{Manufacturing} \\ \text{Agriculture} \end{array} \begin{bmatrix} 4.6 \\ 5.0 \\ 4.0 \end{bmatrix}.$$

6. An economy consisting of four sectors (transportation, manufacturing, agriculture, and services) has the consumption matrix (in millions of dollars worth of products)

		Trans.	Manu.	Agri.	Serv.
	Transportation	0.25	0.28	0.22	0.20
$C =$	Manufacturing	0.15	0.15	0.17	0.23
	Agriculture	0.19	0.20	0.21	0.15
	Services	0.20	0.24	0.19	0.25

a. Draw a weighted digraph for this economy.
b. On which sector of the economy is services the most dependent? The least dependent?
c. If the manufacturing sector has an output of $20 million, what is the input in dollars from services? From transportation?
d. A production matrix, P, in millions of dollars, follows. Use the matrix product CP to find the internal consumption of services and products within this economy. Find the external demand matrix D.

$$P = \begin{array}{l} \text{Transportation} \\ \text{Manufacturing} \\ \text{Agriculture} \\ \text{Services} \end{array} \begin{bmatrix} 50 \\ 40 \\ 45 \\ 50 \end{bmatrix}.$$

e. An external demand matrix, D, in millions of dollars, follows. How much must be produced by each sector to meet this demand?

$$D = \begin{matrix} \text{Transportation} \\ \text{Manufacturing} \\ \text{Agriculture} \\ \text{Services} \end{matrix} \begin{bmatrix} 10 \\ 12 \\ 10 \\ 15 \end{bmatrix}.$$

7. A two-industry system consisting of services and manufacturing has the consumption matrix

$$C = \begin{matrix} \text{Services} \\ \text{Manufacturing} \end{matrix} \begin{matrix} \text{Services} & \text{Manufacturing} \\ \begin{bmatrix} 0.5 & 0.5 \\ 0.2 & 0.3 \end{bmatrix} \end{matrix}.$$

a. Compute the total production necessary to satisfy a consumer demand for 15 units of services and 25 units of manufacturing.
b. Comment on the productivity of this system. Explain your answer.
c. If the consumer demand is for 30 units of services and 50 units of manufacturing, find the production needed to fill these demands.
d. On the basis of the results in parts a and b above, predict the total production of services and goods for a consumer demand for 45 units of service and 75 units of manufacturing. Check your prediction by computing the production matrix for this case.

8. A company has two divisions: service and production. The flow of goods and services within this company is described by the consumption matrix

$$C = \begin{matrix} \text{Services} \\ \text{Products} \end{matrix} \begin{matrix} \text{Services} & \text{Products} \\ \begin{bmatrix} 0.10 & 0.25 \\ 0.05 & 0.10 \end{bmatrix} \end{matrix}.$$

a. Draw a weighted digraph for this situation.
b. The total output for the company during 1 year is $50 million in services and $75 million in products. How much of the total output is used internally by each of the company's divisions?
c. What total output is needed to meet an external consumer demand of $15 million dollars in service and $25 million in products?
d. If the consumer demand increases to $22 million for services and to $30 million for products, what will be the effect on the total production of goods and services?

Projects

9. Research and report to the class on the following.
   a. What computer software is available in your school for solving systems of equations?
   b. What is the largest number of variables that the software can handle?
   c. How long does it take the computer to solve a system with the largest number of variables possible using this software?
   d. It took Professor Leontief 56 hours to solve a system of 42 equations in 42 unknowns using the Mark II in the 1940s and 3 minutes to solve a system of 81 equations in 81 unknowns using the IBM 7090 in the 1960s. If the software available in your school can handle systems of 42 and 81 equations, find out:
      i. How long it takes to solve a system of 42 independent equations in 42 unknowns.
      ii. How long it takes to solve a system of 81 independent equations in 81 unknowns.
   e. Investigate parts a and b, for computer software such as *Mathematica, Maple,* or *Theorist* that your school may not own.
10. Research and report to the class on parts a to c of Exercise 9 for the graphing calculators that are available in your school. Try, in particular, to find this information for the TI-92 calculator.

11. Research and report to the class about the size, cost, and capability of the Mark II and IBM 7090 computers. In your report, compare the information you find about these computers with similar information about the personal computers that are available to students in your school.

Computer/Calculator Explorations

12. Write a graphing calculator (or computer) program that uses matrices for solving systems of n equations in n unknowns.

13. Write a graphing calculator (or computer program) designed to solve the various consumption problems presented in this lesson.

Markov Chains

A **Markov chain** is a process that arises naturally in problems that involve a finite number of events or states that change over time. In this lesson, a situation that illustrates the significant characteristics of a Markov chain is introduced.

Mathematician of Note

Russian mathematician A. A. Markov's (1856–1922) studies of linked chains of events led to the modern study of stochastic processes. As a result of his work, one type of a stochastic process is called a Markov chain.

Consider the following: Students at Lincoln High have two choices for lunch. They can either eat in the cafeteria or eat elsewhere. The director of food service is concerned about being able to predict how many students can be expected to eat in the cafeteria over the long run. She has asked the discrete mathematics class to help her out by conducting a survey of the student body during the first two weeks of school. The results of the survey show that if a student eats in the cafeteria on a given day, the probability that he or she will eat there again the next day is 70% and the probability that he or she will not eat there is 30%. If a student does not eat in the cafeteria on a given day, the probability that he or she will eat in the cafeteria the next day is 40% and the probability that he or she will not eat there is 60%. On Monday, 75% of the students ate in the cafeteria and 25% did not. What can be expected to happen on Tuesday?

A good way to organize all these statistics is with a tree diagram, similar to the way you organized probabilities in Chapter 6 (see Figure 7.2).

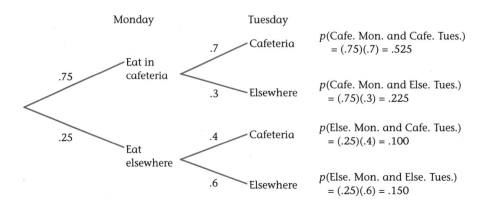

Figure 7.2 Cafeteria statistics organized in a tree diagram.

The director wants to know what portion of the students can be expected to eat in the cafeteria on Tuesday. Look at the tree diagram and notice that this happens if a student eats in the cafeteria on Monday and on Tuesday, or if a student eats elsewhere on Monday and in the cafeteria on Tuesday. This portion is .525 + .100 = .625, or 62.5%. Similarly, the portion of students who will eat elsewhere on Tuesday is .225 + .150 = .375, or 37.5%. Note that this could also be calculated by subtracting .625 from 1.

The tree diagram model is fine if only two stages are required to reach a solution. The director, however, is interested in continuing this process for many days. Because the number of branches of the tree diagram doubles with each additional day, the model soon becomes impractical and so an alternative is needed.

The Monday student data are called the **initial distribution** of the student body and can be represented by a row (or **initial-state**) vector, D_0, where

$$D_0 = \begin{matrix} C & E \\ [.75 & .25] \end{matrix} \quad \begin{matrix} C = \text{eats in the cafeteria} \\ E = \text{eats elsewhere.} \end{matrix}$$

Movement from one state to another is often called a **transition**, so the data about how students choose to eat from one day to the next is written in a matrix called a **transition matrix**, T, where

$$T = \begin{matrix} & C & E \\ C & \begin{bmatrix} .7 & .3 \\ E & .4 & .6 \end{bmatrix} \end{matrix}.$$

Notice that the entries of a transition matrix are probabilities, values between 0 and 1 inclusive. Also notice that the transition matrix is a square matrix and the sum of the probabilities in any row is 1.

Now calculate the product of matrix D_0 and matrix T:

$$D_0T = [.75 \quad .25]\begin{bmatrix} .7 & .3 \\ .4 & .6 \end{bmatrix} = [.75(.7) + .25(.4) \quad .75(.3) + .25(.6)]$$

$$= [.625 \quad .375].$$

Compare these calculations with those made in the tree diagram model. The values in the resulting row vector can be interpreted as the portion of students who eat in the cafeteria and who eat elsewhere on Tuesday. This row vector is called D_1 to indicate that it occurs one day after the initial day. To see what happens on Wednesday, it is only necessary to repeat the process using D_1 in place of D_0:

$$D_1T = [.625 \quad .375]\begin{bmatrix} .7 & .3 \\ .4 & .6 \end{bmatrix} = [.625(.7) + .375(.4) \quad .625(.3) + .375(.6)]$$

$$= [.5875 \quad .4125].$$

The resulting row vector is called D_2 to indicate that it occurs two days after the initial day. Thereafter D_2 shows that approximately 59% of the students will eat in the cafeteria on Wednesday and 41% will eat elsewhere.

Consider how D_2 was calculated.

$$D_2 = D_1T, \text{ but } D_1 = D_0T, \text{ so by substitution, } D_2 = (D_0T)(T).$$

Because matrix multiplication is associative,

$$D_2 = (D_0T)(T) = D_0(T^2).$$

This means that the calculation of the distribution of students on Wednesday can be completed by taking the initial-state vector times the square of the transition matrix.

This observation simplifies additional calculations. If, for example, you want to know the distribution on Friday, four days from Monday, calculate $D_4 = D_0(T^4)$ on a calculator that has matrix features or on a computer equipped with

matrix software:

$$D_0 T^4 = [.75 \quad .25] \begin{bmatrix} .7 & .3 \\ .4 & .6 \end{bmatrix}^4 = [.572875 \quad .427125].$$

About 57% of the students can be expected to eat in the cafeteria on Friday. For a school of 1,000 students, about 573 of them can be expected in the cafeteria on that day.

The movement of students from one state to another can also be shown with a weighted digraph called a **transition digraph** or **state diagram** (see Figure 7.3).

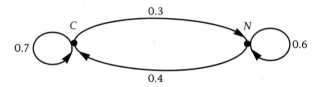

Figure 7.3 Transition digraph for the cafeteria statistics.

Exercises

Use either a calculator or computer software to perform matrix operations in the following exercises.

1. a. Find the distribution of students eating and not eating in the cafeteria each day for the first week of school using the initial distribution $D_0 = [0.75 \; 0.25]$ and the transition matrix T of this lesson.
   b. Find the distribution of students eating and not eating in the cafeteria after 2 weeks (10 school days) have passed. Repeat for 3 weeks (15 days).
   c. What would your report to the director of food services be, based on your computations in parts a and b?
   d. Choose any other initial distribution of students and repeat parts a and b.
   e. Compare the results of parts b and d. Does the initial distribution appear to make a difference in the long run?
   f. Calculate the 15th power of matrix T. Compare the entries in T^{15} to the distribution after the 15th day.

2. When successive applications of a Markov process are made and the rows of powers of the transition matrix converge to a single vector, this

common vector is called the **stable-state vector** for the Markov chain. A sufficient condition, which we will not prove in this text, for a Markov chain to have a stable-state vector is that some power of its transition matrix have only positive entries. Since all the entries in the transition matrix T are nonzero probabilities, this condition is clearly met for the cafeteria Markov chain.

a. What is the stable-state vector for the transition matrix T?

b. Make a conjecture about the relationship between the distribution of students in the long run and the stable-state vector of the transition matrix.

3. Suppose the entire student body eats in the cafeteria on the first day of school. The initial distribution in this case is $D_0 = [1 \quad 0]$. Repeat parts a and b of Exercise 1 for this distribution. After several weeks, what percentage of students will be eating in the cafeteria?

4. Which of the matrices below could be Markov transition matrices? For the matrices that could not be transition matrices, explain why not.

a. $\begin{bmatrix} .7 & .3 \\ .6 & .6 \end{bmatrix}$.

b. $\begin{bmatrix} .1 & .4 & .5 \\ .2 & .6 & .2 \end{bmatrix}$.

c. $\begin{bmatrix} 1.2 & -4 \\ 1 & 0 \end{bmatrix}$.

d. $\begin{bmatrix} .6 & .3 & .1 \\ .3 & .3 & .3 \end{bmatrix}$.

e. $\begin{bmatrix} .75 & .25 \\ 1 & 0 \end{bmatrix}$.

f. $\begin{bmatrix} .45 & .55 \\ .33 & .66 \end{bmatrix}$.

5. There is a 60% chance of rain today. It is known that tomorrow's weather depends on today's according to the probabilities shown in the following tree diagram.

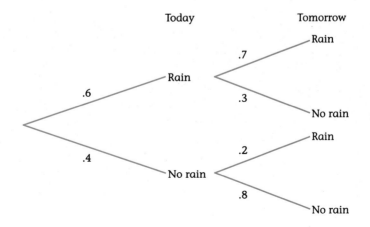

Lincoln Area Forecast

LINCOLN JOURNAL STAR, June 26, 1998

Lincoln, Beatrice, Nebraska City: Today, windy, very hot. Sunny. High: near 100. South wind 20–30 mph. Tonight, partly cloudy, 20% chance of storms. Low: 70–75. Saturday, partly cloudy, 30% chance of storms. High: near 90.

a. What is the probability it will rain tomorrow if it rains today?

b. What is the probability it will rain tomorrow if it doesn't rain today?

c. Write an initial-state matrix that represents the weather forecast for today.

d. Write a transition matrix that represents the transition probabilities shown in the tree diagram.

e. Calculate the forecast for 1 week (7 days) from now.

f. In the long run, for what percentage of days will it rain?

6. A taxi company has divided the city into three districts—Westmarket, Oldmarket, and Eastmarket. By keeping track of pickups and deliveries, the company found that of the fares picked up in the Westmarket district, only 10% are dropped off in that district, 50% are taken to the Oldmarket district, and 40% go to the Eastmarket district. Of the fares picked up in the Oldmarket district, 20% are taken to the Westmarket district, 30% stay in the Oldmarket district, and 50% are dropped off in the Eastmarket district. Of the fares picked up in the Eastmarket district, 30% are delivered to each of the Westmarket and Oldmarket districts, while 40% stay in the Eastmarket district.

a. Draw a transition digraph for this Markov chain.

b. Construct a transition matrix for these data.

c. Write an initial-state matrix for a taxi that starts off by picking up a fare in the Oldmarket district. What is the probability that it will end up in the Oldmarket district after three additional fares?

d. Find and interpret the stable-state vector for this Markov process.

7. Emily, Jon, and Gretchen are tossing a football around. Emily always tosses to Jon, and Jon always tosses to Gretchen, but Gretchen is equally likely to toss the ball to either Emily or Jon.

a. Draw a transition digraph to represent this information.

b. Represent this information as the transition matrix of a Markov chain.

c. What is the probability that Emily will have the ball after three tosses if she was the first one to throw it to one of the others?

d. Find and interpret the stable-state vector for this Markov chain.

e. Explain why there are no zeros in the stable-state matrix even though there were several zeros in the transition matrix.

8. Jim agreed to care for Emily's cat, Ellington, for the weekend. On Friday night Ellington prowled the first floor of Jim's house, randomly moving from room to room, not staying in one room for more than a few minutes. The following floor plan shows the location of the rooms and doorways in Ellington's range. The letters on the floor plan represent Living room, Dining room, Kitchen, Bathroom, and Study.

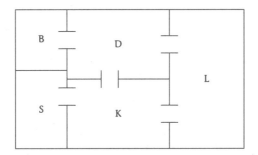

Each of Ellington's movements can be interpreted as a transition in a Markov chain in which a state is identified with the room he is in. The first row of the transition matrix is

$$\begin{array}{ccccc} L & D & K & S & B \end{array}$$
$$L \begin{bmatrix} 0 & \frac{1}{2} & \frac{1}{2} & 0 & 0 \end{bmatrix}.$$

a. Construct the complete transition matrix for this situation.

b. If Ellington starts off in the living room, what is the probability that he will be in the study after two transitions? After three transitions?

c. After a large number of transitions, what is the probability that Ellington will be in the bathroom?

d. In the long run, what percentage of the time will Ellington spend in either the kitchen or the dining room?

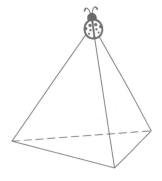

9. A discrete mathematics student observes a bug crawling from vertex to vertex along the edges of a tetrahedron model on the teacher's desk (see figure at left). From any vertex the bug is equally likely to go to any other vertex.
   a. Set up a transition matrix for this situation. (Note: To minimize roundoff errors, if your calculator does not accept fractions, approximate 1/3 to at least four decimal places when you enter the data.)
   b. Determine the probabilities for the location of the bug when the passing bell rings if it moves to a different vertex about 20 times during the class period.

10. Dick's old hound dog, Max, spends much of his time during the day running from corner to corner along the fence surrounding his square-shaped yard. There is a .5 probability that Max will turn in either direction at a corner. The corners of Max's yard point north, east, south, and west.
    a. Draw an initial-state diagram that represents Max's movement.
    b. Construct a transition matrix for this situation.
    c. Look at the behavior of successive powers of the transition matrix. Notice the oscillation of the transition probabilities between the states represented by the rows of the matrices. Does this system appear to stabilize in some way? Explain your answer.
    d. Approximately what percentage of the time will Max spend at each of the corners of his yard? (Note: You need to halve the entries in the matrix to account for the oscillating pattern.)
    e. Max changes his routine one day, and the pattern of his new movements is represented by the following transition matrix. Answer part d for this situation.

$$
\begin{array}{c c}
 & \begin{array}{cccc} N & E & S & W \end{array} \\
\begin{array}{c} N \\ E \\ S \\ W \end{array} &
\left[\begin{array}{cccc}
0 & \frac{1}{2} & 0 & \frac{1}{2} \\
\frac{3}{8} & 0 & \frac{5}{8} & 0 \\
0 & \frac{3}{8} & 0 & \frac{5}{8} \\
\frac{3}{4} & 0 & \frac{1}{4} & 0
\end{array} \right]
\end{array}
$$

11. Using mathematical induction, prove that $D_k = D_0 T^k$ for any original distribution D_0 and transition matrix T, where k is a natural number.

12. A group of researchers are studying the effect of a potent flu vaccine in healthy (well) and infirm (ill) rats. When the rats are injected with the vaccine, three things may occur. The rat may have no reaction where its health status does not change. The rat may have a mild reaction and become ill, or the rat may have a severe reaction and die. The probabilities of each of these reactions are shown in the following matrix.

$$
\begin{array}{c}
 \\
\text{Well} \\
\text{Ill} \\
\text{Dead}
\end{array}
\begin{array}{ccc}
\text{Well} & \text{Ill} & \text{Dead} \\
\left[\begin{array}{ccc}
.8 & .2 & 0 \\
.1 & .6 & 0 \\
0 & 0 & 1
\end{array}\right]
\end{array}
$$

a. Write an initial-state vector for a healthy rat who is injected with the vaccine.

b. In this study the scientists check the status of the rats on a daily basis. Use the transition matrix to predict the health of the rat in part a after 4 days.

c. Use the transition matrix to predict the rat's health in the long run.

d. This Markov chain has a state that is called an **absorbing state**. Which state do you think it is? Why?

13. A hospital categorizes its patients as well (in which case they are discharged), good, critical, and deceased. Data show that the hospital's patients move from one category to another according to the probabilities shown in this transition matrix:

$$
\begin{array}{c}
 \\
\text{Well} \\
\text{Good} \\
\text{Critical} \\
\text{Dead}
\end{array}
\begin{array}{cccc}
\text{Well} & \text{Good} & \text{Critical} & \text{Dead} \\
\left[\begin{array}{cccc}
1 & 0 & 0 & 0 \\
.5 & .3 & .2 & 0 \\
0 & .3 & .6 & .1 \\
0 & 0 & 0 & 1
\end{array}\right]
\end{array}.
$$

a. Write an initial-state matrix for a patient who enters the hospital in critical condition.

b. If patients are reclassified daily, predict the patient's future after 1 week in the hospital.

c. Predict the future of any patient in the long run.

d. Does this Markov chain have any absorbing states? Which states do you think are absorbing? Why?

Project

14. Research and report to the class on the life and work of A. A. Markov.

Computer/Calculator Exploration

15. Write a graphing calculator (or computer) program that can be used to find the stable-state vector for a Markov chain.

Game Theory, Part 1

The basic ideas of game theory were first researched by John von Neumann in the 1920s. But, it was not until the 1940s during World War II that game theory was recognized as a legitimate branch of mathematics. Thus, most of the work in this area has been done over the last 60 years.

Mathematician of Note

This photograph, taken in 1952, shows John von Neumann (1903–1957) standing in front of the EDVAC computer.

You probably tend to think of games as being fun or relaxing ways to spend your time. There are, however, many decision-making situations in fields such as economics or politics that can also be thought of as games. In such games, there are two or more players who have conflicting interests. These players may be individuals, teams of people, whole countries, or even forces of nature. Each player (or side) has a set of alternative courses of action called **strategies** that can be used in making decisions. Mathematical game theory deals with selecting the best strategies for a player to follow in order to achieve the most favorable outcomes.

In this lesson, you will explore some examples of games with two players and use matrices to determine the best strategy for each player to choose. As the first example, consider a simple coin-matching game that Sol and Tina are playing. Each conceals a penny with either heads or tails turned upward. They display their pennies simultaneously. Sol will win

three pennies from Tina if both are heads. Tina will win two pennies from Sol if both are tails, and one penny from Sol if the coins don't match. What is the best strategy for each player?

If you think carefully about the game, you will probably decide that it isn't such a good deal for Sol. As long as Tina displays tails, she cannot lose. If Sol knows that Tina is going to play tails, he should display heads because he will lose more if he doesn't.

You probably think this is a rather boring game. In a sense it is, because both players will do the same thing every time. A game in which the best strategy for both players is to pursue the same strategy every time is called **strictly determined.** Although strictly determined games are fairly boring, there are situations in life in which they cannot be avoided and knowing how to analyze them properly can be beneficial. Although strictly determined games are often very simple, they can be difficult to analyze without an organizational scheme. Matrices offer a way of doing this.

The following matrix presents Sol's view of the game. It is customary to write a game matrix from the viewpoint of the player associated with the matrix rows rather than the player associated with the columns. Such a matrix is called a **payoff matrix**. The entries are the payoffs to Sol for each outcome of the game.

$$\begin{array}{c} \qquad\qquad\text{Tina} \\ \qquad\quad \text{Heads} \quad \text{Tails} \\ \text{Sol} \begin{array}{c} \text{Heads} \\ \text{Tails} \end{array} \left[\begin{array}{cc} 3 & -1 \\ -1 & -2 \end{array} \right]. \end{array}$$

Game Theory Captures a Nobel

New York Times, October 12, 1994

The 1994 Nobel Memorial Prize in Economic Science, a $930,000 award to be divided among three pioneers in the field of game theory, celebrates achievements in building the foundations for analyzing interactions among businesses, nations, and even biological species.

But just as important, the prize awarded to John F. Nash of Princeton University, John C. Harsanyi of the University of California at Berkeley and Reinhard Selton of the University of Bonn acknowledges a sea of changes in economics that has occurred in the last decade.

John von Neumann and Oskar Morgenstern, economists at Princeton, invented the field. Their

continues

This matrix is easy to follow if you are Sol, but the entries are just the opposite if you are Tina. If you find it difficult to think of all the numbers as their opposites, you may find it preferable to write a second matrix from Tina's point of view:

$$\begin{array}{c} & \text{Tina} \\ & \begin{array}{cc} \text{Heads} & \text{Tails} \end{array} \\ \text{Sol} \begin{array}{c} \text{Heads} \\ \text{Tails} \end{array} \left[\begin{array}{cc} -3 & 1 \\ 1 & 2 \end{array} \right]. \end{array}$$

Consider the game from Sol's point of view. Sol does not want to lose any more money than necessary, so he analyzes his strategies from the standpoint of his losses. If he displays heads, the worst he can do is to lose 1 cent. If he displays tails, the worst he could do is lose 2 cents. Since it is better to lose 1 cent than lose 2 cents, Sol decides to display heads.

Sol's analysis can be related to the payoff matrix by writing the worst possible outcome of each strategy to the right of the row that represents it. The worst possible outcome of each strategy is the smallest value of each row, often referred to as the **row minimum**. Sol's best strategy is to select the option that produces the largest of these minimums or, in other words, to select the "best of the worst." Because this value is the largest of the smallest row values, it is called the **maximin** (the maximum of the row minimums).

Game Theory Captures a Nobel (continued)

book, published in 1944, "The Theory of Games and Economic Behavior," was the first to delve deeply into the likely consequences of strategic interactions, where all the actors must consider the potential for reaction.

The great bulk of the work by economists in game theory has been in an area where its insights had been most sorely missed: the organization of industry.

An example: Intel, the microprocessor giant, gave up an effective monopoly on the 86-series chip by allowing Advanced Micro Devices to share the technology. Intel, it seems, decided that computer makers would not lock themselves into a new microprocessor technology unless they were protected from future price-gouging by a monop-olist. So by licensing another manufacturer, Intel successfully increased the demand for its own product.

Here, game theory explained corporate behavior that made no sense in nonstrategic terms.

Game theorists have also been hired to create corporate strategy from scratch, most notably in the case of the Federal Communications Commission's auction in December 1994 of bands on the radio spectrum for use in wireless communications. Every major bidder hired academic game theorists as consultants. The F.C.C.'s goal was to raise the maximum amount of money, at least $10 billion and it used game theory in attempting to reach that target.

Tina

		Heads	Tails	Row minimums
Sol	Heads	3	−1	①
	Tails	−1	−2	−2

In general, the best strategy for the row player in a strictly determined game is to select the strategy associated with the largest of the row minimums.

Because Tina's point of view is exactly the opposite of Sol's, she views the minimums as maximums and vice-versa. Therefore, her best strategy is the one associated with the smallest of the largest column values, the **minimax** (the minimum of the column maximums).

Tina

		Heads	Tails
Sol	Heads	3	−1
	Tails	−1	−2
Column maximums		3	①

In general, the best strategy for the column player in a strictly determined game is to select the strategy associated with the smallest of the column maximums.

Remember, if you find it confusing to reverse your thinking when analyzing the columns, change the sign of all matrix entries and use the same reasoning you used for the rows.

In this game, the value selected by both Sol and Tina is the same one, that is, the −1 that appears in the upper right-hand corner of the matrix. This is the identifying characteristic of strictly determined games. If the value selected by the two players is not the same, then the game is not strictly determined and is much less boring. Games that are not strictly determined are considered in the next lesson.

A strictly determined game is one in which the maximin (the maximum of the row minimums) and the minimax (the minimum of the column

maximums) are the same value. This value is called the **saddle point** of the game. The saddle point can be interpreted as the amount won per play by the row player.

When players have more than two strategies, a game is somewhat harder to analyze. It is often helpful to eliminate strategies that are **dominated** by other strategies. For example, in a competition between two pizza restaurants, Dino's and Sal's, both are considering four strategies: running no special, offering a free minipizza with the purchase of a large pizza, offering a free medium pizza with the purchase of a large one, and offering a free drink with any pizza purchase.

A market study estimates the gain in dollars per week to Dino's over Sal's according to the following matrix.

Sal's

		No special	Mini	Medium	Drink
	No special	200	−400	−300	−600
Dino's	Mini	500	100	200	600
	Medium	400	−100	−200	−300
	Drink	300	0	400	−200

What should the managers of Dino's and Sal's do?

Suppose you are the manager of Dino's and examine the first two rows carefully. You notice that no matter what your competitor does, you always achieve a larger payoff by offering the free mini. It would make no sense, therefore, to run no special. The first row of the matrix is dominated by the second and can be eliminated by drawing a line through it. Similarly, the second row dominates the third, and so the third row can be eliminated.

Sal's

		No special	Mini	Medium	Drink
	No special	~~200~~	~~−400~~	~~−300~~	~~−600~~
Dino's	Mini	500	100	200	600
	Medium	~~400~~	~~−100~~	~~−200~~	~~−300~~
	Drink	300	0	400	−200

Now think of the matrix from the point of view of Sal's manager. Because all the payoffs to Sal's are opposites of the payoffs to Dino's, a column is dominated if all its entries are larger, rather than smaller, than those of another column. Notice that all the values in the first column are larger than the corresponding values in the second column. Because the first column is dominated by the second, it is unwise for Sal's to run no special, and so this strategy can be eliminated. Similarly, the second column dominates the third, and so the third column can be eliminated.

Sal's

	No special	Mini	Medium	Drink
No special	~~200~~	−400	~~300~~	~~600~~
Mini	500	100	200	600
Medium	~~400~~	~~100~~	~~200~~	~~300~~
Drink	300	0	400	−200

(Dino's labels rows: No special, Mini, Medium, Drink)

Once these strategies are eliminated, the game is easier to examine for a minimax and a maximin:

Sal's

	No special	Mini	Medium	Drink	Row minimums
No special	~~200~~	~~400~~	~~300~~	~~600~~	
Mini	500	100	200	600	(100)
Medium	~~400~~	~~100~~	~~200~~	~~300~~	
Drink	300	0	400	−200	−200
Column maximums		(100)		600	

(Dino's labels rows: No special, Mini, Medium, Drink)

The game is strictly determined with a saddle point of 100. Dino's best strategy is to offer the free mini, and Sal's best strategy is to do the same. By pursuing this strategy, Dino's will gain about $100 a week over Sal's.

Exercises

1. Each of the following matrices represents a payoff matrix for a game. Determine the best strategies for the row and column players. If the game is strictly determined, find the saddle point of the game.

a. $\begin{bmatrix} 16 & 8 \\ 12 & 4 \end{bmatrix}$.

b. $\begin{bmatrix} 0 & 4 \\ -1 & 2 \end{bmatrix}$.

c. $\begin{bmatrix} 2 & -3 \\ -3 & 4 \end{bmatrix}$.

d. $\begin{bmatrix} 0 & 1 & 2 \\ 3 & -2 & 0 \end{bmatrix}$.   e. $\begin{bmatrix} 0 & -6 & 1 \\ -4 & 8 & 2 \\ 6 & 5 & 4 \end{bmatrix}$.   f. $\begin{bmatrix} 0 & 3 & 1 \\ -3 & 0 & 2 \\ -1 & -4 & 0 \end{bmatrix}$.

2. a. For the game defined by the following matrix, determine the best strategies for the row and column players and the saddle point of the game.

$$\begin{bmatrix} -4 & 2 \\ 5 & 3 \end{bmatrix}$$

b. Add 4 to each element in the matrix given in part a. How does this affect the best strategies and the saddle point of the game?

c. Multiply each element in the matrix in part a by 2. How does this affect the saddle point of the game and the best strategies?

d. Make a conjecture based on the results of parts b and c.

3. Discuss what would happen in the game given in this lesson if Sol decided to depart from his best strategy. Suppose he switches to displaying tails occasionally. Do you think Tina should still play tails every time? Explain your answer.

4. Use the concept of dominance to solve each of the following games. Give the best row and column strategies and the saddle point of each game.

a.

	E	F	G
A	3	1	7
B	0	1	3
C	4	3	4
D	1	3	6

b.

	E	F	G
A	4	-1	-2
B	0	1	1
C	0	-2	5
D	3	2	4

5. The Democrats and Republicans are engaged in a political campaign for mayor in a small midwestern community. Both parties are planning their strategies for winning votes for their candidate in the final days. The Democrats have settled on two strategies, A and B, and the Republicans plan to counter with strategies C and D. A local newspaper got wind of their plans and conducted a survey of eligible voters. The results of the survey show that if the Democrats choose plan A and the Republicans choose plan C, then the Democrats will gain 150 votes. If the Democrats choose A and the Republicans choose D, the Democrats will lose 50 votes. If the Democrats choose B and the Republicans choose

C, the Democrats will gain 200 votes. If the Democrats choose B and the Republicans choose D, the Democrats will lose 75 votes. Write this information as a matrix game. Find the best strategies and the saddle point of the game.

6. Two major discount companies, Salemart and Bestdeal, are planning to locate stores in Nebraska. If Salemart locates in city A and Bestdeal in city B, then Salemart can expect an annual profit of $50,000 more than Bestdeal's annual profit. If both locate in city A, they expect equal profits. If Salemart locates in city B and Bestdeal in city A, then Bestdeal's profits will exceed Salemart's by $25,000. If both companies locate in city B, then Salemart's profits will exceed Bestdeal's by $10,000. What are the best strategies in this situation and what is the saddle point of the game?

7. Jon and Gretchen each have three dimes. They both hold either one, two, or three coins in a clenched fist and open their fists together. If they both are holding the same number of coins, Jon will take the coins that Gretchen is holding. If they are holding different numbers of coins, then Gretchen will take the coins that Jon is holding.
   a. Write the payoff matrix from Jon's point of view.
   b. Does this game have a saddle point? If so, what are the best strategies for Jon and Gretchen?

8. Mike is going over to see his girlfriend, Nancy, after track practice, when he suddenly remembers that today may be a special anniversary for Nancy and him, and he always brings her a single red rose on this occasion. But he's not sure. Maybe the anniversary is next week. What should he do? If it is their anniversary and he doesn't bring a rose, then he'll be in bad trouble. On a scale from 0 to 10, he'd score a -10. If he doesn't bring a rose and it isn't their anniversary, Nancy won't know anything about his frustration and he'll score a 0. If he brings a rose and it is not their anniversary, then Nancy will be suspicious that something funny is going on but he'll score about a 2. If it is their special anniversary and he brings a rose, then Nancy will be expecting it and he'll score a 5. Write a payoff matrix for this situation. What is Mike's best strategy?

9. School board and Teacher Education Association representatives are meeting to negotiate a contract. Each side can either threaten (reduction in staff or strike), refuse to negotiate, or negotiate willingly. Each side decides its strategy prior to coming to the negotiating table. The

following payoff matrix gives the percentage pay increases for the teachers that would result from each combination of strategies. Find the best strategies for each side.

		School Board		
		Threaten	Refuse	Negotiate
	Threaten	5	4	3
Teachers	Refuse	3	0	2
	Negotiate	4	3	2

Projects

10. Research and report to the class on the life and work of John von Neumann.

11. Research and report to the class on the development of game theory during World War II.

12. Find and report to the class on applications of game theory in foreign policy, political science, economics, or business.

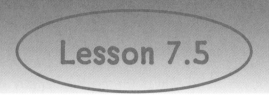
Game Theory, Part 2

The games considered in the previous lesson were strictly determined. In this lesson, games in which there is not a single best strategy for each player are introduced.

Look again at the game of the previous lesson. Suppose that Sol, knowing that he can lose only if Tina plays rationally, proposes changing the game. He now will win four pennies if both coins are heads and one penny if both coins are tails. He will lose two pennies if he shows heads and Tina shows tails, and three pennies if he shows tails and Tina shows heads. The new payoff matrix is:

$$
\begin{array}{c} \\ \text{Sol} \end{array}
\begin{array}{cc} & \text{Tina} \\ & \begin{array}{cc} \text{Heads} & \text{Tails} \end{array} \\ \begin{array}{c} \text{Heads} \\ \text{Tails} \end{array} & \begin{bmatrix} 4 & -2 \\ -3 & 1 \end{bmatrix}. \end{array}
$$

Here is the same matrix with the row minimums and column maximums:

		Tina			
		Heads	Tails		Row minimums
Sol	Heads	4	−2		(−2)
	Tails	−3	1		−3
Column maximums		4	(1)		

The maximin is −2 and the minimax is 1. Since the maximin does not agree with the minimax, the game is not strictly determined. The best strategy for either player is to display a mixture of heads and tails and keep the other player guessing. One way to do this would be to flip the coin and allow it to appear heads or tails at random. But such a strategy would cause heads and tails to appear in roughly equal portions, and it is not clear that this would be best for either player. Another strategy Sol could try is to roll a die and show heads if one, two, three, or four spots appeared, and tails otherwise. He might reason that this would benefit him because he would show heads two-thirds of the time and he wins the most if two heads appear.

Consider what will happen if Sol and Tina each decide to flip their coins. The probability of heads is .5, as is the probability of tails. Because Sol's flip and Tina's flip are made independently, the probability of both showing heads or both showing tails is $.5 \times .5 = .25$. The same is true for the cases in which one shows a head and the other shows a tail (see Figure 7.4).

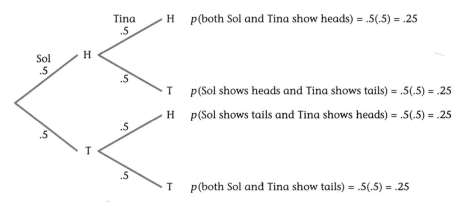

Figure 7.4 Probabilities when Sol and Tina flip their coins.

The probability distribution for Sol's winnings for this case are shown in the following table.

Outcome	HH	HT	TH	TT
Probability	.25	.25	.25	.25
Amount won	4	−2	−3	1

The expected payoff of the game for Sol is .25(4) + .25(−2) + .25(−3) + .25(1) = 1.00 − .50 − .75 + .25 = 0.

Since Tina's payoffs are the opposite of Sol's, her expectation is .25(−4) + .25(2) + .25(3) + .25(−1) = −1.00 + .50 + .75 − .25 = 0. If both players display heads and tails in equal proportions in this way, the game is **fair** because their expectations are equal.

But suppose that Tina decides to play heads 40% of the time, while Sol continues flipping his coin. The probability of both heads is now .5 × .4 = .2, while the probability of both tails is .5 × .6 = .3. The probability that Sol shows heads and Tina shows tails is .5 × .6 = .3 and that Sol shows tails and Tina shows heads is .5 × .4 = .2 (see Figure 7.5).

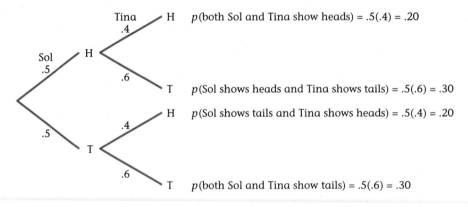

Figure 7.5 Probabilities when Sol flips his coin and Tina shows heads 40% of the time.

The distribution for Sol's winnings now looks like this:

Outcome	HH	HT	TH	TT
Probability	.2	.3	.2	.3
Amount won	4	−2	−3	1

The expected payoff for Sol is now .2(4) + .3(−2) + .2(−3) + .3(1) = .8 − .6 − .6 + .3 = −.1. This means he will lose 0.1 pennies per play, or 1 penny every 10 plays. Tina has an advantage and the game is no longer fair!

You have seen that Tina can gain an advantage over Sol if she knows he will display heads and tails in equal proportions. She does not know that Sol is going to do this, however, so how can she decide her best mixture of strategies? How can Sol decide what is best for him?

Reconsider the game from Sol's point of view, and suppose that Tina plays heads every time while Sol continues to flip his coin. The outcomes for this combination is shown in Figure 7.6. Sol's expected payoff is now $.5(4) + .5(-3) = .20 - 1.5 = -1.3$.

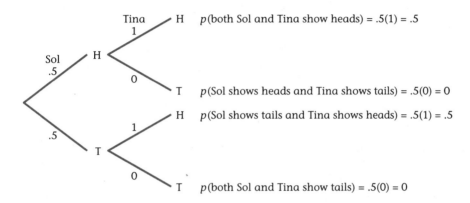

Figure 7.6 Probabilities when Sol flips his coin and Tina always plays heads.

If Tina decides to play tails each time while Sol continues to flip his coin (see Figure 7.7), Sol's expectation is $.5(-2) + .5(1) = -1.0 + .5 = -.5$.

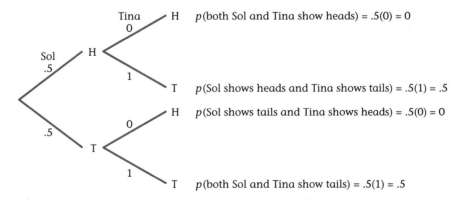

Figure 7.7 Probabilities when Sol flips his coin and Tina always plays tails.

Another way to show these calculations is to write the probabilities of Sol's displaying heads and tails in a row matrix and find the matrix product:

$$[.5 \quad .5] \begin{bmatrix} 4 & -2 \\ -3 & 1 \end{bmatrix} = [.5(4) + .5(-3) \quad .5(-2) + .5(1)]$$

$$= [.2 - 1.5 \quad -1 + .5] = [-1.3 \quad -.5].$$

Suppose Sol switches to displaying heads 60% of the time. Then this matrix product is

$$[.6 \quad .4] \begin{bmatrix} 4 & -2 \\ -3 & 1 \end{bmatrix} = [.6(4) + .4(-3) \quad .6(-2) + .4(1)]$$

$$= [2.4 - 1.2 \quad -1.2 + .4] = [1.2 \quad -.8].$$

This means that if Sol displays heads 60% of the time, he will gain 1.2 pennies per play if Tina always displays heads and lose 0.8 pennies per play if Tina always displays tails.

In general, if the probability Sol will display heads is p, his expected winnings per play, if Tina displays all heads or all tails, are

$$[p \quad 1-p] \begin{bmatrix} 4 & -2 \\ -3 & 1 \end{bmatrix} = [4p - 3(1-p) \quad -2p + 1(1-p)].$$

Because it is not very likely that Tina will display all heads or all tails, Sol's best strategy is to act in such a way that the two expectations are balanced or equalized. To find the value of p that does this, set the two expectations equal to each other and solve the resulting equation.

$$4p - 3(1-p) = -2p + 1(1-p)$$
$$4p - 3 + 3p = -2p + 1 - p$$
$$7p - 3 = -3p + 1$$
$$10p = 4$$
$$p = .4$$
$$1 - p = .6.$$

Sol's best strategy is to display heads four-tenths of the time and tails six-tenths of the time. One way he could accomplish this is to generate a random number on a calculator and display heads if the number that comes up is less than or equal to 0.4.

Tina's best strategy can be determined in a similar way. Call the probability that she displays heads q. Because she is the column player, multiply the payoff matrix times a column matrix to obtain her expectations if Sol plays either all heads or all tails:

$$\begin{bmatrix} 4 & -2 \\ -3 & 1 \end{bmatrix}\begin{bmatrix} q \\ 1-q \end{bmatrix} = \begin{bmatrix} 4q - 2 + 2q \\ -3q + 1 - q \end{bmatrix}.$$

Equate the two entries in the resulting matrix and solve to find Tina's best strategy.

$$
\begin{aligned}
4q - 2 + 2q &= -3q + 1 - q \\
6q - 2 &= -4q + 1 \\
10q &= 3 \\
q &= .3 \\
1 - q &= .7.
\end{aligned}
$$

In this case, Tina's best strategy is to display heads three-tenths of the time and tails seven-tenths of the time.

If both players pursue these strategies, the probability that a pair of heads will appear is $.4(.3) = .12$, or 12% of the time, and that a pair of tails will appear is $.6(.7) = .42$, or 42% of the time. The probability that Sol shows heads and Tina shows tails is $.4(.7) = .28$ or 28% of the time, and that Sol shows tails and Tina shows heads is $.6(.3) = .18$, or 18% of the time (see Figure 7.8).

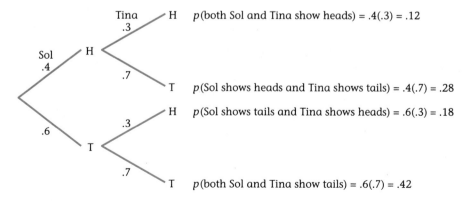

Figure 7.8 Probabilities when both Sol and Tina play their best strategies.

The resulting probability distribution from Sol's point of view is:

Outcome	HH	HT	TH	TT
Probability	.12	.28	.18	.42
Amount won	4	−2	−3	1

Sol's expected payoff for the game is .12(4) + .28(−2) + .18(−3) + .42(1) = .48 − .56 − .54 + .42 = −.2. If both players pursue their best strategy, the game is in Tina's favor and she can expect to win 0.2 of a penny per play, or 2 pennies every 10 plays, from Sol.

Using Matrices to Find the Expected Payoff

The expected payoff for a game can be found very easily by using matrices and a calculator. To do this, we form a row matrix A using the probabilities for the row player and a column matrix C using the probabilities of the column player. Then the expected payoff for the game for the row player equals the matrix product ABC, where B is the payoff matrix for the game.

In the preceding example, in which both Sol and Tina play their best strategies,

$$A = [.4 \quad .6], C = \begin{bmatrix} .3 \\ .7 \end{bmatrix}, \text{ and}$$

$$ABC = [.4 \quad .6] \begin{bmatrix} 4 & -2 \\ -3 & 1 \end{bmatrix} \begin{bmatrix} .3 \\ .7 \end{bmatrix} = -.2,$$

as shown on the following calculator screens.

```
[A]
        [[.40 .60]]
[B]
 [[4.00   -2.00]
  [-3.00 1.00 ]]
```

```
[C]
              [[.30]
               [.70]]
[A]*[B]*[C]
              [[-.20]]
```

A Graphical Solution for Sol's Best Strategy

Sol's search for a best strategy can be visualized graphically:

1. Draw a horizontal line to represent the probability of Sol's displaying heads. Scale this axis in tenths from 0 to 1 (see Figure 7.9).

2. Draw vertical axes at each end of the horizontal axis and scale them from the minimum amount Sol can win (—3 in this case) to the maximum amount (4 in this case).

3. Draw a diagonal line to represent what happens if Tina always displays heads. To do this, notice that if Tina displays heads and Sol displays tails, Sol will lose 3 cents. Place a dot at −3 on the vertical axis on the left, where the probability of Sol's displaying heads is 0. Similarly, if Sol displays heads and Tina displays heads, he will win 4 cents. Place a dot at the 4 on the vertical axis on the right where the probability of Sol's displaying heads is 1. Connect the two dots with a diagonal line. Sol's expected winnings for his various strategies for displaying heads when Tina always displays heads can be read from this line (see line 1 in Figure 7.9).

4. Repeat the procedure in step 3 to draw a diagonal line that shows what happens if Tina always displays tails. Place a dot at 1 on the

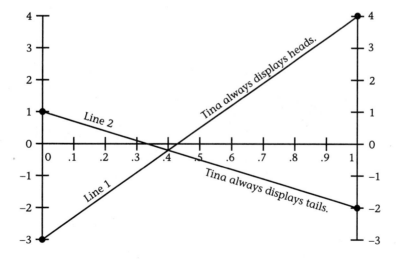

Figure 7.9 Sol's best strategy for displaying heads.

left vertical axis since Sol wins 1 cent when both he and Tina show tails. Similarly, place a dot at -2 on the vertical axis on the right since Sol loses 2 cents if he displays heads and Tina displays tails. Connect the two dots. Sol's expected winnings for his various strategies when Tina always displays tails can be read from this line (see line 2 in Figure 7.9).

Since Sol's best strategy is to act in such a way that the two expectations are balanced or equalized, the intersection of the two lines lies directly below Sol's best strategy for displaying heads. His expected payoff for playing this strategy can be read from the vertical axes.

To calculate the exact values for Sol's best strategy and his expected payoff from the graph, let x represent the probability of Sol's displaying heads and P represent the payoff. Then the equations of line 1 and line 2 in slope intercept form are $P = 7x - 3$ and $P = -3x + 1$, respectively. Setting these two equations equal and solving for x, we get

$$7x - 3 = -3x + 1$$
$$10x = 4$$
$$x = .4$$

Substituting .4 for x in the first equation, we get

$$P = 7x - 3 = 7(.4) - 3 = 2.8 - 3 = -.2$$

Exercises

1. Suppose that in the example of this lesson, Sol decides to return to flipping his coin while Tina continues to pursue her best strategy of playing heads three-tenths of the time.

   a. Set up a tree diagram to compute the probabilities of each of the four outcomes for this game.

   b. What is the probability that both Sol and Tina will show heads?

   c. What is the probability that Tina will show tails and Sol will show heads?

   d. What is the probability that Tina will show heads and Sol will show tails?

   e. What is the probability that both Sol and Tina will show tails?

   f. Write a probability distribution chart for Sol's winnings.

   g. Calculate Sol's expected payoff for this game. Explain what this means in terms of pennies won or lost.

   h. How does this payoff compare with Sol's expectation if he plays his best strategy as computed in this lesson?

2. Use matrices and a calculator as shown on page 388 to find the expected payoffs for Sol in this exercise.

   a. Suppose that in the example of this lesson, Sol decides to play heads three-fourth of the time while Tina continues to pursue her best strategy of playing heads three-tenths of the time. Find Sol's expectation for this situation.

   b. Choose two or three other strategies for Sol to play while Tina continues to pursue her best strategy of playing heads three-tenths of the time. Compute Sol's expected payoff for these strategies.

   c. Suppose now that Sol returns to using his best strategy of playing heads four-tenths of the time while Tina plays a variety of strategies. Choose three or four different strategies for Tina to play while Sol plays his best strategy and find Sol's expected payoff in each case.

   d. Compare your results with your classmates' results for Sol's expectation in parts a to c. Make a conjecture based on your observations.

3. Suppose that Sol and Tina change their game so that the payoffs to Sol are

$$
\begin{array}{cc}
 & \text{Tina} \\
 & \begin{array}{cc} \text{Heads} & \text{Tails} \end{array} \\
\text{Sol} \begin{array}{c} \text{Heads} \\ \text{Tails} \end{array} & \begin{bmatrix} 3 & -2 \\ -2 & 1 \end{bmatrix}.
\end{array}
$$

   a. Use the row matrix $[p \quad 1-p]$ to find Sol's best strategy for this game.

   b. Use the column matrix $\begin{bmatrix} q \\ 1-q \end{bmatrix}$ to find Tina's best strategy for this game.

   c. Set up a tree diagram to compute the probabilities of each of the four outcomes for this game.

   d. Prepare a probability distribution chart for Sol's winnings.

   e. Find Sol's expectation for this game.

   f. Interpret your answer in part e in terms of how many pennies Sol can expect to win or lose over a number of games.

   g. Construct a graph showing Sol's best strategy for playing heads in this game.

   h. Find the equations of the lines in part g. Set these equations equal and find Sol's best strategy and his expected payoff for this game.

4. The procedure outlined in this lesson is designed to determine the best mixture of strategies when a game is not strictly determined. Therefore,

you should always inspect a game to see whether it is strictly determined and apply the saddle point technique of the previous lesson if it is. It is, however, easy to forget to do this. To see what will happen if you attempt to determine a mixture of strategies for a game that is strictly determined, apply the techniques of this lesson to the strictly determined game that Sol and Tina were playing in the last lesson and try to find the best mixture of strategies for each of them. The payoff matrix for this game is reprinted here.

$$
\begin{array}{cc}
 & \text{Tina} \\
 & \begin{array}{cc} \text{Heads} & \text{Tails} \end{array} \\
\text{Sol} \quad \begin{array}{c} \text{Heads} \\ \text{Tails} \end{array} & \begin{bmatrix} 3 & -1 \\ -1 & -2 \end{bmatrix}.
\end{array}
$$

5. a. For the game defined by the following matrix, determine the best strategies for both players.

$$
\begin{bmatrix} 1 & 3 \\ 4 & 2 \end{bmatrix}
$$

   b. Add 5 to each element in the matrix given in part a. How does this effect the best strategies?
   c. Multiply each element in the matrix in part a by 3. How does this effect the best strategies?
   d. Make a conjecture based on the results of parts b and c.
   e. Challenge: Use algebra to prove your conjecture.

6. In a game known as Two-Finger Morra, two players simultaneously hold up either one or two fingers. If they hold up the same number of fingers, player 1 will win the sum (in pennies) of the digits from player 2. If they hold up different numbers, then player 2 will win the sum from player 1. Write the payoff matrix for this game. Find the best strategy for each player and the expectation for the row player. Is this a fair game? Explain your answer.

7. In another version of the game in Exercise 6, if the sum of the fingers held out by each player is even, player 1 will win 5 cents. If the sum is odd, player 2 will win 5 cents. Write the payoff matrix for this version. Find the best strategy for each player and the payoff expectation for the row player. Is this a fair game? Explain your answer.

8. A group of parents in a small town in the Midwest are in an uproar about a new social studies program that the school district has adopted. They are seeking to have the program removed from the curriculum. A second group of parents believe the new program is a solid choice and are organizing in favor of keeping it. In order to bring the issue before the voters in the town, the opposing group must collect 400 supporting signatures from registered voters. Both sides are campaigning vigorously by making telephone calls, sending out mailings, and going door to door to contact voters. The local newspaper has estimated the number of signatures that the opposing group is expected to collect with each combination of strategies. What are the best strategies for both groups of parents? If both follow their best strategies, can the opposing group expect to gather enough signatures to get the issue on the ballot? (Hint: Use the concept of dominance to eliminate a row and column.)

$$
\begin{array}{c}
\text{Group in favor} \\
\begin{array}{ccc}
\text{Phone} & \text{Mail} & \text{Door}
\end{array} \\
\text{Group against}
\begin{array}{c}
\text{Phone} \\
\text{Mail} \\
\text{Door}
\end{array}
\left[
\begin{array}{ccc}
150 & 75 & 100 \\
350 & 300 & 200 \\
500 & 100 & 400
\end{array}
\right]
\end{array}
$$

9. Two rival TV networks compete for prime time audiences by showing comedy, drama, and sports. The following matrix gives the payoffs for network A in terms of percentages of regular viewers who watch its channel for various combinations of programs. Find the best strategy for each network and the expectation for network A.

$$
\begin{array}{c}
\text{Network B} \\
\begin{array}{ccc}
\text{Comedy} & \text{Drama} & \text{Sports}
\end{array} \\
\text{Network A}
\begin{array}{c}
\text{Comedy} \\
\text{Drama} \\
\text{Sports}
\end{array}
\left[
\begin{array}{ccc}
10 & 50 & 20 \\
40 & 30 & 50 \\
30 & 20 & 60
\end{array}
\right]
\end{array}
$$

10. In a campaign for student council president at Northeast High the top two candidates, Betty and Bob, are each making two promises about what they will do if they are elected. The payoff matrix in terms of the number of votes Betty will gain follows. What is the best strategy for each candidate and what is Betty's expectation?

$$
\begin{array}{cc}
 & \text{Bob} \\
 & \begin{array}{cc} A & \quad B \end{array} \\
\text{Betty} \begin{array}{c} 1 \\ 2 \end{array} & \begin{bmatrix} 200 & 100 \\ 50 & 180 \end{bmatrix}
\end{array}
$$

Project

11. The games you studied in this and the previous lesson are known as *zero-sum games*, because one person's loss is the other's gain. If, for example, Sol wins $2, then Tina loses the same amount. In some games, a particular outcome may be worth 2 to one player, but not -2 to the other. Examples of such games include the prisoner's dilemma, chicken, and arms races between countries. Research and report on games that are not zero sum.

Computer/Calculator Exploration

12. Write a graphing calculator (or computer) program using matrices to find the expected payoff for the row player in a two-person game that is not strictly determined.

A Look at a Dominance Matrix

In this example a dominance matrix is used to examine "pecking order" behavior among five neighborhood cats (Bruiser, Wraque, Wruin, Pebbles, and Boy) who enjoy hunting for mice in an open field near their homes. The purpose of using this simple struggle for top cat in the field is to give you an understanding about how this application can be used to measure and compare the dominance of one person over another in political or business situations.

Close observation of the behavior of the cats reveals that there is a definite sense of who is allowed to hunt near the choicest mouse holes when more than one cat is in the field. Bruiser lives up to his name and chases away Wruin, Boy, and Pebbles. Feisty little Wraque stands up to Bruiser and let's Boy know, in no uncertain terms, that she is the boss cat when their paths cross. Wruin dominates her little sister, Wraque, and will not tolerate Pebbles. Pebbles always picks on Wraque and Boy. Finally, Boy, even though he is the smallest, somehow manages to intimidate fat, fuzzy Wruin.

The directed graph on the next page illustrates this furry dominance. The direction indicates who is the dominant cat for each possible pair. Letters B, Q, W, P, and Y represent Bruiser, Wraque, Wruin, Pebbles, and Boy, respectively.

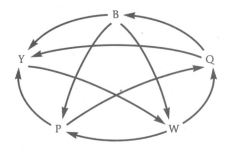

A 5 by 5 **dominance matrix** D can be used to represent this situation where a 1 represents the dominance of one cat over another and 0s other-wise. The relationship expressed in the dominance matrix is the same as that represented in the digraph.

$$D = \begin{array}{c} \\ B \\ Q \\ W \\ P \\ Y \end{array} \begin{array}{ccccc} B & Q & W & P & Y \\ \left[\begin{array}{ccccc} 0 & 0 & 1 & 1 & 1 \\ 1 & 0 & 0 & 0 & 1 \\ 0 & 1 & 0 & 1 & 0 \\ 0 & 1 & 0 & 0 & 1 \\ 0 & 0 & 1 & 0 & 0 \end{array}\right] \end{array}$$

Notice that the entries along the main diagonal in the dominance matrix are all zero. This represents the fact that a cat cannot dominate itself. Notice, also, that if entry D_{ij} is a 1, then entry D_{ji} is a 0. This is true since only one cat can dominate the other. They cannot be mutually dominant. It can also be seen by looking at the matrix that the relationship among the cats is not transitive. For example, Bruiser dominates Wruin, and Wruin dominates Wraque, but Bruiser does not dominate Wraque.

Since the relationship among the cats is not transitive, it is not immedi-ately apparent who is the most powerful. One way to decide this issue would be to look at the dominance matrix and sum the numbers in each row. If we do this, the result is called an **authority vector**:

$$A = \begin{array}{c} \text{Bruiser} \\ \text{Wraque} \\ \text{Wruin} \\ \text{Pebbles} \\ \text{Boy} \end{array} \left[\begin{array}{c} 3 \\ 2 \\ 2 \\ 2 \\ 1 \end{array}\right].$$

This places Bruiser at the top of the pecking order in dominating three of the other cats. Wraque, Wruin, and Pebbles tie with dominance over two each, and poor Boy lands at the bottom with the least amount of power. This *direct* measure may be all right for a determining the top cat in the neighborhood. However, if we were using this system to examine the relative power or influence over others among a group of people, we would probably be interested in indirect measures of influence as well. For example, if B has direct influence over P and P has direct influence over Q, then there is a sense that B may have some *indirect* influence over Q. This indirect, or second-order, influence should be taken into consideration in determining the degree of B's power.

Return now to the dominance matrix D for this group of cats. Suppose you square this matrix:

$$D^2 = \begin{array}{c} \\ B \\ Q \\ W \\ P \\ Y \end{array} \begin{array}{c} \begin{array}{ccccc} B & Q & W & P & Y \end{array} \\ \left[\begin{array}{ccccc} 0 & 2 & 1 & 1 & 1 \\ 0 & 0 & 2 & 1 & 1 \\ 1 & 1 & 0 & 0 & 2 \\ 1 & 0 & 1 & 0 & 1 \\ 0 & 1 & 0 & 1 & 0 \end{array}\right] \end{array}.$$

What do the entries in this matrix represent? Look at the entry in row 1 column 2. This value is found by multiplying the components of row 1 of matrix D by the corresponding components in the second column and summing the five products:

$$(0 \times 0) + (0 \times 0) + (1 \times 1) + (1 \times 1) + (1 \times 0) = 1 + 1 = 2.$$

Notice that the entries in D^2 can only be nonzero if both factors in at least one product are equal to 1. In the first nonzero product, the first 1 comes from the fact that Bruiser dominates Wruin, and the second 1 comes from the fact that Wruin dominates Wraque. And since this product is nonzero, we can deduce that Bruiser has an indirect second-order influence over Wraque through Wruin. Similarly, the second nonzero product indicates that Bruiser dominates Pebbles, and Pebbles dominates Wraque. This shows that Bruiser also has an indirect second-order influence over Wraque through Pebbles. Thus the 2 in row 1, column 2 of D^2 indicates that Bruiser has two second-order influences over Wraque. We can verify this conclusion by looking at the digraph. One edge points from B to W and another points from W to Q. A second indirect path points from B to P and from P to Q.

You can use a similar argument to show that D^3 will represent the number of third-order influences that exist for each cat.

$$D^3 = \begin{array}{c} \\ B \\ Q \\ W \\ P \\ Y \end{array} \begin{array}{ccccc} B & Q & W & P & Y \\ \left[\begin{array}{ccccc} 2 & 2 & 1 & 1 & 3 \\ 0 & 3 & 1 & 2 & 1 \\ 1 & 0 & 3 & 1 & 2 \\ 0 & 1 & 2 & 2 & 1 \\ 1 & 1 & 0 & 0 & 2 \end{array}\right] \end{array}$$

Now if you add all the entries in corresponding rows from the three matrices D, D^2, and D^3 (omitting the diagonal of matrix D^3), the result will be an influence vector showing the total number of direct first-order and indirect second- and third-order influences exercised by each cat over the others:

$$I = \begin{array}{c} Bruiser \\ Wraque \\ Wruin \\ Pebbles \\ Boy \end{array} \left[\begin{array}{c} 15 \\ 10 \\ 10 \\ 9 \\ 5 \end{array}\right]$$

The resulting vector indicates that Bruiser is still top cat and Boy remains at the bottom, but at least one of the ties has been eliminated. The question now is where do you stop when you are adding up dominance matrices in this way? Also, is it perhaps more reasonable to give less weight to the second-, third-, and higher-order influence matrices than we give to the direct-order matrix? One way to accomplish this is by dividing D^2 by 2 and D^3 by 3, and so on, if higher-order influence matrices are used in determining the power of each individual member of the group.

The case of the dominant cat may not be of great interest to persons other than the cat owners. However, the method developed in the cat problem can be applied to groups of people such as legislators or corporate leaders to determine who is the most influential or who has the most power. In these situations, one person may show support for another by voting for a bill or merger that he or she has sponsored. A 1 in the dominance matrix is assigned to the person who receives the most support from the other.

Reprinted with special permission of King Features Syndicate.

1. Write a summary of what you think are the important points of this chapter.

2. Suppose that a three-sector economy has the consumption matrix

$$C = \begin{array}{c} \\ A \\ B \\ C \end{array} \begin{array}{ccc} A & B & C \\ \left[\begin{array}{ccc} 0.1 & 0.2 & 0.3 \\ 0.1 & 0.3 & 0.2 \\ 0.2 & 0.1 & 0.2 \end{array} \right] \end{array}.$$

a. Draw a weighted digraph for this economy.

b. A production matrix, P, follows. Find the internal consumption matrix product CP. Find the external demand matrix D, where $D = P - CP$.

$$P = \begin{bmatrix} 8 \\ 12 \\ 15 \end{bmatrix}.$$

c. An external demand matrix, D, follows. Find the production matrix P for this economy. Recall that $P = (I - C)^{-1}D$.

$$D = \begin{bmatrix} 6 \\ 8 \\ 12 \end{bmatrix}.$$

3. Mike and Nancy are playing poker for pennies to kill time during lunch. Mike is holding a very poor hand and is considering bluffing or not bluffing. Nancy can either call or not call the bluff. The payoff matrix for this situation is shown on the next page. Over the course of

several games in which Mike comes up with a poor hand, what should his strategy be?

$$\begin{array}{c} & & \text{Nancy} \\ & & \text{Call} \qquad \text{Not call} \\ \text{Mike} \begin{array}{c} \text{Bluff} \\ \text{Not bluff} \end{array} & \begin{bmatrix} -10 & 10 \\ -2 & 0 \end{bmatrix} \end{array}$$

4. Mike and Nancy soon get bored playing the game described in Exercise 3. They each draw two cards from the deck. Mike draws a 4 of spades and an ace of hearts. Nancy draws a 3 of clubs and a 2 of diamonds. They make up a new game to play with the following rules. Each player will show one card. If both cards shown are the same color, Nancy will pay Mike the sum of the face value of the cards in pennies. If the cards shown are of different colors, Mike will pay Nancy the sum of the face values shown.
   a. Find the best strategies for both Mike and Nancy.
   b. Use a probability tree to calculate the probability of each of the four possible outcomes when Mike and Nancy play their best strategies.
   c. Set up a probability distribution for this game.
   d. Find the expected value of the game for Mike.
   e. Explain what is meant by the expected value of the game for Mike in this situation.

5. The discrete mathematics teacher has three class starter activities, one of which she uses to begin class every day: a pop quiz, a quickie review, and a small-group problem-solving activity. She never uses the same activity two days in a row. If she gave a pop quiz yesterday, she will toss a coin, and do a quickie review if it comes up heads. If she used a review, she will toss two coins and switch to problem solving if two heads come up. If she did a problem-solving activity, then she will toss three coins, and if three heads come up, she gives a pop quiz again. The transition matrix for this scheme is

$$\begin{array}{c} & \begin{array}{ccc} Q & R & P \end{array} \\ \begin{array}{c} Q \\ R \\ P \end{array} & \begin{bmatrix} 0 & \frac{1}{2} & \frac{1}{2} \\ \frac{3}{4} & 0 & \frac{1}{4} \\ \frac{1}{8} & \frac{7}{8} & 0 \end{bmatrix} \end{array}$$

   a. If the teacher gives a quiz on Monday, what is the probability that she will give another quiz on Friday?
   b. In the long run, how often should the students expect that the teacher will start class with a quiz?

c. What activity will the teacher use most often to begin class, and how frequently will she use it?

6. The Super X sells three kinds of sandwiches that many of the students at Southeast High especially like for lunch—Super X Original, Italian Special, and Barbecue Beef. The Super X clerk observed that the same students were coming in for sandwiches for lunch every school day and that the kind of sandwich that each student purchased depended on what he or she had ordered on the previous visit. He conducted a survey and found that of the students who ordered the Original on their last visit, 20% ordered it again the next time, whereas 25% switched to Italian and 55% switched to Barbecue Beef. Of the students who ordered the Italian sandwich the last time, 35% did so again the next time, but 45% switched to the Original and 20% switched to the Barbecue Beef. Of the students who got the Barbecue Beef the last time, 55% ordered it the next time, 20% switched to the Original, and 25% switched to Italian.
   a. Set up the transition matrix for this Markov chain.
   b. If the same students tend to buy Super X sandwiches for lunch every day, what is the probability that a student who buys the Italian sandwich on Monday will have it again on Wednesday.
   c. In the long run, what percentage of the orders will be for the Original? For the Italian? For the Barbecue Beef?
   d. How will access to this information help the Super X clerk?

7. A certain economy consists of three industries: transportation, petroleum, and agriculture. The production of $1 million worth of transportation requires an internal consumption of $0.2 million worth of transportation, $0.4 million worth of petroleum, and no agriculture. The production of $1 million worth of petroleum requires an internal consumption of $0.3 million worth of transportation, $0.2 million worth of petroleum, and $0.3 million worth of agriculture. The production of $1 million worth of agriculture requires an internal consumption of $0.3 million worth of transportation, $0.2 million worth of petroleum, and $0.25 million worth of agriculture.
   a. Draw a weighted digraph for this economy.
   b. Write a consumption matrix, C, representing this information.
   c. On what sector of the economy is transportation the most dependent? The least dependent?
   d. If the agriculture sector has an output of $5.4 million dollars, what is the input in dollars from petroleum? From agriculture?

e. A production matrix, P, in millions of dollars follows. Find the internal consumption matrix product CP and the external demand matrix D.

$$P = \begin{matrix} \text{Transportation} \\ \text{Petroleum} \\ \text{Agriculture} \end{matrix} \begin{bmatrix} 20 \\ 25 \\ 15 \end{bmatrix}$$

f. An external demand matrix, D, in millions of dollars, follows. How much must each sector produce to meet this demand?

$$D = \begin{matrix} \text{Transportation} \\ \text{Petroleum} \\ \text{Agriculture} \end{matrix} \begin{bmatrix} 4.6 \\ 5.2 \\ 3.0 \end{bmatrix}$$

8. Two computer companies (1 and 2) are competing for sales in two large school districts (A and B). The following payoff matrix shows the differences in sales for companies 1 and 2 in hundreds of thousands of dollars if they focus their full sales force on either school district. Find the best strategy for each company.

Computer company 2

$$\text{Computer company 1} \quad \begin{matrix} A \\ B \end{matrix} \begin{bmatrix} \overset{\displaystyle A}{3} & \overset{\displaystyle B}{7} \\ -7 & -3 \end{bmatrix}$$

9. Suppose that in the final days of a political campaign for mayor in a small midwestern city, the Democrats and Republicans are planning their strategies for winning undecided voters to their political camps. The Democrats have decided on two strategies, plan A and plan B. The Republicans plan to counter with plans C and D. The following matrix gives the payoff for the Democrats of the various combinations of strategies. The numbers represent the percentage of the undecided voters joining the Democrats in each case. Find the best strategies for both parties and the expectation for the Democrats.

Republicans

$$\text{Democrats} \quad \begin{matrix} \text{Plan A} \\ \text{Plan B} \end{matrix} \begin{bmatrix} \overset{\displaystyle \text{Plan C}}{30} & \overset{\displaystyle \text{Plan D}}{60} \\ 50 & 40 \end{bmatrix}$$

10. A manufacturing company has divisions in Massachusetts, Nebraska, and California. The company divisions use goods and services from each other as shown in the following consumption matrix C.

$$C = \begin{matrix} & \begin{matrix} \text{Mass.} & \quad \text{Neb.} & \quad \text{Calif.} \end{matrix} \\ \begin{matrix} \text{Mass.} \\ \text{Neb.} \\ \text{Calif.} \end{matrix} & \begin{bmatrix} 0.04 & 0.02 & 0.03 \\ 0.03 & 0.01 & 0.05 \\ 0.01 & 0.02 & 0.04 \end{bmatrix} \end{matrix}$$

a. Draw a weighted digraph for this situation.
b. Find the total production needed to meet a final consumer demand of $50,000 from Massachusetts, $30,000 from Nebraska, and $40,000 from California.
c. What will the internal consumption be for each division to meet the demands in part b?
d. Suppose there is a increase in consumer demand of $10,000 from Massachusetts, $8,000 from Nebraska, and $12,000 from California. What will be the change in internal consumption and in the total production of goods and services for each division?

11. Two competing dairy stores choose daily strategies of raising, not changing, or lowering their milk prices. The following payoff matrix shows the percentage of customers who go from store A to store B for each combination of strategies. What should each store do?

		Store B	
	Raise	No change	Lower
Raise	4	−1	−4
Store A No change	2	1	−2
Lower	5	2	3

Bibliography

Bittinger, M. L., and J. C. Crown. 1989. *Finite Mathematics*. Reading, MA: Addison-Wesley.

Bogart, Kenneth P. 1988. *Discrete Mathematics*. Lexington, MA: Heath.

Brandenburger, A. M., A. A. Brandenburger, and B. J. Nalebuff. 1996. *Co-opetition: The Competitive Edge in Business, Politics, and Everyday Life*. Garden City, NY: Doubleday.

Brams, S. J. 1985. *Rational Politics: Decisions, Games, and Strategy.* Washington, D.C.: C Q Press.

Case, J. H. 1979. *Economics and the Competitive Process.* New York: New York University Press.

COMAP. 1997. *For All Practical Purposes: Introduction to Contemporary Mathematics.* 4th ed. New York: W. H. Freeman.

Cozzens, M. B., and R. D. Porter. 1987. *Mathematics and Its Applications.* Lexington, MA: Heath.

Dixit, A. K., and B. J. Nalebuff. 1991. *Thinking Strategically.* New York: Norton.

Keller, M. K. 1983a. *Food Service Management and Applications of Matrix Methods.* Lexington, MA: COMAP.

Keller, M. K. 1983b. *Markov Chains and Applications of Matrix Methods: Fixed Point and Absorbing Markov Chains.* Lexington, MA: COMAP.

Kemeny, J. G., J. L. Snell, and G. L. Thompson. 1957. *Finite Mathematics.* Englewood Cliffs, NJ: Prentice-Hall.

Leontief, L. 1986. *Input-Output Economics.* New York: Oxford University Press.

Mauer, S. B., and A. Ralston. 1991. *Discrete Algorithmic Mathematics.* Reading, MA: Addison-Wesley.

National Council of Teachers of Mathematics. 1988. *Discrete Mathematics Across the Curriculum K-12.* Reston, VA: National Council of Teachers of Mathematics.

North Carolina School of Science and Mathematics. *New Topics for Secondary School Mathematics: Matrices.* 1988. Reston, VA: National Council of Teachers of Mathematics.

Poundstone, William. 1993. *Prisoner's Dilemma.* Garden City, NY: Doubleday.

Rapoport, A. 1966. *Two-Person Game Theory: The Essential Ideas.* Ann Arbor: University of Michigan Press.

Rapoport, A., and A. M. Chammah. 1965. *Prisoner's Dilemma: A Study in Conflict and Cooperation.* Ann Arbor: University of Michigan Press.

Ross, K. A., and C. R. B. Wright. 1985. *Discrete Mathematics.* Englewood Cliffs, NJ: Prentice-Hall.

Straffin, P. 1993. *Game Theory and Strategy*. Washington, D.C.: Mathematical Association of America.

Tuchinsky, P. M. 1989. *Management of a Buffalo Herd*. Lexington, MA: COMAP.

Wheeler, R. E., and W. D. Peebles. 1987. *Finite Mathematics with Applications to Business and the Social Sciences*. Monterey, CA: Brooks/ Cole.

Williams, J. D. 1982. *The Compleat Strategyst*. New York: Dover.

Zagare, F. C. 1985. *The Mathematics of Conflict*. Lexington, MA: COMAP.

Recursion

Recursion is a process that creates new objects from existing ones that were created by the same process. The recurrence relations you wrote in previous chapters are an example: they enable you to calculate new numbers from existing ones that were calculated with the same formula.

Recursive processes can be geometric, although the computer programs used to implement geometric recursion do so by performing numerical calculations. Fractal images are among the best-known examples of geometric recursion. For example, fractal techniques have been used to create artificial landscapes for science fiction movies.

How can recursion be used to create appealing images? How can recursion help people plan their financial futures? The mathematics of recursion can answer these and other important questions.

Introduction to Recursive Thinking

The simplest recursive processes are numerical: one number in a list is determined by applying simple mathematical calculations to one or more of the preceding numbers. This is the type of recursion you have considered in previous chapters and is also the type with which this chapter begins.

Reconsider a problem that you first saw in Lesson 2.6. Luis and Britt were examining the number of handshakes that occur when every person in a group shakes hands with every other person. The following is a table similar to the one you made in Lesson 2.6.

Number of People in the Group	Number of Handshakes
1	0
2	1
3	3
4	6
5	10

When a new person entered a group in which everyone had shaken hands, the new person had to shake hands with each of the people who were already in the group. Thus, the number of handshakes in a group of n people is $n - 1$ more than the number of handshakes in a group of

$n - 1$ people. If H_n represents the number of handshakes in a group of n people, this recurrence relation can be expressed symbolically as $H_n = H_{n-1} + (n - 1)$.

Your work with this recurrence relation included writing a formula called a **solution to the recurrence relation** and using mathematical induction to prove the formula correct. In this case, the solution, also called a **closed-form solution**, is $H_n = \dfrac{n(n - 1)}{2}$.

Closed-form solutions are useful because, unlike recurrence relations, they calculate a value directly. You can, for example, find the number of handshakes in a group of ten people without knowing the number of handshakes in a group of nine people. However, closed-form solutions can be difficult to find. If such solutions are found by trial and error, mathematical induction can be used to prove their validity.

There are techniques other than trial and error that can be used to find closed-form solutions. For example, the counting techniques discussed in Chapter 6 are useful for certain kinds of recurrence relations. The handshake problem requires that every pair of people shake hands. In a group of n people, there are $C(n, 2)$ ways of selecting a pair, and so there are $C(n, 2) = \dfrac{n!}{(n - 2)!2!}$ handshakes. But $n! = n(n - 1)(n - 2)!$, so the counting solution is equivalent to the solution you hypothesized and proved in Chapter 2.

The closed-form calculation of the number of handshakes in a group of, say, 100 people is a simple one: $100 \times \dfrac{99}{2} = 4{,}950$. Obtaining this solution with a recurrence relation requires extending the table on page 408 to 100 rows.

Extending a table to 100 or more rows is a tedious task when done by hand. Fortunately, there are several ways to apply technology to the problem. Indeed, the speed of computer and calculator technology has made recursive methods much more useful today than they were only a few decades ago.

One type of technology that is very useful when working with recurrence relations is a computer *spreadsheet*. A spreadsheet is a matrix consisting of columns labeled with the letters A, B, C, . . . and rows labeled with the numerals 1, 2, 3, ... A particular location in the spreadsheet is called a *cell* and is denoted by its column letter and row number, such as A1 or C5. Cells may contain verbal information, numeric values, or formulas based on references to other cells. Spreadsheets have copy features that allow formulas to be copied into other cells so that tables can be generated rapidly.

Another way technology can be applied to recurrence relations is by developing a calculator or computer program that generates values from the relation. Programming requires that an appropriate algorithm be adapted to the language used by the calculator or computer. The following is an algorithm for the handshake problem that can be adapted to a calculator or computer. The variable N represents the number of people in the group, and H represents the number of handshakes.

1. Store the number 1 for variable N and the number 0 for variable H.

2. Display N and H.

3. Add 1 to N and store the result as the new value of N.

4. Add $N - 1$ to H and store the result as the new value of H.

5. Repeat steps 2 through 4.

Step 4 of this algorithm used the recurrence relation to calculate the number of handshakes. The closed form could also be used in this step. To do so, replace step 4 with "store $\dfrac{N(N-1)}{2}$ as the new value of H."

Finally, some calculators have special functions designed to generate values from recurrence relations.

Using Computer Spreadsheets with Recurrence Relations

To create a spreadsheet for the handshake problem, type suitable labels in the first row. Type initial values of 1 for the number of people in cell A2 and 0 for the number of handshakes in cell B2. In cell A3, type the formula A2 + 1. In cell B3, type the formula B2 + A2, which is equivalent to the recurrence relation $H_n = H_{n-1} + (n - 1)$. The remaining rows are filled by copying row 3. Note that most spreadsheets require the initial character of a formula to be either + or =. The completed spreadsheet is shown here in two ways: with formulas and with numeric results.

	A	B	C
1	Number of people	Number of handshakes	Closed form
2	1	0	=A2*(A2−1)/2
3	=A2+1	=A2+B2	=A3*(A3−1)/2
4	=A3+1	=A3+B3	=A4*(A4−1)/2
5	=A4+1	=A4+B4	=A5*(A5−1)/2
6	=A5+1	=A5+B5	=A6*(A6−1)/2
7	=A6+1	=A6+B6	=A7*(A7−1)/2
8	=A7+1	=A7+B7	=A8*(A8−1)/2
9	=A8+1	=A8+B8	=A9*(A9−1)/2
10	=A9+1	=A9+B9	=A10*(A10−1)/2

	A	B	C
1	Number of people	Number of handshakes	Closed form
2	1	0	0
3	2	1	1
4	3	3	3
5	4	6	6
6	5	10	10
7	6	15	15
8	7	21	21
9	8	28	28
10	9	36	36

Writing Programs for Recurrence Relations

On the left is a computer algorithm written in BASIC that generates a table for the handshake problem. On the right is a similar calculator algorithm for Texas Instruments graphing calculators.

```
10 N = 1:H = 0                1→N:0→H

20 PRINT N, H                 Lbl A

30 N = N + 1                  Disp N, H

40 H = H + N − 1              N+1→N:H+N−1→H

50 GO TO 20                   Goto A
```

Because these algorithms do not end, a statement should be added to terminate the table at some value. For example, to stop the program after calculation of the number of handshakes in a group of ten people, add the line 45 IF N>10 THEN STOP to the BASIC algorithm or the two lines If N>10 and Stop before the last line of the calculator algorithm. Note that the calculator algorithm does not display paired values of N and H on a single line. One way to remedy this inconvenience is store the values in a 1 × 2 matrix and display the matrix.

Calculators with Recursion Features

The following screens demonstrate the recursion features of one type of graphing calculator. The left screen shows the entry of the handshake recurrence relation, which is done after the calculator is placed in its sequence mode. The right screen shows the resulting table, which appears after the table's initial value and increment have been set.

```
Plot1  Plot2  Plot3
 nMin=1
\..u(n)∎u(n-1)+(n-
1)
 u(nMin)∎{0}
\..v(n)=
 v(nMin)=
\..w(n)=
```

n	u(n)	
1	0	
2	1	
3	3	
4	6	
5	10	
6	15	
7	21	

u(n)∎u(n-1)+(n-...

Exercises

1. Consider a variation of this lesson's handshake problem. There are an equal number of men and women in Luis and Britt's group and each person shakes hands with all members of the opposite sex.

    a. Draw a graph for a group of four couples in which the vertices represent the men and women in the group and the edges represent the handshakes. Recall your work in graph theory. What kind of a graph is this?

    b. If there are only one man and one woman in the group, how many handshakes will be made? With two couples? With three couples?

    c. Complete the following table to investigate the number of handshakes that are made.

Number of Couples	Number of Handshakes	Recurrence Relation
1		
2		
3		
4		
5		

d. Assume that there are H_{n-1} handshakes for $n - 1$ couples and that another couple joins the party. How many additional handshakes are now possible?

e. Write a recurrence relation that describes the relationship between the number of handshakes (H_n) for n couples and the number of handshakes (H_{n-1}) for $n - 1$ couples.

2. Consider another variation of the handshake problem in which each man shakes hands with each of the women *except* his date.
   a. Make a table showing the number of handshakes that occur when there is one couple. Two couples. Three couples. Four couples.
   b. Assuming you know the number of handshakes with $n - 1$ couples, how many additional handshakes are made when the nth couple arrives?
   c. Write a recurrence relation for the total number of handshakes (H_n) when there are n couples.

3. a. Write a recurrence relation to describe each of the following patterns. (Note: Do not give closed-form formulas.)

      i. 1, 4, 7, 10, 13, ...   ii. 1, 2, 4, 8, 16, 32, ...

     iii. 1, 3, 6, 10, 15, 21, 28, ...   iv. 1, 2, 6, 24, 120, 720, ...

   b. What is the next term in each of the patterns in part a?

4. The ability to recognize patterns is considered a mark of intelligence. Therefore, most IQ tests include questions about numerical patterns. For example, a question on an IQ test gives the sequence 1, 2, 3, 5, 8, 11 and asks which of the numbers does not belong.
   a. Explain why the correct answer is that the last number, 11, does not fit the pattern.
   b. Write a recurrence relation that describes the pattern.

5. You cannot use a recurrence relation to generate terms unless you have an initial value. For example, $t_n = 2t_{n-1} - 3$ has terms 5, 7, 11, 19, . . . if the initial value t_1 is 5. But if the initial value is 6, then the terms

are 6, 9, 15, 27, ... Notice, however, if the initial value is 3, then all the terms are 3. An initial value for which all the terms of the recurrence relation are the same is called a **fixed point**.

Find the fixed point for each of the following recurrence relations if one exists.

a. $t_n = 2t_{n-1} - 4.$ b. $t_n = 3t_{n-1} + 2.$

c. $t_n = 2t_{n-1}.$ d. $t_n = t_{n-1} + 3.$

6. If two rays have a common endpoint, one angle is formed. If a third ray is added, three angles are formed. See the following figure.

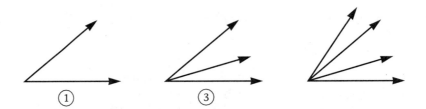

a. How many angles are formed if a fourth ray is added? A fifth ray?
b. Write a recurrence relation for the number of angles formed with *n* rays.
c. Write a closed-form solution.
d. Use your closed-form solution to find the number of angles formed by ten rays.

7. For the original handshake problem in which everyone shakes hands with everyone else, construct a table for one through eight people in the following manner.

First column: term number

Second column: number of handshakes

Third column: differences of successive numbers from column 2

Fourth column: differences of successive numbers from column 3

a. What do you notice about the last column?
b. What degree is the polynomial that was obtained for the closed-form solution of the handshake problem? Compare your answer with the number of difference columns in your table.

8. Consider the closed-form polynomial $S_n = 4n^3 - 3n + 2$.

a. Make a table, as in Exercise 7, for $n = 1, \ldots, 8$. Include difference columns until the numbers in the last difference column are the same.

b. How many difference columns did you need?

c. How does the number of difference columns compare with the degree of the closed-form polynomial?

9. Let V_n be the number of vertices in a complete binary tree. (A binary tree is complete if each vertex of the tree has either two or no children.) Level 0 is the root of the tree. The first three trees follow.

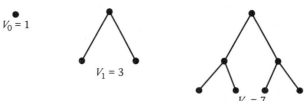

a. Make a table for $V_0, \ldots, V_6$.

b. Write a recurrence relation to describe V_n.

In Exercise 9, it was convenient to begin the table with V_0 because the root of the tree is considered level 0. There are other cases in which the initial value of the recurrence relation is labeled with the subscript 0. It is, for example, a useful practice when working with time intervals as in Exercises 10 and 11.

10. In the 1980s, residents of the southwestern United States became concerned about an influx of African "killer" bees. In 1987, the number of African bees was estimated at 5,000. It was also estimated that the population would increase at a rate of 12% annually. Let B_n be the number of African bees in

The Buzz About Africanized Bees Proves Mostly Bad Publicity

CHRISTIAN SCIENCE
MONITOR,
October 10, 1996

Killer bees, known for defending their home by attacking trespassers in swarms, caused quite a buzz in the 1970s and '80s with the threat of an imminent United States invasion—even spawning the low-budget disaster movies "The Swarm" and "The Bees" in 1978. But Africanized bees have been living in

continues

The Buzz About Africanized Bees Proves Mostly Bad Publicity
(*continued*)

the Southwest since 1990 without major incident.

Eric Mussen, extension apiculturist at the University of California at Davis, says people are aware to steer clear of the bees and so avoid confrontations. Just recently, a "killer" bee incident was reported in California when tree trimmers were stung by bees that had nested in the tree they were pruning. But despite their bad press, there have been no human or livestock deaths attributed to Africanized bees in the U.S. A 1994 article in *U.S. News and World Report* says people are more likely to be struck by lightning than attacked by Africanized bees.

The Africanized bee arrived in the Western Hemisphere

in 1956, when geneticist Warwick Kerr took 75 African queen bees to an apiary near Rio Carlo, Brazil, with the intent to cross breed them with European bees. Africanized bees were known for producing a lot of honey and thriving during rough conditions.

After 26 colonies were accidentally let loose in 1957, the bees spread throughout South and North America, traveling 100 to 200 miles per year.

Nussen says the behavior of pure Africanized bees hasn't changed a lot, but they are inbreeding with the European bee—developing a larger more aggressive hybrid population that is negatively affecting bee colonies in Texas.

Texas each year, where $n = 0$ corresponds to the beginning of the year 1987. Then B_1 would indicate "the end of year 1."

a. Make a table with entries for B_0, . . . , B_4.

b. Write a recurrence relation for B_n.

c. Use a spreadsheet or calculator to determine when the population of African bees is predicted to exceed 100,000. In what year is this predicted to occur?

11. Susie put $500 in a bank account that pays 5% interest per year, compounded yearly. Let n be the number of years she leaves the money in the bank, let A_0 be the initial amount of money ($500), and let A_n be the amount of money in the bank at the end of n years.

a. By creating a table, find out how many years it will take for Susie's money to double.

b. Write a recurrence relation for this situation.

12. The term *fractal* was invented in the 1970s by Benoit Mandelbrot to describe a class of geometric objects with unusual properties. One of those properties is that a fractal contains parts that resemble the whole. The following sequence of figures demonstrates the construction of such a "self-similar" object.

A short algorithm describes the process used to create the figures:
1. Draw a line segment.
2. For each line segment in the current figure, erase the middle third and draw a "peak" by constructing an equilateral triangle having the erased portion as a base.
3. Repeat step 2.
    a. How does the length of a segment in one figure of the sequence compare with the length of a segment in the preceding figure?
    b. How does the number of segments in one figure of the sequence compare with the number of segments in the preceding figure?
    c. Write a recurrence relation that describes the relationship between the total length of the segments in figures n and $n - 1$.

Computer/Calculator Explorations

13. Some computer drawing utilities such as Geometer's Sketchpad support recursion. Use a utility with recursion features to construct figures like those in Exercise 12.

14. The Logo computer language supports recursion. Use Logo to construct figures like those in Exercise 12.

15. Perform this spreadsheet experiment: Type the number 1 in cell A1. In cell A2 type the formula A1, and in cell B2 type the formula A1 + B1. Copy the formula in cell A2 into cell A3; copy the formula in cell B2 into cells B3, C3, D3, ... (Stop at a convenient cell in row 3.) Copy row 3 into several of the rows that follow it. Identify the results and explain why they are produced by this procedure.

Projects

16. Research and report on several current applications of recursion not discussed in this chapter. (You may want to begin your research now and continue it as this chapter progresses.)

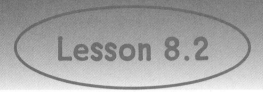
Finite Differences

Previous lessons discussed two approaches to the problem of finding a closed-form solution to the handshake problem:

1. Trial and error followed by an induction proof of the hypothesized formula

2. Counting techniques

This lesson considers a technique known as the method of finite differences, which can be used to find a closed-form solution to the handshake problem and a variety of related problems.

   Recall that the handshake problem is described recursively by $H_1 - 0$, $H_n = H_{n-1} + (n - 1)$. The following is a table generated by this recurrence relation. The third column contains the differences between successive values in the second column; the fourth column contains the differences between successive values in the third column.

Number of People	Number of Handshakes	Differences First	Second
1	0	—	—
2	1	1	—
3	3	2	1
4	6	3	1
5	10	4	1
6	15	5	1
7	21	6	1
8	28	7	1

The constant second differences indicate that the closed-form solution for this recurrence relation is a second-degree polynomial: $an^2 + bn + c$.

Consider what happens when the general second-degree polynomial is evaluated for consecutive integral values of n, and first and second differences are found. The following table shows the results.

	Value of	Differences	
Value of n	Polynomial	First	Second
1	$a + b + c$	—	—
2	$4a + 2b + c$	$3a + b$	—
3	$9a + 3b + c$	$5a + b$	$2a$
4	$16a + 4b + c$	$7a + b$	$2a$
5	$25a + 5b + c$	$9a + b$	$2a$

Notice that not only are the second differences constant, the value of the difference is twice the value of the coefficient of n^2. In the case of the handshake problem, this result means that the constant difference of 1 indicates that one term of the closed-form solution is $\frac{1}{2}n^2$.

The remaining terms of the closed-form solution can be found by substituting values from the table into the polynomial $H_n = \frac{1}{2}n^2 + bn + c$.

Although the method just described works well when the closed-form solution is second degree, it is much more tedious for degrees higher than 2. The following alternative method uses technology and is therefore much easier to extend to higher degrees.

Reconsider the handshake problem, a situation in which you know the solution is second degree: $H_n = an^2 + bn + c$. Since there are three values you need to know (a, b, and c), select any three pairs of values from your table. Although any three will do, the first three are the most convenient because of their relatively small values. Form three equations by substituting these three pairs into the general second-degree polynomial $H_n = an^2 + bn + c$.

When $n = 1$, $0 = a + b + c$.

When $n = 2$, $1 = 4a + 2b + c$.

When $n = 3$, $3 = 9a + 3b + c$.

Solve this system using the matrix techniques developed in Chapter 7:

$$\begin{bmatrix} 1 & 1 & 1 \\ 4 & 2 & 1 \\ 9 & 3 & 1 \end{bmatrix}^{-1} \times \begin{bmatrix} 0 \\ 1 \\ 3 \end{bmatrix} = \begin{bmatrix} 0.5 \\ -0.5 \\ 0 \end{bmatrix}$$

The finite differences method can be used whenever the differences in consecutive values of the recurrence relation become constant in a finite number of columns. The degree of the closed-form solution is the same as the number of columns needed to achieve the constant differences. The number of equations in the system needed to find the closed-form solution is 1 more than its degree.

A Finite Differences Example

Consider a stack of cannonballs at Fort Recurrence (see Figure 8.1).

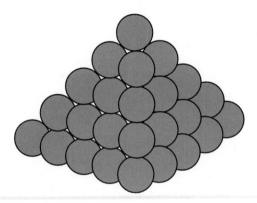

Figure 8.1 Cannonballs at Fort Recurrence.

The following table displays the number of cannonballs in a pyramid of *n* layers.

Number of Layers	Number of Canonballs	Differences		
		First	Second	Third
1	1	—	—	—
2	5	4	—	—
3	14	9	5	—
4	30	16	7	2
5	55	25	9	2
6	91	36	11	2

The recurrence relation that describes the number of cannonballs in a stack of n layers is $C_n = C_{n-1} + n^2$. The constant differences in the third column indicate that the closed-form solution is third degree: $C_n = an^3 + bn^2 + cn + d$. The system created by this general third-degree polynomial and the first four values in the table is:

When $n = 1$, $1 = a + b + c + d$.

When $n = 2$, $5 = 8a + 4b + 2c + d$.

When $n = 3$, $14 = 27a + 9b + 3c + d$.

When $n = 4$, $30 = 64a + 16b + 4c + d$.

The matrix solution is

$$\begin{bmatrix} 1 & 1 & 1 & 1 \\ 8 & 4 & 2 & 1 \\ 27 & 9 & 3 & 1 \\ 64 & 16 & 4 & 1 \end{bmatrix}^{-1} \times \begin{bmatrix} 1 \\ 5 \\ 14 \\ 30 \end{bmatrix} = \begin{bmatrix} 0.3333 \\ 0.5 \\ 0.1667 \\ 0 \end{bmatrix}$$

The closed-form solution, therefore, is $C_n = \frac{1}{3}n^3 + \frac{1}{2}n^2 + \frac{1}{6}n$. Note that unlike the case in which the solution is second degree, the coefficient of the first term is not one-half the constant difference.

Unfortunately, the finite difference method does not apply to recurrence relations that never achieve constant differences. In such cases, other methods that are described in later lessons of this chapter are often successful. The following exercises investigate several situations that can be described with recurrence relations in which the differences eventually become constant.

Including Differences in a Spreadsheet

If your spreadsheet already contains the number of people and the number of handshakes in columns A and B, then adding columns for differences requires very little effort. A difference column can be added by typing only one new formula and then copying it into as many additional cells as necessary. If, for example, the spreadsheet has the number of handshakes for a group of 1 in cell B2, for a group of 2 in cell B3, and so forth, the first difference is placed in cell C3 by typing the formula B3 − B2. This formula is then copied into other cells of column C. Since the values in column C are not constant, the same formula is copied into the cells of column D starting in cell D4. Because the values in column D are constant, the process stops.

The first spreadsheet following shows the formulas; the second shows the values that result.

	A	B	C	D
1	Number of people	Number of handshakes	First differences	Second differences
2	1	0		
3	=A2+1	=B2+A2	=B3−B2	
4	=A3+1	=B3+A3	=B4−B3	=C4−C3
5	=A4+1	=B4+A4	=B5−B4	=C5−C4
6	=A5+1	=B5+A5	=B6−B5	=C6−C5
7	=A6+1	=B6+A6	=B7−B6	=C7−C6
8	=A7+1	=B7+A7	=B8−B7	=C8−C7
9	=A8+1	=B8+A8	=B9−B8	=C9−C8
10	=A9+1	=B9+A9	=B10−B9	=C10−C9
11				

	A	B	C	D
1	Number of people	Number of handshakes	First differences	Second differences
2	1	0		
3	2	1	1	
4	3	3	2	1
5	4	6	3	1
6	5	10	4	1
7	6	15	5	1
8	7	21	6	1
9	8	28	7	1
10	9	36	8	1
11				

Difference Columns on a Graphing Calculator

Some graphing calculators have a function that calculates the differences between successive pairs of values in a list. Note that the calculator used to create the following screens places a given difference opposite the first member of the pair rather than the second.

L1	L2	▪3	3
1	0	------	
2	1		
3	3		
4	6		
5	10		
6	15		
7	21		

L3 =ₐList(L2)

L1	L2	L3	3
1	0	1	
2	1	2	
3	3	3	
4	6	4	
5	10	5	
6	15	6	
7	21	------	

L3(1)=1

Exercises

1. Use finite differences to determine the degree of the closed-form formula that was used to generate the given sequence.
   a. $-3, -2, 3, 12, 25, 42, 63, 88, 117, 150, 187, 228, 273, 322, \ldots$.
   b. $0.29, 0.52, 0.75, 0.98, 1.21, 1.44, 1.67, 1.90, 2.13, 2.36, 2.59, \ldots$.
   c. $0, -2, -2, 0, 4, 10, 18, 28, 40, 54, 70, 88, 108, 130, 154, \ldots$.
   d. $1, 3, 9, 27, 81, 243, 729, 2187, 6561, 19683, 59049, 111147, \ldots$.

2. For each part of Exercise 1, determine the closed-form formula that will generate the sequence.

3. a. Write a recurrence relation for the number of edges T_n in a complete graph with n vertices, K_n.
   b. For your recurrence relation in part a, what is the initial condition? (Hint: How many edges are in a graph with one vertex?)
   c. Use finite difference techniques to determine a closed-form formula for the number of edges in a K_n graph.

4. $a_1 = 1$ and $a_n = 3a_{n-1} - 5$.
   a. Find the first few (six to eight) terms.
   b. Find the fixed point for this recurrence relation. (Hint: When a recurrence relation has a fixed point, all the terms are the same. Replace a_n and a_{n-1} with a single variable such as x, then solve. Check your solution by using it as an initial value in the recurrence relation.)

5. A triangle has no diagonal, a quadrilateral has two diagonals, and a pentagon has five diagonals.
   a. Write a recurrence relation for the number of diagonals in an n-sided polygon.
   b. Use finite difference techniques to find a closed-form formula for the number of diagonals in an n-sided polygon.

6. An auditorium has 24 seats in the front row. Each successive row, moving toward the back of the auditorium, has 2 additional seats. The last row has 96 seats.
   a. Create a table with a column for the number of the row and a column for the number of seats in that row. Complete at least the first six entries in the table.
   b. Write a recurrence relation for the number of seats in the nth row.
   c. Find a closed-form solution for the number of seats in the nth row. (One way to do this is to use finite differences.)
   d. How many rows are in the auditorium? Explain.
   e. Add a third column, "Total seats," to your table from part a. Complete at least the first six sums in this column.
   f. Write a recurrence relation for the number of seats in the first n rows of the auditorium.
   g. Write a closed-form formula for the number of seats in the first n rows of the auditorium.

7. A house purchased in 1985 increased in value at the rate of 8% per year.
   a. If the original cost of the house was $38,000, calculate the value of the house each year from 1985 to 1998. (A spreadsheet might be nice to use here.)
   b. Write a recurrence relation for the value of the house at the end of the nth year since 1985.
   c. Calculate the finite differences for your numbers in part a. Do you eventually obtain constant differences?

8. In 1998, a herd of 50 deer is increasing at the rate of approximately 4% per year.
   a. Make a table that gives the number of deer at the end of each year ($T_0 = 50$).
   b. If the herd's habitat can provide food for a maximum of 325 deer, in what year will there not be enough food?
   c. Write a recurrence relation for the number of deer at the end of the nth year.
   d. Calculate the finite differences for your table in part a. Do you eventually obtain constant differences?

9. This lesson includes an analysis of second-degree polynomials that discovered a connection between the leading coefficient of a second-degree closed-form solution and the constant difference. Perform a similar analysis for the third-degree polynomial. How is the leading coefficient related to the constant difference?

From Endangered to Dangerous: Once Headed for Extinction, Deer Now are Overpopulated

KANSAS CITY STAR, May 18, 1996

White-tailed deer are the nation's most deadly animal. They cause 120 deaths a year in car crashes—no other animal causes more human deaths.

No animal causes more property damage.

The average damage is more than $600 to each of the 300,000 vehicles a year in deer collisions, says George Harrison in an article he wrote for the current issue of *Sports Afield*. Harrison cites figures from the Insurance Information Institute. The total cost: more than $180 million.

Sixty years ago people were genuinely concerned that deer soon would be extinct. Today, more than 25 million white-tailed deer and 5 million mule deer roam the United States.

And the populations continue to climb. It is estimated the nation has more deer now than at the time of European settlement.

Computer/Calculator Explorations

10. Graphing calculators have statistical functions that fit various kinds of mathematical functions to a set of data. Many of these calculators include several kinds of polynomials in this collection of functions. Prepare a report on the polynomial-fitting capabilities in your calculator and show how they can be used to find closed-form polynomial solutions to recurrence relations for which differences become constant.

Arithmetic and Geometric Recursion

Counting techniques and finite differences are two methods that can be used to find closed-form solutions for recurrence relations. However, there are many kinds of recurrence relations, and no method is capable of finding a closed-form solution for all of them.

Two of the most common types of recurrence relations are those in which a term is generated by either adding a constant to the previous term or multiplying the previous term by a constant. The first type is called **arithmetic**, and the constant is called the **common difference**. The second type is called **geometric**, and the constant is called the **common ratio**. This lesson considers arithmetic and geometric recurrence relations and a few of their many applications.

A surprising fact about these two types of recurrence relations is that a geometric recurrence relation with a common ratio larger than 1 and positive first term will eventually grow to a larger value than an arithmetic recurrence relation, even if the latter's first term and common difference are relatively large. For example, consider a job that employs you for 30 days and in which you have a choice of two methods of payment. Method 1 pays $5,000 the first day and includes a $10,000 raise each day after that. Method 2 pays only $.01 the first day but doubles the amount you are paid each successive day. The questions are, of course, Which salary is better? How much better?

The first method of payment is arithmetic, and the common difference is $10,000. The second method is geometric, and the common ratio is 2. If

P_n is the payment on the nth day, the arithmetic recurrence relation is $P_n = P_{n-1} + 10,000$; the geometric recurrence relation is $P_n = 2P_{n-1}$.

To determine the amount by which one salary is better than the other, you must find the total amount each pays over the 30-day period. If T_n represents the total, then a recurrence relation that describes the total is $T_n = T_{n-1} + P_n$.

Of course, all these questions can be answered by using a computer or calculator to generate a comparative table for the entire 30-day period. However, there are formulas that can find all the relevant information directly. Since these formulas have many applications, this lesson next discusses how the formulas are developed.

Formulas for Arithmetic and Geometric Terms

In method 1, the pay on the first day is $5,000, the pay on the second day is $5,000 + $10,000, and the pay on the third day is $5,000 + $10,000 + $10,000 = $5,000 + 2($10,000). Therefore, the pay on the nth day is $5,000 plus $n - 1$ raises of $10,000 each. Thus, the formula for the pay on the nth day is $5,000 + (n - 1)$10,000.

> In general, the nth term of an arithmetic recurrence relation is found by adding $n - 1$ common differences to the first term: $t_n = t_1 + (n - 1)d$.

In method 2, the pay on the first day is $.01, the pay on the second day is $.01(2), and the pay on the third day is $.01(2)(2) = $.01(2^2)$. Therefore, the pay on the nth day is $.01 doubled $n - 1$ times. Thus, the formula for the pay on the nth day is $.01(2^{n-1})$.

> In general, the nth term of a geometric recurrence relation is found by multiplying the first term by the common ratio $n - 1$ times: $t_n = t_1(r^{n-1})$.

With these formulas, the comparison of wages on the 30th day is a simple matter:

Method 1: $t_{30} = \$5,000 + (30 - 1)\$10,000 = \$295,000$.

Method 2: $t_{30} = \$.01(2^{30-1}) = \$5,368,709.12$.

The trend lines in Figure 8.2 show the wages for each method over the 30-day period.

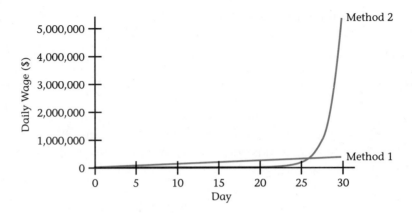

Figure 8.2 Daily wages by two methods.

Sums of Arithmetic and Geometric Terms

A complete comparison of the two methods requires the total of the 30 daily wages for each method. To examine the total pay for method 1, consider the general arithmetic recurrence relation with first term t_1 and common difference d.

Term Number	Term	Sum of First n Terms	Differences First	Second
1	t_1	t_1	—	—
2	$t_1 + d$	$t_1 + (t_1 + d) = 2t_1 + d$	$t_1 + d$	—
3	$t_1 + 2d$	$(t_1 + 2d) + (2t_1 + d) = 3t_1 + 3d$	$t_1 + 2d$	d
4	$t_1 + 3d$	$(t_1 + 3d) + (3t_1 + 3d) = (4t_1 + 6d)$	$t_1 + 3d$	d

The constant second differences indicate that the closed-form solution is a second-degree polynomial; $t_n = an^2 + bn + c$. The related system of equations created from the first three pairs in the table is:

When $n = 1$, $t_1 = a + b + c$.

When $n = 2$, $2t_1 + d = 4a + 2b + c$.

When $n = 3$, $3t_1 + 3d = 9a + 3b + c$.

Matrices can be used to solve the system:

$$\begin{bmatrix} 1 & 1 & 1 \\ 4 & 2 & 1 \\ 9 & 3 & 1 \end{bmatrix}^{-1} \begin{bmatrix} t_1 \\ 2t_1 + d \\ 3t_1 + 3d \end{bmatrix} =$$

$$\begin{bmatrix} 1 & 1 & 1 \\ 4 & 2 & 1 \\ 9 & 3 & 1 \end{bmatrix}^{-1} \begin{bmatrix} 1 & 0 \\ 2 & 1 \\ 3 & 3 \end{bmatrix} \begin{bmatrix} t_1 \\ d \end{bmatrix} =$$

$$\begin{bmatrix} 0 & 0.5 \\ 1 & -0.5 \\ 0 & 0 \end{bmatrix} \begin{bmatrix} t_1 \\ d \end{bmatrix} = \begin{bmatrix} 0.5d \\ t_1 - 0.5d \\ 0 \end{bmatrix}.$$

In general, the sum of the first n terms of an arithmetic recurrence relation is $0.5dn^2 + (t_1 - 0.5d)n$.

The sums in method 2 do not generate constant differences in a finite number of steps, and, therefore, a closed-from solution cannot be found by the finite difference method. It can, however, be found by algebraic means.

Consider the general geometric recurrence relation with first term t_1 and common ratio r. The sum of the first n terms, S_n, is

$$S_n = t_1 + t_1r + t_1r^2 + \cdots + t_1r^{n-1}.$$

Multiply this equation by r:

$$rS_n = r(t_1 + t_1r + t_1r^2 + \cdots + t_1r^{n-1}).$$

Distribute r on the right side of the equation, and subtract this equation from the original equation:

$$S_n = t_1 + t_1r + t_1r^2 + \cdots + t_1r^{n-1}$$

$$rS_n = \qquad t_1r + t_1r^2 + t_1r^3 + \cdots + t_1r^{n-1}$$

$$S_n - rS_n = t_1 - t_1r^n.$$

Now factor both sides and divide by $(1 - r)$:

$$S_n(1 - r) = t_1(1 - r^n),$$

or

$$S_n = \frac{t_1(1 - r^n)}{(1 - r)}.$$

In general, the sum S_n of the first n terms of a geometric recurrence relation is

$$\frac{t_1(1 - r^n)}{(1 - r)}.$$

Multilingual Software Starts to Open Doors in Cyberspace

London,
(CBS MARKETWATCH),
May 13, 1998

The multilingual jungle of the Internet has created demand for translation software that allows users to quickly read web pages in their own tongues, and that has the companies behind the software among the most popular exhibits at Internet World 98 this week in London.

Orange, California-based newcomer LanguageForce launched its new Deluxe Universal Translator at the London conference. LanguageForce uses word-for-word technology developed in Russia and says its software can translate documents, e-mail, and WWW pages in 33 different languages, includ-

ing everything from Arabic and Chinese to Japanese and Swahili.

In 1997, only its second year, the company had revenues of more than $6 million, said president Ian Simpson. "But we're reaching for $11 million in 1998." And he says that $16 million–$20 million is likely in 1999. The overall translation market, including the old-fashioned kind that uses humans with dictionaries, will be worth about $11.2 million in 1998; and that is projected to grow to around $17.2 million in 2003, Simpson says, meaning that his company should see "geometric growth," as translation software grabs an increasing share of the pie.

With the arithmetic and geometric sum formulas, comparison of the total wages is a simple matter:

Method 1: $S_{30} = 0.5 \times 10{,}000 \times 30^2 + (5{,}000 - 0.5 \times 10{,}000) \times 30 = \$4{,}500{,}000.$

Method 2: $S_{30} = 0.01(1 - 2^{30})/(1 - 2) = \$10{,}737{,}418.23.$

Including Sums in Spreadsheets

If your spreadsheet has the term numbers and the terms of a recurrence relation in columns A and B, respectively, adding a column for the sum requires entering a single formula, then copying it into as many cells as you need. For example, if the first term of the recurrence relation is in cell B1, the second term is in B2, and so forth, type B1 in cell C1 and the formula C1 + B2 in cell C2. Copy the formula into cells C3, C4, and so on.

Including Sums in Programs

Lesson 8.1 gave an algorithm for generating terms of the recurrence relation $H_n = H_{n-1} + (n - 1)$ and implementations for the BASIC language and a calculator language. To include sums in the algorithm, introduce a variable (S) for the sum, give it an initial value equal to the first term, and have it accumulate values of the terms as the terms are generated. Following are Lesson 8.1's implementations with instructions to accumulate sums.

10 N = 1:H = 0:S = H	$1 \rightarrow N:0 \rightarrow H:H \rightarrow S$
20 PRINT N, H, S	Lbl A
30 N = N + 1	Disp N, H, S
40 H = H + N − 1	$N + 1 \rightarrow N:H + N − 1 \rightarrow H:S + H \rightarrow S$
50 S = S + H	Goto A
60 GO TO 20	

Keep in mind that neither algorithm terminates unless additional instructions are added.

Sums on a Graphing Calculator

Some graphing calculators have functions that automatically calculate sums. The screen on the left shows the closed-form solution $t_n = 0.01 \times 2^{n-1}$ as it is being used to calculate the 30 daily wages by method 2. The results are stored in one of the calculator's lists. The second screen shows the sum function as it is being used to find the sum of the 30 daily wages.

```
seq(.01*2^(X-1),
X,1,30)→L1
{.01 .02 .04 .0…
```

```
seq(.01*2^(X-1),
X,1,30)→L1
{.01 .02 .04 .0…
sum(L1)
         10737418.23
```

In this lesson's exercises you have the opportunity to apply your knowledge of arithmetic and geometric terms and sums in a variety of situations.

Exercises

1. Consider the following sequences:
   i. 2, 5, 8, 11, 14, ...
   ii. 64, 32, 16, 8, 4, 2, 1, ...

iii. 10, 12, 14.4, 17.28, 20.736, ...

iv. 2, 3, 5, 8, 13, 21, ...

v. $\dfrac{3}{10}, \dfrac{3}{100}, \dfrac{3}{1,000}, \dfrac{3}{10,000}, \cdots$

vi. 3, 4, 6, 9, 13, 18, ...

    a. Which of the sequences are geometric, which are arithmetic, and which are neither?

    b. Write a recurrence relation for each sequence.

    c. For those sequences that are arithmetic or geometric, write a closed-form formula for the nth term.

2. Consider the general recurrence relation for a geometric sequence, $t_n = rt_{n-1}$. Find the fixed point for this recurrence relation if one exists. (Recall the hint in Exercise 4b, page 424, of Lesson 8.2.)

3. Consider the general recurrence relation for an arithmetic sequence, $t_n = t_{n-1} + d$. Find the fixed point for this recurrence relation if one exists.

4. A teen chat line is a 900 telephone number line that charges $2 for the first minute and $.95 for each additional minute.

    a. What recurrence relation does the statement of the problem suggest? (Make a table if necessary.)

    b. Find a closed-form formula to describe the cost of the call.

    c. Assume that 5,000 teens use the line each week and talk for an average of 15 minutes. How much income is produced? Suppose also that the bulk of the cost to the company operating the line is the cost of the long distance line which averages $3 for a 15-minute call. How much profit does the company make each week?

5. George deposits $5,000 in the bank at an interest rate of 4.8% compounded yearly.

    a. Write a recurrence relation for the amount of money in the account at the end of n years.

    b. Write a closed-form formula for the amount of money in the account at the end of n years.

    c. How much money will be in George's account at the end of 3 years?

    d. Suppose the bank decides to become competitive with other banking institutions in the city by offering an interest rate of 4.8% that compounds monthly. What monthly interest rate will George receive?

    e. Write a recurrence relation for the amount of money in the account at the end of n months.

    f. Write a closed-form formula for the amount of money in the account at the end of n months.

    g. Find the amount in George's account at the end of 3 years (36 months).

    h. Compare the amount of money in George's account at the end of 3 years when the interest is compounded yearly with the amount at the end of 3 years when the interest is compounded monthly.

6. In Exercise 5, suppose George is given the choice of investing his $5,000 at 4.8% compounded monthly or at 5.0% compounded yearly. Compare the two methods of investment at the end of 1 year. At the end of 2 years. At the end of 3 years.

7. Find the fifteenth term and the sum of the first 15 terms for the sequences in parts a to c.

    a. 4, 9, 14, 19, ...

    b. 45.75, 47, 48.25, 49.5, ...

    c. 3650, 3623, 3596, 3569, 3542, ...

    d. From your work in other mathematics courses, you may be familiar with the use of the Greek letter Σ (sigma) to indicate sums. For example, if you are summing the first five terms of the sequence whose terms are generated by $2n - 3$, the sum can be indicated in this way: $\sum_{1}^{5}(2n - 3)$. The term numbers of the first and last terms that are included in the sum are written at the bottom and at the top of sigma, respectively, and the formula for the nth term is written in parentheses to the right. Use this type summation notation to represent each of the sums you found in parts a to c.

8. The number of deer on Fawn Island is currently estimated to be 500 and is increasing at a rate of 8% per year. At the present time, the island can support 4,000 deer, but acid rain is destroying the vegetation on the island and the number of deer that can be fed is decreasing by 100 per year. In how many years will there not be

Because of the demise of natural predators, deer populations sometimes threaten to multiply out of control. Discrete mathematical models play an important role in managing deer populations.

enough food for the deer on Fawn Island? Explain how you obtained your answer.

9. The cost of n shirts selling for $14.95 each is given by $C_n = 14.95n$.
   a. What is the equivalent recurrence relation?
   b. Is the recurrence relation arithmetic, geometric, or neither? If it is arithmetic or geometric, what are the first term and the common difference or ratio?

10. At 5.5% compounded yearly, what amount must Bill's parents invest for Bill at age 10 if they want Bill to be a millionaire when he reaches age 50?

11. To double your investment at the end of 11 years, what annual interest rate do you need to receive?

12. The following table gives data relating temperature in degrees Fahrenheit to the number of times a cricket chirps in one minute. Although the data are not quite a perfect arithmetic sequence, they are close to one.

Temperature (°F)	50	52	55	58	60	64	68
Chirps per minute	40	48	60	73	80	98	114

   a. Assuming that the data form an arithmetic sequence, with the temperature as the term number and the chirps per minute as the term value, what is the common difference per degree rise in temperature?
   b. What is the temperature when a cricket is chirping 110 times per minute?
   c. At what temperature does a cricket stop chirping?

B.C. by permission of Johnny Hart and Field Enterprises, Inc.

13. A ball is dropped from a height of 8 feet and rebounds on each bounce to 75% of its height on the previous bounce.
   a. How high does the ball reach after the sixth bounce?
   b. What is the total distance that the ball has traveled just before the seventh bounce?

14. The tenth term of an arithmetic sequence is 4, and the twenty-fifth term is 20. Find the first term and the common difference for this arithmetic sequence.

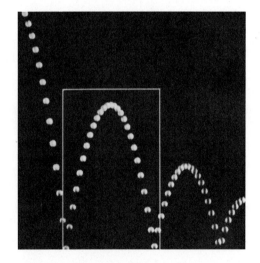

A rubber ball rebounds to a fraction of the height of the previous bounce, as shown in this time-lapse photograph.

CHRISTIAN SCIENCE
MONITOR,
September 29, 1997

Parents, Legislators Ask Why College Tuition Keeps Rising

Question: What do these items have in common? Hair dryers, restaurant-quality dorm food, faculty salaries, designer student services, and penny-pinching legislatures.

Answer: They're all reasons colleges and universities cite for again hiking tuition and fees above the inflation rate.

American undergraduates will pay 5 percent more this year than last for tuition and fees at a four-year college, according to the College Board's Annual Survey of Colleges, released last week. Inflation is sputtering along at 2.2 percent.

Adjusting for inflation, college tuition has jumped a whopping 90 percent over the past 15 years—higher than health care. Meanwhile, family income increased by only 9 percent over the same period.

The average four-year private college costs $19,213; the average four-year public university, $7,472.

15. The average cost of a year of college education in a public university is about $7,500. The cost is increasing an average of 5% per year.
    a. If the current rate continues, what will be the cost of a year of college education in 30 years?
    b. Ten years from now you start saving for the college education of a child. The best interest rate you can get is 6% compounded annually. What amount would you have to put in the bank in order to pay for a year of college 20 years later?

16. Over time, the number of bacteria in a culture grows geometrically. There are 600 bacteria at $t = 0$ and 950 bacteria at $t = 3$ hours.
    a. What is the approximate common ratio for this sequence?
    b. If the growth rate continues, approximately how many bacteria will there be in 10 hours?
    c. At what time will there first be 50,000 bacteria in the culture?

17. A 50-meter-long pool is constructed with the shallow end 0.85 meter deep. For each meter of length, starting at the shallow end, the pool deepens by 0.06 meter.
    a. How deep is the deepest part of the pool? Explain.
    b. A rope is to be placed across the pool where the pool depth is 1.6 meters in order to mark the end of the shallow section. How far from the shallow end of the pool should the rope be placed?

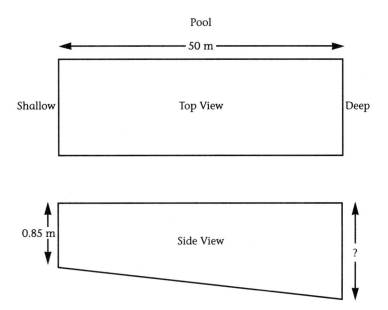

Pool

18. A company has sales of $6 million one year and $8 million the next year. The company expects to experience "geometric growth" at a similar rate in succeeding years. (See the news article on page 430.)
    a. Write a recurrence relation that describes the company's growth and find a solution to the recurrence relation.
    b. Predict the company's sales 5 years from now and 10 years from now.
    c. Do you think the predictions are realistic? Explain.

Computer/Calculator Explorations

19. The fractal curve construction described in Exercise 12 of Lesson 8.1 on pages 416 and 417 can be applied to the sides of an equilateral triangle to produce a fractal that is often called a snowflake curve. At right is one stage of such a construction. Adapt the procedure you developed in either Exercise 13 or Exercise 14 in Lesson 8.1 to produce several snowflake curves.

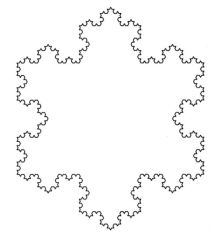

Projects

20. Some geometric recurrence relations can be used to generate infinite sequences with finite properties. For example, the recurrence relation $t_n = \frac{1}{2} t_{n-1}$ with $t_1 = 1$ generates the sequence $1, \frac{1}{2}, \frac{1}{4}, \frac{1}{8}, \ldots$. The terms of this sequence approach 0, and the sequence of sums: $1, 1 + \frac{1}{2}, 1 + \frac{1}{2} + \frac{1}{4}, \ldots$ approaches 2. Research and report on conditions for which the sequence of terms and the sequence of sums of a geometric recurrence relation approach a finite value. Apply the results of your investigation to the perimeter and area of a snowflake curve based on an equilateral triangle with sides of length 1 (see Exercise 19).

21. In this lesson, the closed-form formulas for terms and sums of arithmetic and geometric recurrence relations are proved by algebraic means. Prepare a report showing how these formulas can be proved by mathematical induction.

Mixed Recursion, Part 1

The previous lesson considered arithmetic and geometric recurrence relations. This lesson examines recurrence relations that have both a geometric and arithmetic component.

To begin, consider a game called Towers of Hanoi. The game involves three pegs and disks of varying sizes stacked from largest to smallest on one of the pegs (see Figure 8.3).

The object of the game is to move the disks from the first peg to the third peg in as few moves as possible. Disks may be placed temporarily on the middle peg or moved back to the first peg, but only one disk may be moved at a time and a disk may never be placed on top of one that is smaller than it.

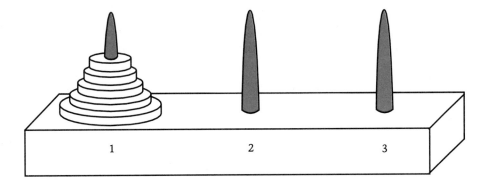

Figure 8.3 Towers of Hanoi.

Artwork appearing on the box for the original Towers of Hanoi puzzle. The puzzle was first marketed in 1883 by the French mathematician Edouard Lucas under the pseudonym "Professor N. Claus (of Siam)." The text translates as "The Tower of Hanoi, Authentic Brain Teaser of the Annamites, A Game Brought Back from Tonkin by Professor N. Claus (of Siam), Mandarin of the College of Li-Sou-Stian!"

Explore This

Cut several round or square pieces of different sizes from paper or poster board. Use the pieces as disks. Number three spots on a large piece of paper and stack several of the pieces you cut from largest to smallest on spot number one. You now have a rudimentary version of the Towers of Hanoi puzzle.

At the direction of your instructor, team up with one or more members of your class and play the game. Keep a record of the fewest moves in which games with different numbers of disks can be completed in a table like the following.

Number of Disks	Fewest Moves to Complete Puzzle
1	1
2	
3	
4	
5	
⋮	

As you progress, look for patterns in the data as well as in the way you are completing the puzzle. If M_n represents the fewest moves in which a game of n disks can be completed, look for a recurrence relation that describes the relationship between M_n and M_{n-1}. If time permits, conjecture a closed-form formula for M_n.

Your investigation of the Towers of Hanoi puzzle should have concluded that neither an arithmetic nor a geometric recurrence relation described the number of moves: you cannot generate a term of the sequence by adding a constant or by multiplying by a constant; you need to do both.

A **mixed recurrence relation** is one in which both multiplication by a constant and addition of a constant are required. The general form of a mixed recurrence relation is $t_n = at_{n-1} + b$. Many common situations can be modeled with mixed recurrence relations. Examples include financial applications such as annuities and loan repayment, the way an object cools or heats, and the way in which diseases spread in a population.

To use mixed recurrence relations, you should be able to recognize that a given set of data can be modeled with a mixed recurrence relation. Consider a table of data generated by the relation $t_n = 2t_{n-1} + 3$ with $t_1 = 4$.

n	t_n	Differences
1	4	—
2	$2(4) + 3 = 11$	$11 - 4 = 7$
3	$2(11) + 3 = 25$	$25 - 11 = 14$
4	$2(25) + 3 = 53$	$53 - 25 = 28$
5	$2(53) + 3 = 109$	$109 - 53 = 56$

Note that the values in the difference column grow by a factor of $a = 2$. This characteristic of data generated by a mixed recurrence relation makes it possible to identify situations in which a mixed recurrence relation is an appropriate model and to find the values of a and b.

> **When data are generated by a mixed recurrence relation, the ratio of any difference to the one preceding it is the same as the value of a in $t_n = at_{n-1} + b$.**

For example, consider the following table showing the increase in the value of a house over several years after its purchase.

Year	Value	Differences
0	80,000	—
1	82,000	2,000
2	84,200	2,200
3	86,620	2,420

The ratio of the second difference to the first is $\dfrac{2,200}{2,000} = 1.1$. The ratio of the third difference to the second is $\dfrac{2,420}{2,200} = 1.1$. A mixed recurrence relation is therefore an appropriate model, and the value of a is 1.1.

The recurrence relation that models the value of the house after n years is $V_n = 1.1V_{n-1} + b$. The value of b can be found by substituting any two successive values from the second column of the table into this equation. For example, if the first two values are used, the result is $82,000 = 1.1(80,000) + b$. Solving for b gives $b = 82,000 - 1.1(80,000)$, or $b = -6,000$.

The completed model for the value of the house after n years is $V_n = 1.1V_{n-1} - 6{,}000$.

A Mixed Recursion Example

Many people save money by making regular deposits—from each paycheck, for example. Often such accounts are part of a retirement plan that shelters current income from taxes. An account to which you make regular contributions is called an **annuity**. Annuities can be modeled with mixed recurrence relations.

Consider an account that pays 6% annual interest compounded monthly, to which monthly additions of $200 are made. Since the interest is compounded monthly, the model is based on a monthly interest rate of $\frac{0.06}{12} = 0.005$. The following table shows the growth of the account during the first few months.

Month	Balance ($)
0	200
1	$1.005(200) \quad + 200 = 401$
2	$1.005(401) \quad + 200 = 603.01$
3	$1.005(603.01) + 200 = 806.03$

If B_n represents the balance at the end of the nth month, then $B_n = 1.005B_{n-1} + 200$. This mixed recurrence relation can be used to create a table that tracks the growth of the account over many months or years.

Annuities on Spreadsheets

Spreadsheet techniques discussed in previous lessons can be used to create annuity models. However, it is useful to create the spreadsheet in a way that allows monthly payments and interest rates to be changed easily so you can explore various options.

Create a spreadsheet in which a monthly payment of $200 and an annual interest rate of 0.06 are stored in cells A1 and B1. Type column headings of "Month" and "Balance" in the next row. In cell A3 type 0 and in cell B3 type the simple formula A1. In cell A4 type the formula A3 + 1. In cell B4 type the formula (1 + $B$1/12)*B3 + $A$1. This formula is the key to the spreadsheet model. The dollar sign causes the referenced columns and rows to be fixed, which means that the spreadsheet will not change them when the formula is copied into other cells.

	A	B
1	200	0.06
2	Month	Balance
3	0	= A1
4	= A3 + 1	= (1 + $B$1/12)*B3 + $A$1
5		
6		

Complete the spreadsheet by copying the formulas in row 4 into as many rows as you like.

Annuities on a Graphing Calculator

The screens above show how the annuity example in this lesson is modeled on one type of graphing calculator and the table that results. The table can be scrolled to any desired value. However, annuities are often active for many years, and scrolling to, say, the tenth year (120th month) can be tedious. This calculator allows calculation of any term on the home screen (below).

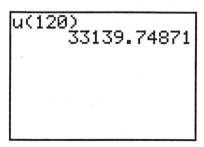

Exercises

1. Use the annuity example of this lesson with a monthly addition of $200 and an annual rate of 6% compounded monthly.
    a. Extend the table to 20 years (240 months). Record the balance at the end of 20 years.
    b. Determine the portion of the balance after 20 years that was paid into the account by its owner.
    c. Determine the portion of the balance after 20 years that was paid in interest.
    d. Compare the amount in the account after 30 years with the amount after 20 years.
    e. Change the interest rate to 8% compounded monthly and determine the amount in the account after 20 years. How does it compare with the balance after 20 years when the interest is 6%?
    f. Change the monthly addition to $300 and determine the amount in the account paying 6% after 20 years. How does the balance compare with the balance after 20 years in the 6% account with $200 monthly additions?

2. Following is a table similar to the one you made for the Towers of Hanoi puzzle.

Number of Disks	Number of Moves
1	1
2	3
3	7
4	15
5	31

The recurrence relation that describes these data is $M_n = 2M_{n-1} + 1$. You may have noticed that the number of moves always seems to be 1 smaller than a power of 2 and conjectured the closed-form formula $M_n = 2^n - 1$. In this exercise, assume that this is the correct formula.
    a. The original version of the puzzle was accompanied by a legend that at the beginning of time God created a temple in which a group of priests are working tirelessly to complete a 64-disk version of the puzzle. Find the number of moves necessary to complete the 64-disk version.
    b. The legend says that the priests move one disk a second and work nonstop in shifts; when the puzzle is completed, God will end the world. Find the number of years from the beginning of creation until the end of the world according to this legend.

3. The closed-formula formula for the Towers of Hanoi problem, $M_n = 2^n - 1$, can be proved by mathematical induction.
   a. What initial step is necessary?
   b. The proof assumes that the formula works for a puzzle with k disks and then attempts to prove the formula must also work for a puzzle with $k + 1$ disks. Rewrite the assumption and goal of the induction proof in terms of the formula.
   c. Show how the proof is completed by applying the recurrence relation.

4. Marty had $2,000 in her savings account when she graduated from college and began work. She converted the savings account to an annuity to which she will deposit $100 each month. She expects the annuity to earn 0.5% monthly.
   a. Complete this table.

n (in Months)	t_n
0	2,000
1	$1.005(2,000) + 100 = 2,110$
2	
3	

   b. Find the recurrence relation for the amount of money at the end of n months.
   c. Use your recurrence relation to find the amount in Marty's account at the end of the fourth month. At the end of the fifth month.

5. The Hamadas have borrowed $12,000 to buy a boat. The yearly interest on the loan is 8.4% and the monthly payment is $286.
   a. Complete the table.

t (in Months)	t_n
0	12,000
1	$1.007(12,000) - 286 = 11,798$
2	
3	

   b. Write a recurrence relation for the loan balance at the end of n months.
   c. Use a spreadsheet, calculator, or computer program to extend the table in part a until the loan is paid off. The values that you are calculating form what is known as an *amortization schedule*.
   d. Explore what happens to the Hamadas' loan if the interest rate and monthly payment remain the same but they borrow $34,000. What will be the amount in the loan at the end of 1 month? At the end of 2 months? At the end of n months? What is the $34,000 called?

6. Newton's law of cooling says that over a fixed time period the change in temperature of an object is proportional to the difference between the temperature of the object and its surrounding environment. If t is the temperature of the object, s is the temperature of the surroundings, and a is the constant of proportion, then $t_n - t_{n-1} = a(t_{n-1} - s)$.

   A cup of cocoa was brewed to a temperature of 170°F. When set in a room whose temperature is 70°F, the temperature of the cocoa dropped to 162°F in 1 minute.
   a. Write the recurrence relation for the temperature of the cocoa after n minutes.
   b. Simplify the recurrence relation you wrote in part a to the form $t_n = at_{n-1} + b$.
   c. Use the recurrence relation to find the temperature of the hot chocolate after 2 minutes.
   d. Find the fixed point for this recurrence relation. What is the significance of the fixed point in this situation?

7. To help him finish his final year of college, Sam took out a loan of $5,000. At the end of the first year after he graduated, there was a $4,500 balance, and at the end of the second year, $3,950 remained. The amount of money left at the end of n years can be modeled by the mixed recurrence relation $t_n = at_{n-1} + b$.
   a. The information stated above is summarized in the following table:

n	t_n
0	5,000
1	4,500
2	3,900

   Use the general form of a mixed recurrence relation and the data in the table to write a system of equations. Solve for a and b. What is the recurrence relation for the amount of money in Sam's account after n years?
   b. What will be the balance owed on the loan at the end of the third year? At the end of the fourth year?
   c. What is the rate of interest on this loan?
   d. Find the fixed point for this recurrence relation. What is the significance of this amount of money?

8. Suppose a college's tuition over the past 3 years has risen from $8,000 to $8,700 to the present cost of $9,435. Use a mixed recurrence relation of the form $t_n = at_{n-1} + b$ to predict next year's tuition.

9. A virus is spreading through Central High. In the following table, n represents a given period of time, and t_n represents the number of people exposed to the virus at the end of the time period.

n	t_n
5	500
6	750
7	900
8	990
9	1,044

a. Write a recurrence relation that models these data.

b. Use the recurrence relation to find the number of people exposed to the virus at the end of time period 10. At the end of time period 4.

10. Models for exposure to disease often assume that the number of people exposed during a given time interval is directly proportional to the number of people not yet exposed at the beginning of the time interval. In other words, if t_n represents the number of people exposed during time period n, P represents the total population, and k is the constant of proportion, then $t_n - t_{n-1} = k(P - t_{n-1})$.

a. Find values of k and P that show this recurrence relation is equivalent to the one you found in Exercise 9.

b. What is the fixed point for your recurrence relation in Exercise 9? What is the significance of the fixed point in this situation?

Flu Epidemic Is Latest Plague to Hit Olympics

Milwaukee Journal Sentinel, February 18, 1998

In an Olympic Games plagued for a week by bad weather, the latest enemy is a vicious flu that has swept through Japan and landed squarely in the lungs of dozens of athletes, officials and journalists in Nagano.

Nearly 900,000 people have taken ill and at least four children have died in Japan this winter in one of the worst influenza outbreaks in years. At least 20 people, including 17 school-children and three elderly people, have died from complications caused by the flu.

Among the athletes affected by the flu are: German figure skater Tanja Szewczenko, who was forced to withdraw from the women's competition; Canadian pairs figure skaters Marie-Claude Savard-Gagnon and Luc Bradet, who finished 16th; Norwegian speedskater Aadne Sondral, who won the gold medal in the 1,500-meter race but had to withdraw from the 1,000 meters; and Canadian silver medalist Elvis Stojko, who blamed a "brutal flu" for his failure to win the gold in figure skating.

Some of the worst-hit people in Nagano have been the 8,000 journalists covering the Games. Most of the journalists work together in the Main Press Center, a huge convention center, and live in apartment complexes constructed for the Games. They travel back and forth on crowded shuttle buses, which seem to be perfect incubators for the flu.

11. The terms of a recurrence relation can be graphed on a rectangular coordinate system in which term numbers are placed on the x-axis and the values of the terms on the y-axis.
   a. Prepare such a graph for the mixed recurrence relation $t_n = 0.5t_{n-1} + 1$ with $t_1 = 4$. Show at least the first ten terms.
   b. What does the graph tell you about the long-term behavior of the recurrence relation?

Computer/Calculator Explorations

12. People often use annuities to save for a goal. Consider someone who wants to have $10,000 at the end of 5 years and expects to earn 6% annual interest compounded monthly. What monthly deposit should this person make? Use a spreadsheet to experiment with different monthly deposits until you find one the does the job and report to your class on the results.

13. Automobile manufacturers occasionally have promotions in which the customer is offered a choice between low-interest financing and a cash rebate. Select a promotion of that type that has recently been offered in your area (you may have to call a few dealerships for the information) and a car to which the promotion applies.
   a. Construct a spreadsheet that compares the two options. In other words, track both the low-interest loan on the car's value after the required down payment and a regular loan on the car's value after the rebate and the required down payment. Which is the better choice?
   b. Construct a spreadsheet that compares taking the low-interest loan with paying cash less the rebate for the car. Would it have been better to take the loan and put the cash in an account, such as a certificate of deposit, that pays a relatively high interest rate or to pay cash for the car to get the rebate? (You will need to check with a bank to get current interest rates.)

14. Conduct an investigation into the advantages and disadvantages of short-term versus long-term loans for a car or a house. For example, compare the payments and total cost of a 30-year home loan versus a 15-year home loan.

15. Mortgage companies sometimes offer a home-loan repayment plan with payments every four weeks instead of monthly, resulting in a shorter loan term. Contact mortgage companies in your area for details of such a plan. Compare the total cost of both plans for a typical home loan.

Mixed Recursion, Part 2

The previous lesson considered several real-world situations that can be modeled with a mixed recurrence relation. The long duration of some applications such as annuities often means that tables of several hundred rows must be created. Although calculators and computers make construction of large tables feasible, a general closed-form solution for mixed recurrence relations would further decrease the time required to answer many questions. This lesson develops the closed-form solution for mixed recurrence relations.

As an example, consider the mixed recurrence relation $t_n = 2t_{n-1} - 3$ with $t_1 = 4$. Recall that the fixed point of a recurrence relation can be found by equating t_n and t_{n-1}. In this case, solving the equation $x = 2x - 3$ gives a fixed point of 3. Now consider a table showing the first four terms of this recurrence relation. Notice, as shown in the third column, that subtracting the fixed point from each term produces a sequence of powers of 2.

n	t_n	$t_n - 3$
1	4	1
2	5	2
3	7	4
4	11	8

Therefore, the closed-form solution is $t_n = 2^{n-1} + 3$.

Grab Retirement $$ Early

CNN, May 18, 1998

Mary Wilson is 27 years old and hasn't saved anything for retirement. For that reason alone, she speaks for a large part of her generation.

Wilson, a production assistant at a New York book publisher, said she's still early in her career and doesn't have the luxury of looking that far ahead.

She's not alone in her reluctance to begin planning for her golden years. According to a 1997 study done by the Employee Benefit Research Institute, 25 percent of all people between the ages of 25 and 33 have no retirement savings.

The incentives to start saving in your 20s are nearly overwhelming. Time is certainly on your side, since the power of compounding can make a large sum out of a small investment.

If at age 21 you invest $2,000 in an individual retirement account each year for five years, assuming 9 percent annual return, you'd have $400,000 by age 65 though your investment was only $10,000.

If you waited until age 40, you could invest $2,000 for 20 years (a total of $40,000) and still only end up with $130,000.

One of the best places to start investing is with a 401(k) plan. Usually you'll have to be at your job for one year before you're eligible, but once you're eligible, your employer will often match a portion of your own contributions.

Drawing a conclusion from a single example is unwise, so consider the same recurrence relation, but this time with $t_1 = 8$. The following table shows the first four terms.

n	t_n	$t_n - 3$
1	8	5
2	13	10
3	23	20
4	43	40

Compare the third column of this table with the third column of the previous table. The powers of 2 are still there, but each has been multiplied by 5, which is the difference between the first term and the fixed point. Therefore, the closed-form solution is now $t_n = 5(2^{n-1}) + 3$. Keep in mind that 5 is the difference between the first term and the fixed point, 2 is the multiplier in the original recurrence relation, and 3 is the fixed point.

If there is uncertainty about the validity of a closed-form formula, mathematical induction can be used to prove that the formula is correct. Consider the closed-form formula $t_n = 5(2^{n-1}) + 3$.

First, be sure the formula produces the correct first term: $5(2^{1-1}) + 3 = 5(1) + 3 = 8$.

Now you must show that if $t_n = 5(2^{n-1}) + 3$ generates the nth term, then $t_{n+1} = 5(2^{(n+1)-1}) + 3 = 5(2^n) + 3$ generates the $n + 1$th term.

The term t_{n+1} is generated by multiplying the previous term by 2 and subtracting 3: $t_{n+1} = 2t_n - 3$, but $t_n = 5(2^{n-1}) + 3$, so by substitution $t_{n+1} = 2[5(2^{n-1}) + 3] - 3$. Simplify this expression:

$$2[5(2^{n-1}) + 3] - 3$$
$$= 5(2)(2^{n-1}) + (3)2 - 3$$
$$= 5(2^n) + 6 - 3$$
$$= 5(2^n) + 3.$$

Thus, mathematical induction guarantees that $t_n = 5(2^{n-1}) + 3$ will always generate the terms of $t_n = 2t_{n-1} - 3$ with $t_1 = 8$.

It is tempting to base a general conclusion on the previous example: a closed-form solution for a recurrence relation of the type $t_n = at_{n-1} + b$ was $t_n = (t_1 - p)(a^{n-1}) + p$, where p, the fixed point, is $\dfrac{b}{1 - a}$, which is obtained by solving the equation $x = ax + b$. However, the previous induction proof applies only to one specific recurrence relation and, therefore, does not establish a general result. You will do that in this lesson's exercises.

An Annuity Example

Consider the annuity example of Lesson 8.4 (see "A Mixed Recursion Example" on page 442). The account paid 6% annual interest compounded monthly and included monthly deposits of $200. The recurrence relation for the balance at the end of the nth month is

$$B_n = \left(1 + \frac{0.06}{12}\right) B_{n-1} + 200 \quad \text{or} \quad B_n = 1.005 \, B_{n-1} + 200.$$

The following table tracks the account for the first few months.

Month	Balance
0	200
1	$1.005(200) + 200 = 401$
2	$1.005(401) + 200 = 603.01$
3	$1.005(603.01) + 200 = 806.03$

Note that $t_0 = 200$, $t_1 = 401$, $a = 1.005$, and $b = 200$. Therefore, the fixed point is $\dfrac{200}{1 - 1.005} = -40{,}000$

Substituting for t_1, a, and b in the general closed-form solution for mixed recurrence relations, $t_n = (t_1 - p)(a^{n-1}) + p$, gives

$t_n = (401 + 40{,}000)(1.005^{n-1}) - 40{,}000$, or $t_n = 40{,}401(1.005^{n-1}) - 40{,}000$.

Determining the amount in the account after, say, 20 years requires evaluating the closed-form solution for $n = 20$:

$$40{,}401(1.005^{20-1}) - 40{,}000 = \$93{,}070.22.$$

A commonly asked question about an annuity involves the time required for it to reach a certain amount. For example, to determine when this annuity will reach \$200,000, solve the equation $200{,}000 = 40{,}401(1.005^{n-1}) - 40{,}000$. Doing so requires a little algebra:

$$40{,}401(1.005^{n-1}) - 40{,}000 = 200{,}000$$

$$40{,}401(1.005^{n-1}) = 200{,}000 + 40{,}000$$

$$(1.005^{n-1}) = \frac{240{,}000}{40{,}401}$$

$$(n-1)\log(1.005) = \log\left(\frac{240{,}000}{40{,}401}\right)$$

$$n - 1 = \frac{\log\left(\dfrac{240{,}000}{40{,}401}\right)}{\log(1.005)}$$

$$n = \frac{\log\left(\dfrac{240{,}000}{40{,}401}\right)}{\log(1.005)} + 1.$$

The solution can be evaluated on a calculator to obtain about 358 months, or just under 30 years.

Point of Interest

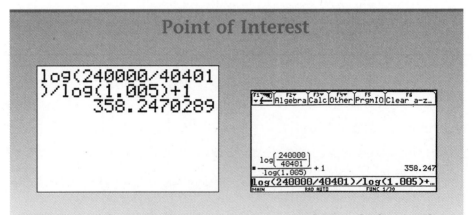

Evaluation of the example solution on two graphing calculators. The expression is typed the same way on both calculators, but the second displays it the way it is usually written.

In the following exercises, the general closed form for mixed recurrence relations is applied to situations similar to those you considered in Lesson 8.4. Use the closed form to solve the problems, and use the recurrence relation together with either a spreadsheet or calculator to check your answers.

Exercises

1. Find the fixed point and the closed form for each of these recurrence relations. Use the closed form to find the 100th term.
   a. $t_1 = 1$, $t_n = 2t_{n-1} + 3$.　　　　　b. $t_1 = 5$, $t_n = 3t_{n-1} - 7$.
   c. $t_1 = 2$, $t_n = 4t_{n-1} - 5$.

2. An annuity pays an annual rate of 8% compounded monthly and includes monthly additions of $150.
   a. Write the recurrence relation for B_n, the balance at the end of the nth month.
   b. Use your recurrence relation to build a table showing B_n for $n = 0$, 1, 2, 3, 4.
   c. Find the fixed point for the recurrence relation.
   d. Use the fixed point and the value of B_1 to write the closed form.
   e. Use the closed form to find the account balance at the end of 30 years.
   f. Use the closed form to determine the amount of time that it will take the account to grow to $500,000.
   g. Suppose the owner of the account would like to it to reach $500,000 in 30 years. What monthly additions are required? (Hint: You must find the value of b in $B_n = \left(1 + \dfrac{0.08}{12}\right) B_{n-1} + b$. Write an expression for the fixed point, leaving b as an unknown. Write the appropriate closed form, set it equal to 500,000 when $n = 360$, and solve for b.)

3. Jilian has borrowed $10,000 to buy a car. The annual interest rate is 12% compounded monthly and the monthly payments are $220.
   a. Write a recurrence relation for the unpaid balance at the end of the nth month.
   b. Use the recurrence relation to tabulate the unpaid balance at the end of months 0, 1, 2, 3, and 4.
   c. Find the fixed point for the recurrence relation.
   d. Find the closed form.
   e. Use the closed form to determine the unpaid balance at the end of 2 years.

f. Use the closed form to determine the amount of time needed to repay the loan. That is, set the closed form equal to 0 and solve for n. Round your answer to the nearest whole number of months.

g. Multiply your previous answer by the monthly payment and determine the amount Jilian really paid for her car. What is the total amount of interest that she paid?

h. Suppose Jilian wanted to pay for the car in 3 years. What would her monthly payment be? (See the hint in part g of Exercise 2.)

i. Interpret the fixed point you found in part c. That is, what is the meaning of the fixed point's dollar value in Jilian's situation?

4. In this lesson the closed-form solution for $t_n = at_{n-1} + b$ was given as $t_n = (t_1 - p)(a^{n-1}) + p$, where p is the fixed point $\dfrac{b}{1 - a}$. Mathematical induction was used to prove this formula, but only for a specific case. This exercise uses mathematical induction to prove that this closed form is correct for all mixed recurrence relations.

a. To begin, verify that the closed form works for t_1: Replace n with 1 in $(t_1 - p)(a^{n-1}) + p$ and show that this really is t_1.

b. The next step is to show that whenever the closed form generates the correct value of t_n, it will also generate the correct value of t_{n+1}. To begin this process, write the closed form for t_n and for t_{n+1}.

c. In a mixed recurrence relation, a term is generated by multiplying the previous term by a and adding b. Generate the $(n + 1)$th term by multiplying the closed form for the nth term by a and adding b.

d. Algebraically simplify the previous expression until it matches the closed form for the $(n + 1)$th term. (Hint: You might find it helpful to replace the second occurrence of the fixed point with $\dfrac{b}{1 - a}$, but not the first.)

5. In Exercise 6 from Lesson 8.4 (page 446) you applied Newton's law of cooling to a cup of cocoa that dropped from 170°F to 162°F in 1 minute in a room whose temperature was 70°F.
   a. Rewrite the recurrence relation and recalculate the fixed point for the recurrence relation.
   b. Write the closed form for the temperature of the cocoa after n minutes.
   c. Use the closed form to find the temperature of the cocoa after 5 minutes.
   d. Use the recurrence relation to create a table showing the temperature each minute through the first 5 minutes. Compare the last entry of the table with your answer to part c.
   e. Would it make sense to try to use the closed form to determine the time needed for the cocoa to freeze? Try to do so. What happens?

6. The following data represent the number of people at Central High who have heard a rumor.

Number of Hours after Rumor Began	Number of People Who Have Heard It
1	80
2	240
3	320
4	360

   a. Write a recurrence relation for the number of people who have heard the rumor after n hours.
   b. Find the fixed point for the recurrence relation.
   c. Use the fixed point to write the closed form.
   d. Use the closed form to determine the number of people who have heard the rumor after 10 hours.
   e. The way in which rumors and other information spread through a population is similar to the way in which disease spreads through a population. Compare this exercise with Exercises 9 and 10 of Lesson 8.4 (page 447) and explain the significance of the fixed point you found in part c.

7. People often express surprise at figures like those in the news article on page 450.
   a. Check the claim in the article for a 21-year-old person who saves $2,000 a year for 5 years, then stops. Find the account balance when the person is 65 and explain how you used your knowledge of recurrence relations.
   b. Check the claim for the person who starts saving at age 40. Be sure to explain how you used your knowledge of recurrence relations.
   c. If a person has saved $400,000 for retirement, what can the person withdraw each month without decreasing the account balance if the account is earning 8% annual interest compounded monthly when the person retires?

8. The fixed point of a mixed recurrence relation cannot be calculated if $a = 1$ because the denominator of $\dfrac{b}{1 - a}$ is 0. Discuss what to do when this happens.

Projects

9. Design your own savings plan. State the age at which you plan to start saving, the age at which you would like to retire, and the amount of money you would like to have in your account on retirement. Contact investment specialists in your area to determine an approximate rate of return on current annuities. Determine the amount you need to pay into the account on a regular basis to achieve your goal. Discuss the amount you will be able to withdraw from the account to meet regular expenses without depleting the account and if the account is depleted gradually over a reasonable life expectancy. (Keep a copy of the results. One day you will be glad you took this course.)

Cobweb Diagrams

In the last two lessons of this chapter you studied mixed recurrence relations primarily from a numerical viewpoint. Your understanding of these recurrence relations will benefit from a visual technique for representing their behavior.

As an example, consider the recurrence relation $t_n = 2t_{n-1} - 1$ with $t_1 = 3$. The first four terms are shown in the following table.

n	t_n
1	3
2	5
3	9
4	17

Although it may seem obvious, it is important to realize that a given term can be thought of as t_n or t_{n-1}. The first term 3, for example, is t_n if you are thinking of n as 1, but it is t_{n-1} if you are thinking of n as 2. As you build a table of values for a recurrence relation, each term first takes on the role of t_n, then the role of t_{n-1}, and then fades into history, something like an officer of an organization who serves a term as president, then a term as past president, and finally disappears from office. This succession of terms can be visualized by graphing the recurrence relation $t_n = 2t_{n-1}$ as separate functions $y = t_n$ and $y = 2t_{n-1} - 1$, or $y = x$ and $y = 2x - 1$. Associate the x axis with t_n and the y axis with t_{n-1}.

The following algorithm describes the process of graphing the generation of successive terms.

1. Graph the lines $y = x$ and $y = 2x - 1$.

2. Locate the first term of the recurrence relation on the x axis.

3. Draw a vertical line from the first term to the line $y = 2x - 1$. You can think of this step as representing the substitution of the first term into $2t_{n-1}$ in order to generate the second term.

4. Draw a horizontal line from $y = 2x - 1$ to $y = x$. You can think of this line as representing the term's transition from the role of t_n to the role of t_{n-1} in preparation for the generation of the next term.

5. Draw a vertical line upward from $y = x$ to $y = 2x - 1$.

6. Repeat steps 4 and 5 for as many terms as desired.

The graph that results is sometimes called a cobweb diagram (see Figure 8.4).

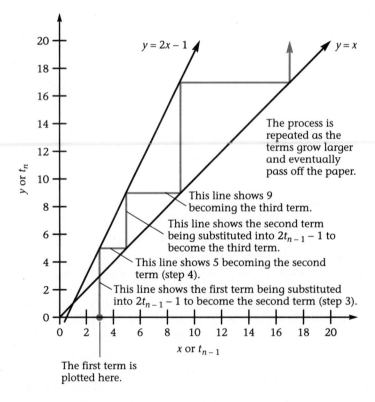

Figure 8.4 Cobweb diagram for the recurrence relation $t_n = 2t_{n-1} - 1$ with $t_1 = 3$.

Note that the intersection of the lines $y = x$ and $y = 2x - 1$ represents the fixed point. If the first term is 1, the first vertical line segment hits the point of intersection and the process can go nowhere from there.

Cobwebs on Graphing Calculators Without Recursion Features

The following is a generic cobweb diagram algorithm that can be adapted to any graphing calculator.

1. Set a suitable graphing range.

2. Graph the line $y = x$ as the calculator's first function ($Y1$) and the line whose equation is found by replacing t_n with y and t_{n-1} with x in the recurrence relation as the calculator's second function ($Y2$). (Be sure all other functions and plots are turned off.)

3. Input the value of the initial point for variable A.

4. Replace X with the value of A.

5. Draw a line from the point $(X, 0)$ to $(X, Y2)$

6. Pause and wait for the user to press ENTER.

7. Draw a line from $(X, Y2)$ to $(Y2, Y2)$.

8. Pause and wait for the user to press ENTER.

9. Replace X with the current value of $Y2$.

10. Draw a line from (X, X) to $(X, Y2)$.

11. Pause and wait for the user to press ENTER.

12. Go back to Step 7.

The first three steps can be included in the program or performed before the program is run. The specific commands to implement the algorithm vary with the calculator model. Consult your calculator manual or talk with someone knowledgeable about your calculator's programming features if you are unsure of what to do.

Cobwebs on Graphing Calculators with Recursion Features

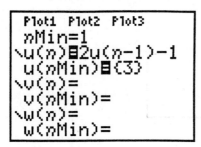 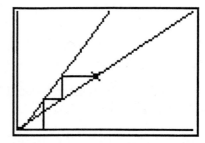

Some graphing calculators have a special graphing mode for cobweb diagrams. With the calculator in the sequence mode, the recurrence relation is entered (left screen). After the calculator is placed in a special web graphing mode, it draws the line $y = x$ and a line representing the recurrence relation (right screen). A segment of the cobweb is drawn each time the user presses a designated key.

The behavior exhibited in this lesson's cobweb diagram example is not the only kind of behavior that can occur. The following exercises explore some other kinds of behavior.

Exercises

1. Construct a cobweb diagram for the indicated number of terms of each recurrence relation. Find all the terms before beginning the graph so that you can choose a suitable scale for the axes.
   a. $t_n = 3t_{n-1} - 8$ with $t_1 = 5$, four terms.
   b. $t_n = 5 - t_{n-1}$ with $t_1 = 3$, four terms.
   c. $t_n = 9 - 0.5t_{n-1}$ with $t_1 = 2$, four terms.
   d. $t_n = 0.5t_{n-1} + 6$ with $t_1 = 2$, four terms.

2. Find the fixed point for each of the recurrence relations in Exercise 1 and mark the fixed point on your cobweb diagram. Mathematicians categorize some fixed points as *repelling*, and others as *attracting*. Which fixed points in Exercise 1 do you think are attracting. Which are repelling? Which appear to be neither?

3. Experiment with mixed recurrence relations $t_n = at_{n-1} + b$. Try various values of a and b. When is a fixed point attracting and when is it repelling? When is it neither?

4. The following figure is a cobweb diagram for a recurrence relation.

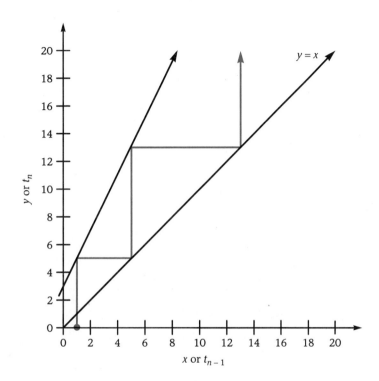

a. What is the first term of the recurrence relation?
b. How many terms can be determined from the diagram?
c. Make a table showing all the terms that can be read from the diagram.
d. Write the recurrence relation.
e. Find the fixed point of the recurrence relation from the graph. Show how algebra can be used to find the fixed point.

5. The deer population, estimated at 12,000, in a region has grown so large that the deer are becoming pests. Wildlife biologists estimate that the population is growing at a rate of 4% a year. Officials want to issue a sufficient number of hunting permits so that the population is decreased to 10,000 over the next 10 years.
a. Recommend a number of permits that will accomplish this goal.
b. If officials want to hold the population constant when it reaches 10,000, how many permits should they issue?
c. Do you think this is a reasonable plan? Explain.

6. Consider the recurrence relation $t_n = 4 - 0.5(t_{n-1})^2$ with $t_1 = 0$. This recurrence relation is called *second degree* because a term is generated by squaring the previous term.
   a. Make a table showing the first 6 terms.
   b. Draw a cobweb diagram displaying the first 6 terms.
   c. Change the first term to 1. Use a spreadsheet or calculator to explore the behavior of the first 20 terms. Also explore the related cobweb diagram.
   d. Change the first term to 2. What happens?
   e. Change the first term to 5. What happens?
   f. Use algebra to find the fixed points. (Because this recurrence relation is second degree, it has two.)
   g. You have tried this recurrence relation with first terms of 0, 1, 2, and 5. In which cases were the terms attracted to a fixed point and in which cases were they repelled? Were there any cases in which neither seemed to happen or in which it wasn't possible to tell?

7. A model that is sometimes used to represent the growth of a population in an environment that is capable of supporting only a limited number of the species says that if the uninhibited growth rate of the population is r (in decimal form) and the maximum number the environment can support is m, then the recurrence relation that describes the total number of the species t_n in a given time period is

   $$t_n = \left[1 + r\left(1 - \frac{t_{n-1}}{m}\right)\right]t_{n-1}.$$

   An important idea reflected in this model is that the uninhibited growth rate is reduced as the population approaches the maximum number the environment can support.

   The population of a particular species of animal has an annual growth rate of 10%, and the environment is capable of supporting 10 (in thousands) of the animals.
   a. Write the recurrence relation for the number of animals after n years.
   b. Use a table or a cobweb diagram to explore the growth of the population if it currently is 5,000. Describe the results.
   c. Use your table or cobweb diagram to experiment with different initial populations. Be sure to include one that is over 10. Describe the results.
   d. What are the fixed points of the recurrence relation? Explain their significance in this situation.

Computer/Calculator Explorations

8. A fractal related to the snowflake curve (see Exercise 19 in Lesson 8.3 on page 437) is called a Sierpinski gasket. Like the snowflake curve, it is based on an equilateral triangle. It is constructed by connecting the midpoints of the sides of the triangle and removing the triangle that results. The same process is applied to the remaining triangles, and so forth. The first three stages follow.

    Use a computer drawing utility or Logo to develop a recursive procedure that draws several of these figures. Investigate their properties. What, for example, can you conclude about the areas and perimeters of these figures?

Projects

9. The *order* of a recurrence relation is the difference between the highest and lowest subscripts. Nearly all the recurrence relations in this chapter are first order. The recurrence relation $t_n = t_{n-1} + t_{n-2}$ is second order because the difference between n and $n - 2$ is 2. Investigate some recurrence relations of order higher than 1. Find some applications of these recurrence relations to include in your report.

10. A recent mathematical topic related to mixed recurrence relations is called *chaos*. Prepare a report on this topic. Include the role of mixed recurrence relations and a few applications in your report.

Fractal Dimensions

Fractal art is familiar to almost everyone. Its appealing images adorn tee shirts, posters, company logos, and advertisements. But fractals often do their work more discreetly: they are used to create realistic landscapes for movies, to compress data for storage in computers, to estimate the lengths of coastlines, and to create models of biological structures.

A fractal is a figure whose dimension is not a whole number. This fact seems strange to anyone who has never thought about the possibility of a fractional dimension. To understand the concept, it's necessary to think about some familiar objects in a slightly different way.

Consider a line segment, a square, and a cube. Their dimensions are 1, 2, and 3, respectively. Subdivide their sides into, say, 3 equal parts (see Figure 8.5).

Each division forms several objects similar in shape to the original: the segment is divided into 3 segments, the square is divided into 9 squares, and the cube is divided into 27 cubes. The number of similar pieces can be counted by raising 3 to a power: $3^1 = 3$ segments, $3^2 = 9$ squares, and $3^3 = 27$ cubes. The exponent is the same as the figure's dimension.

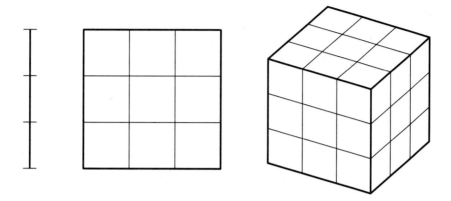

Figure 8.5 Dividing a segment, square, and cube.

Consider a similar analysis of the snowflake curve. The first three steps of its construction are shown in Figure 8.6.

Figure 8.6 The first three steps of the snowflake curve construction.

The second stage is constructed by dividing the first stage into 3 equal pieces, eliminating 1 of them, and adding 2 new ones. The second stage has 4 parts similar to the first stage; the same can be said of the third compared to the second. The fractal dimension is the power d of 3 that equals 4: $3^d = 4$. Solving for d gives the fractal dimension of the snowflake curve:

$$3^d = 4$$

$$d \log 3 = \log 4$$

$$d = \frac{\log 4}{\log 3}.$$

The snowflake curve's dimension is approximately 1.26.

This procedure works for the snowflake curve because it contains copies of itself, in other words, because it is self-similar. To find the fractal dimension of an irregular object such as a coastline, a different procedure is used.

Place a large square over a map of the coastline. Bisect the sides of the square and draw lines dividing the square into 4 smaller squares. Count the number of smaller squares that the coastline intersects. Find the quotient of the log of the number of small squares the coastline intersects and the log of the number of segments into which the sides of the original square are divided (2 in this case).

The process is repeated with a larger number of divisions of the sides of the original square. The ratio

$$\frac{\log(\text{number of squares containing coastline})}{\log(\text{number of segments})}$$

approaches the fractal dimension of the coastline as the number of divisions increases. (Coastline dimensions are usually between 1.15 and 1.25.) Figure 8.7 shows one step in the application of the method.

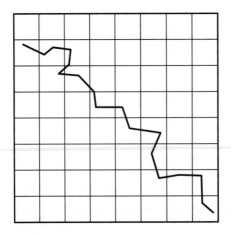

The large square is divided into $8 \times 8 = 64$ smaller squares. The shape passes through 15 of the smaller squares. Therefore, the quotient

is $\dfrac{\log 15}{\log 8} \approx 1.30$.

Figure 8.7 A step in the calculation of a fractal dimension.

1. Write a summary of what you think are the important points of this chapter.

2. For each of the following, write a recurrence relation to describe the pattern, find a closed-form solution, and find the 100th term.
   a. 2, 6, 10, 14, . . .   b. 3, 8, 23, 68, . . .   c. 3, 6, 12, 24, . . .

3. Which of the sequences in Exercise 2 are arithmetic? Which are geometric?

4. Find the fixed point for each of the following recurrence relations if one exists. Check your fixed point by using it and the recurrence relation to write the first four terms.
   a. $t_n = 5t_{n-1}$.   b. $t_n = 5t_{n-1} + 3$.   c. $t_n = t_{n-1} - 3$.

5. In Exercise 4, find a closed-form formula for each recurrence relation if the first term is 2. Use the closed form to find the 100th term.

6. Which of the recurrence relations in Exercise 4 are arithmetic? Which are geometric?

7. Consider the following sequence of squares.

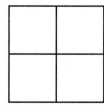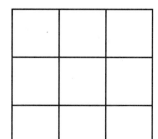

a. Let S_n represent the total number of squares of all sizes in the figure whose sides are n units long. For example, $S_2 = 5$ because there are 4 small squares and 1 large one in the second figure. Complete the following table by counting squares and drawing additional figures in the sequence.

n	S_n
1	1
2	5
3	
4	
5	

b. Add difference columns to your table.
c. Find a closed-form formula for S_n. Use your formula to find the total number of squares of all sizes on a checkerboard.

8. Use finite differences to determine the degree of the closed-form polynomial that was used to generate this sequence and show how to find the closed form by solving an appropriate system.

$$-5, -2, 3, 10, 19, 30, 43, \ldots$$

9. The following table shows the number of gifts given on the nth day of Christmas and the total number of gifts given through the first n days as described in the song, "The Twelve Days of Christmas."

Day	Gifts on That Day	Total Number of Gifts
1	1	1
2	$1 + 2 = 3$	$1 + 3 = 4$
3	$1 + 2 + 3 = 6$	$4 + 6 = 10$
4		
5		
6		

a. Complete the table through the sixth day.
b. Write a recurrence relation for the number of gifts given on the nth day, G_n, and a recurrence relation for the total number of gifts given through the nth day, T_n.
c. Find a closed-form formula for G_n and for T_n.

10. In 1999, the cost of a first-class letter was \$.33 for the first ounce and \$.22 for each additional ounce or fraction thereof.
   a. Write a recurrence relation to describe the amount of postage P_n on a letter that weighs between $n - 1$ and n ounces.
   b. Find a closed-form formula for P_n.

11. Roberto deposited \$1,000 in an account paying 4.8% annual interest compounded monthly.
   a. Complete the following table showing the balance in Roberto's account at the end of the first few months.

Month	Balance
0	\$1,000
1	
2	
3	

   b. Write a recurrence relation for the balance in Roberto's account at the end of the nth month.
   c. Find a closed-form formula for the balance at the end of the nth month.
   d. Determine the number of years it will take for the amount in Roberto's account to double.

12. Joan deposited \$5,000 in an annuity account to which she will make \$100 monthly additions. The account pays 6.4% annual interest compounded monthly.
   a. Complete the following table showing the balance in Joan's account at the end of the nth month.

Month	Balance
0	\$5,000
1	
2	
3	

   b. Write a recurrence relation for the amount in Joan's account at the end of the nth month.
   c. Find a closed-form formula for the amount in Joan's account at the end of the nth month.

d. Find the balance in Joan's account at the end of 5 years.

e. How long will it take Joan's account to grow to $50,000?

13. Martha has borrowed $11,000 to buy a car. Her loan carries an annual interest rate of 9.6% compounded monthly and her monthly payments are $230.

a. Write a recurrence relation for the unpaid loan balance at the end of the nth month.

b. Write a closed-form formula for the unpaid balance at the end of the nth month.

c. How long will it take Martha to pay for her car?

d. What is the total amount of interest that Martha will have paid?

e. If Martha wants to pay off the loan in 3 years, what monthly payments would she make?

14. The following data show the change in value of a painting over a period of years.

Year	Value
1	$8,000
2	$16,000
3	$28,000
4	$46,000

a. Write a recurrence relation to describe the value of the painting after n years.

b. Find a closed-form formula for the value of the painting after n years.

15. Create a cobweb diagram that displays the first four terms of $t_n = 0.5\, t_{n-1} + 12$ with $t_1 = 12$.

16. The absorption and elimination of medicine by the human body can be modeled with a mixed recurrence relation. Suppose a medication has a recommended dosage of 500 milligrams every 4 hours and that a person's body eliminate's about 60% of the medication in a 4-hour period. (Actual rates vary with the medication and the individual's weight and metabolism.)

a. Write a recurrence relation to model the amount of the medication in the person's body after n 4-hour periods.

b. Explore the behavior of the medication over several days. Describe your findings.

c. On the basis of your answer to part b, describe what you would expect to find in a cobweb diagram. Construct one to confirm your answer.

d. Doctors sometimes advise a patient to take a double dose the first time. On the basis of your analysis in parts a to c, why do you think that is?

Bibliography

Arganbright, Deane. 1985. *Mathematical Applications of Electronic Spreadsheets*. New York: McGraw-Hill.

Cannon, Lawrence O., and Joe Elich. 1993. "Some Pleasures and Perils of Iteration." *Mathematics Teacher*, March, pp. 233–239.

Devaney, Robert L. 1990. *Chaos, Fractals, and Dynamics: Computer Experiments in Mathematics*. Menlo Park, CA: Addison-Wesley.

Gleick, James. 1987. *Chaos: Making a New Science*. New York: Viking Penguin.

Mandelbrot, Benoit B. 1983. *The Fractal Geometry of Nature*. New York: Freeman.

Peitgen, Heinz-Otto, Hartmut Jurgens, and Dietmar Saupe. 1992. *Fractals for the Classroom, Part One: Introduction to Fractals and Chaos*. New York: Springer-Verlag.

Pickover, Clifford A, ed. 1996. *Fractal Horizons: The Future Use of Fractals*. New York: St. Martin's.

Sandefur, James T. 1990. *Discrete Dynamical Systems: Theory and Applications*. Oxford: Oxford University Press.

Seymour, Dale, and Margaret Shedd. 1973. *Finite Differences: A Pattern-Discovery Approach to Problem Solving*. Palo Alto, CA: Dale Seymour.

Stewart, Ian. 1989. *Does God Play Dice?: The Mathematics of Chaos*. Cambridge, MA: Blackwell.

Answers to Selected Exercises

Chapter 1

Lesson 1.1

11. a. 120   720

   b. The number of schedules possible when there are n choices is n times the number of schedules possible when there are $n - 1$ choices.

12. 8   12   17

Lesson 1.2

5. a. A 30.8%   69.2%
   B 19.2%   0.0%
   C 23.1%   0.0%
   D 26.9%   30.8%

7. Plurality: B   Borda: D   Runoff: C   Sequential runoff: C

11. $C_n = C_{n-1} - 1$

Lesson 1.3

3. a. A and B

4. a. A

   b. B

8. a. A

   b. C

9. a. From first to last: A, B, C, D.

   b. From first to last, the new ranking is: B, A, D.

11. a. There are two new comparisons. A total of three comparisons must be made.

  b. There are three new comparisons. A total of six comparisons must be made.

  c.
1	0	0
2	1	1
3	2	3
4	3	6
5	4	10
6	5	15

Lesson 1.4

1. Nondictatorship.

2. If the method were repeated, the same ranking might not result. Therefore, condition 5 is violated. Nondictatorship (condition 1) is also violated.

4. Condition 4.

7. None.

12. a. { } {A} {B} {C} {A, B} {A, C} {B, C} {A, B, C}

  b. { } {A} {B} {C} {D} {A, B} {A, C} {A, D} {B, C} {B, D} {C, D} {A, B, C} {A, B, D} {A, C, D} {B, C, D} {A, B, C, D}

  c. $V_n = 2V_{n-1}$

14. 4 5

15. $V1_n = V1_{n-1} + 1$ or $V1_n = \dfrac{n}{n-1} V1_{n-1}$

Lesson 1.5

1. a. The possible coalitions: { ; 0} {A; 3} {B; 2} {C; 1} {A, B; 5} {B, C; 3} {A, C; 4} {A, B, C; 6}.
    The winning coalitions: {A, B; 5} {A, C; 4} {A, B, C; 6}.

  b. A:3 B:1 C:1

  c. A:2 B:2 C:0

6. $C_n = 2C_{n-1}$

7. { } {A} {B} {C} {D} {A, B} {A, C} {A, D} {B, C} {B, D} {C, D}
{A, B, C} {A, B, D} {A, C, D} {B, C, D} {A, B, C, D}.

8. a. The winning coalitions are: {A, B; 51%} {A, C; 51%} {A, B, C; 76%}
{A, B, D; 75%} {A, C, D; 75%} {B, C, D; 74%} {A, B, C, D; 100%}.
Of these, A is essential to 5, B to 3, C to 3, and D to 1.

b. The winning coalitions are: {A, B; 88%} {A, C; 54%} {A, D; 52%}
{A, B, C; 95%} {A, B, D; 93%} {A, C, D; 59%} {B, C, D; 53%}
{A, B, C, D; 100%}.
Of these, A is essential to 6, B to 2, C to 2, and D to 2.

c. In part a, D has 24% of the stock and one-twelfth of the power. In
part b, D has 5% of the stock, but two-twelfths of the power.

Chapter 1 Review

2. a. D

b. B

c. A

d. E

e. C

f. C

3. 19, 42, 89

4. a. It can occur in either the runoff or sequential runoff method.

5. a. Wilson, no.

b. They ranked him last.

c. The voters in the last group could have switched to Roosevelt, their
second choice, and thereby prevented Wilson from winning.

d. Borda, runoff, and Condorcet give the election to Roosevelt.

6. Conditions 2, 3, and 5.

7. Arrow proved that no group ranking method that ranks three or more
choices will always adhere to his five fairness conditions.

8. Yes.

9. a. Clinton: $.43 + .2 \times .38 + .35 \times .19 = .5725$ or about 57%

Bush: $.38 + .15 \times .43 + .3 \times .19 = .5015$, or about 50%

Perot: $.19 + .3 \times .43 + .2 \times .38 = .395$, or about 40%

11. a. {A, B, C, D; 12}, {A, B, C; 10}, {B, C, D; 8}, {A, C, D; 9}, {A, B, D; 9}, {A, B; 7}, {A, C; 7}

b. A: 5, B: 3, C: 3, D: 1

c. No, A's power is disproportionately high, while D's is low.

d. All voters now have equal power.

Chapter 2

Lesson 2.2

2. a. $35,000 $30,000

b. $70,000 − $35,000 = $35,000

c. $35,000

d. $30,000 $37,500 $32,500

f. Marmaduke would receive $35,000.

3. a.

	Amy	Bart	Carl
Final settlement	4,788.89	4,988.89	4,722.23

6. a. $\begin{bmatrix} 10,000 & 94,000 & 0 \\ 8,000 & 100,000 & 0 \\ 9,000 & 96,000 & 0 \end{bmatrix}$

b. The value to Alan of the items that Betty receives.

Lesson 2.3

1. b. 42.857

c. Sophomore quota: 10.83.

d. Sophomore seats: 11; junior seats: 6; senior seats: 4.

2. a. Sophomore adjusted ratio: 42.18; junior adjusted ratio: 40.

b. Sophomore seats: 11; junior seats: 6; senior seats: 4.

6. a.   4.545   4.762
   10.454   10.952
   When the ideal ratio is 22, the decimal part of the 100-member class is larger than the decimal part of the 230-member class. The situation is reversed when the ideal ratio drops to 21.

   b. For a small class.

Lesson 2.4

1. a. 10.83   10   10   11   11
       5.6   5   5   6   6
       4.57   4   4   5   5

   Sophomores: 11; juniors: 6; seniors: 4.

   c. 43.63   43.82
      43.56   43.83

   d. 43.56 (Sr)   43.63 (Jr)   44.19 (So.)
      Sophomore seats: 11; junior seats: 6; senior seats: 4.

   e. 43.82 (Jr)   43.83 (Sr)   44.24 (So).
      Sophomore seats: 11; junior seats: 5; senior seats: 5.

   f. The method favored by a given class can be seen in the following table of final apportionment results:

	Hamilton	Jefferson	Webster	Hill
Sophomore	11	11	11	11
Junior	6	6	6	5
Senior	4	4	4	5

4. a. 50

   b. 3.7   4
      2.6   3
      1.6   2

   c. 52.8571
      52
      53.3333

   d. Freshman: 21; sophomore: 4; junior: 3; senior: 2.

Lesson 2.5

1. Ann will feel she has exactly one-third. Bart and Carl each could feel he received more than one-third.

4. One-sixth or 0.16.

5. b. One-sixth.

    c. Four-sixths or 0.67.

    d. One-third or 0.33.

7. a. $2 \times 3 = 6$

    b. $k(k + 1)$ or $k^2 + k$

8. a. Yes.   No.

    b. Yes.   Yes.

    c. Probably not. Yes.

Lesson 2.6

1. a. $k + 1$ $k - 1$

    b. $k + 2$ k $2k + 1$ $2k - 1$

2. a. New handshakes: 3; total handshakes: 6.

    b. New handshakes: 4; total handshakes: 10.

3. a. 7

    b. k

    c. $H_n = H_{n-1} + (n - 1)$

4. a. 45

    b. $\dfrac{k(k - 1)}{2}$ $\dfrac{2k(2k - 1)}{2}$ $\dfrac{(k + 1)k}{2}$

    c. $\dfrac{k(k - 1)}{2}$

    d. $\dfrac{(k + 1)k}{2}$

e. k

f. $\dfrac{k(k-1)}{2} + k = \dfrac{k^2 - k}{2} = \dfrac{2k}{2} = \dfrac{k^2 + k}{2} = \dfrac{(k+1)k}{2}$

6. a. $V_{k+1} = 2V_k$

   b. $V_n = 2^n$

   c. $V_{k+1} = 2^{k+1}$

Chapter 2 Review

2. Answers are rounded to the nearest dollar.

	Joan	Henry	Sam
Fair share	$8,880	$8,600	$8,433
Items received	Lot	Computer, stereo	Boat
Cash	$ 880	$6,000	$1,733
Final settlement	$9,342	$9,062	$8,895

3. Answers are rounded to the nearest dollar.

	Anne	Beth	Jay
Fair share	$1,800	$1,567	$1,617
Items received	Car, computer		Stereo
Cash	−$2,600	$1,567	$ 417
Final settlement	$2,005	$1,772	$1,822

4. a. 10

   b. 64.7, 24.7, 10.6

   c. 65, 25, 10

   d. 64, 24, 10

   e. 9.9538, 9.8800, 9.6364

   f. 65, 25, 10

   g. 65, 25, 11

   h. 10.0310, 10.0816, 10.0952

   i. 64, 25, 11

   j. 65, 25, 11

   k. 10.0313, 10.0837, 10.1067

   l. 64, 25, 11

   m. 64, 25, 11

   n. State A gained population and State C lost, but A lost a seat to C.

5. Balinski and Young proved that any apportionment method will some-times produce one of three undesirable results: violation of quota, the loss of a seat when the size of the legislative body is increased even if population doesn't decrease, and the loss of a seat by one state whose population has increased to another whose population has decreased.

6. Arnold and Betty.

7. Have each of the original four divide his or her piece into five pieces that he or she considers equal. Have the new person select a piece from each of the others.

8. $\dfrac{k^2 - k}{2}$ or $\dfrac{k(k - 1)}{2}$

Chapter 3

Lesson 3.1

3.

$$\begin{array}{c} & \text{Jackets} & \text{Shirts} & \text{Pants} \\ \begin{array}{l} \text{Boutique 1} \\ \text{Boutique 2} \\ \text{Boutique 3} \end{array} & \left[\begin{array}{ccc} 25 & 75 & 75 \\ 30 & 50 & 50 \\ 20 & 40 & 35 \end{array} \right] \end{array}$$

4. a. $A_{21} = \$1.09$, $A_{12} = \$10.86$, $A_{32} = \$3.89$

b. A_{21} represents the cost of drinks at Vin's.
A_{12} represents the cost of pizza at Toni's.
A_{32} represents the cost of salad at Toni's.

c. S_3 represents the cost of pizza at Sal's.

8. $A + B = \begin{bmatrix} 10.10 + 1.15 & 10.86 + 1.10 & 10.65 + 1.25 \\ 3.69 + 0.00 & 3.89 + 0.45 & 3.85 + 0.50 \end{bmatrix}$

$$C = \begin{array}{c} & \text{Vin's} & \text{Toni's} & \text{Sal's} \\ \begin{array}{l} \text{Pizza} \\ \text{Salad} \end{array} & \left[\begin{array}{ccc} \$11.25 & \$11.96 & \$11.90 \\ \$3.69 & \$4.34 & \$4.35 \end{array} \right] \end{array}$$

12. Decrease in batting average is shown using a negative sign.

Lesson 3.2

1. a. T represents the cost of four pizzas with additional toppings and four salads with a choice of two dressings from each of the three pizza houses.

   b. $47.60

   c. T_{12} represents the cost of four pizzas with two toppings at Toni's.

   d. T_{21} represents the cost of four salads with choice of two dressings at Vin's.

2. a., b.

$$J = \begin{array}{c} \\ \text{Pearl} \\ \text{Jade} \end{array} \begin{array}{cccc} e & p & n & b \\ \begin{bmatrix} 16 & 8 & 12 & 10 \\ 40 & 20 & 24 & 18 \end{bmatrix} \end{array}$$

   c. 24.

   d. J_{21} represents the number of jade earrings that Nancy expects to sell in June.

   e. J_{12} represents the number of pearl pins that Nancy expects to sell in June.

8.

$$\begin{array}{c} \\ \text{Rate} \end{array} \begin{array}{ccc} \text{CD} & \text{CU} & \text{Bond} \\ [0.073 & 0.065 & 0.075] \end{array} \quad \begin{array}{c} \\ \text{CD} \\ \text{CU} \\ \text{Bd} \end{array} \begin{array}{c} \$ \\ \begin{bmatrix} 10,000 \\ 17,000 \\ 12,000 \end{bmatrix} \end{array} = \$2,735$$

9. a. The transpose of a row matrix is a column matrix and the transpose of a column matrix is a row matrix.

   b.

$$M^T = \begin{array}{c} \\ e \\ p \\ n \\ b \end{array} \begin{array}{cc} p & j \\ \begin{bmatrix} 8 & 20 \\ 4 & 10 \\ 6 & 12 \\ 5 & 9 \end{bmatrix} \end{array}$$

   c. Answers will vary. A possible answer: It may be necessary to use the transpose of a matrix when performing matrix multiplication. (See Exercise 10.)

10. a.
$$\begin{array}{cccc} & e & p & n & b \\ \text{hours} & [2 & 1 & 2.5 & 1.5] \end{array}$$

b.
$$\begin{array}{cccc} & e & p & n & b \\ \text{hours} & [2 & 1 & 2.5 & 1.5] \end{array} \quad \begin{array}{c} \\ e \\ p \\ n \\ b \end{array}\begin{array}{cc} p & j \\ \left[\begin{array}{cc} 8 & 20 \\ 4 & 10 \\ 6 & 12 \\ 5 & 9 \end{array}\right] \end{array}$$

c.
$$\begin{array}{cc} & p \quad j \\ \text{hours} & [42.5 \quad 93.5] \end{array}$$

d. It takes Nancy 42.5 hours to make the pearl jewelry and 93.5 hours to make the jade jewelry.

Lesson 3.3

1. a.
$$\begin{array}{c} \\ \text{Mike} \\ \text{Liz} \\ \text{Kate} \end{array}\begin{array}{ccc} \text{Mike} & \text{Liz} & \text{Kate} \\ \left[\begin{array}{ccc} \$261,000 & 0 & \$250 \\ \$235,000 & 0 & \$215 \\ \$255,000 & 0 & \$325 \end{array}\right] \end{array}$$

b. The entries in row 1 represent the value to Mike of the items that he, Liz, and Kate received. Row 2 represents the values to Liz, and row 3 the values to Kate.

2. a.
$$\text{Matrix } Q: \begin{array}{c} \\ \text{Emma} \\ \text{Ken} \end{array}\begin{array}{ccccc} \text{Burger} & \text{Special} & \text{Potato} & \text{Fries} & \text{Shake} \\ \left[\begin{array}{ccccc} 0 & 1 & 0 & 1 & 1 \\ 1 & 0 & 1 & 0 & 1 \end{array}\right] \end{array}$$

b.
$$\text{Matrix } C: \begin{array}{c} \\ \text{Burger} \\ \text{Special} \\ \text{Potato} \\ \text{Fries} \\ \text{Shakes} \end{array}\begin{array}{ccc} \text{Cal.} & \text{Fat} & \text{Chol.} \\ \left[\begin{array}{ccc} 450 & 40 & 50 \\ 570 & 48 & 90 \\ 500 & 45 & 25 \\ 300 & 30 & 0 \\ 400 & 22 & 50 \end{array}\right] \end{array}$$

c. The dimensions of Q are 2 by 5, and those of C are 5 by 3.

d. The dimensions of the product will be 2 by 3.

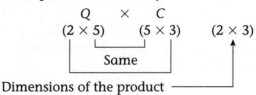

e. The dimensions of C can be described as Foods by Contents and the dimensions of Q times C can be described as Persons by Contents.

Persons by Foods $\times$ Foods by Contents = Persons by Contents

Same

Dimensions of the product

f.

	Cal.	Fat	Chol.
Emma	1,270	100	140
Ken	1,350	107	125

8. The diagram below shows the polygons plotted for parts a through h.

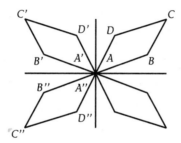

a. $T_1 P = \begin{bmatrix} 0 & -6 & -8 & -2 \\ 0 & 2 & 6 & 4 \end{bmatrix}$

c. Polygon $A'B'C'D'$ is the reflection of polygon $ABCD$ in the y-axis.

d. $T_2 P = \begin{bmatrix} 0 & -6 & -8 & -2 \\ 0 & -2 & -6 & -4 \end{bmatrix}$

e. Polygon $A''B''C''D''$ is the reflection of polygon $A'B'C'D'$ in the x-axis.

f. $RP = \begin{bmatrix} 0 & -6 & -8 & -2 \\ 0 & -2 & -6 & -4 \end{bmatrix}$

The effect of R on P is to rotate $ABCD$ 180 degrees about the origin.

9. $T_3 = \begin{bmatrix} -1 & 0 \\ 0 & 1 \end{bmatrix}$

$T_3 P'' = \begin{bmatrix} 0 & 6 & 8 & 2 \\ 0 & -2 & -6 & -4 \end{bmatrix}$

h.
$$T_4 = \begin{bmatrix} -1 & 0 \\ 0 & -1 \end{bmatrix} = T_3 T_2$$

$$T_4 P' = \begin{bmatrix} 0 & 6 & 8 & 2 \\ 0 & -2 & -6 & -4 \end{bmatrix}$$

Lesson 3.4

1. a. 18.97

b. 9.96, 8.1, 7.29, 9.36, 2.4

c. 18.97, 9.96, 8.1, 7.29, 9.36, 2.4; total 56.

d. 9 months: 18.32, 11.38, 8.96, 7.29, 5.83, 5.62; total 57.
12 months: 18.02, 10.99, 10.24, 8.06, 5.83, 3.5; total 57.

e. Answers may vary. A possible answer: The population continues to grow. The rate of growth seems to have slowed.

f. Answers may vary. A possible answer: The population growth may continue to slow or even become constant.

2. a. 118.4

b. The product is the number of newborn deer after 1 cycle.

c. 30, 24, 21.6, 21.6, 8.4

d. i. $[50 \ 30 \ 24 \ 24 \ 12 \ 8]\begin{bmatrix} 0.6 \\ 0 \\ 0 \\ 0 \\ 0 \\ 0 \end{bmatrix}$ ii. $\begin{bmatrix} 0 \\ 0.8 \\ 0 \\ 0 \\ 0 \\ 0 \end{bmatrix}$

iii. $\begin{bmatrix} 0 \\ 0 \\ 0.9 \\ 0 \\ 0 \\ 0 \end{bmatrix}$ iv. $\begin{bmatrix} 0 \\ 0 \\ 0 \\ 0.9 \\ 0 \\ 0 \end{bmatrix}$ v. $\begin{bmatrix} 0 \\ 0 \\ 0 \\ 0 \\ 0.7 \\ 0 \end{bmatrix}$

3. a. Distribution after 3 months: 0, 21, 0, 0, 0, 0; total 21.

b. Distribution after 3 months: 11, 3, 4.5, 4.5, 4, 3; total 30.

Lesson 3.5

1. a. $P_5 = [19.47 \quad 10.81 \quad 9.89 \quad 9.22 \quad 6.45 \quad 3.50]$

   b. Total: 59.35

   c. $P_7 = [20.47 \quad 12.11 \quad 10.51 \quad 8.76 \quad 7.12 \quad 4.43]$; total: 63.41.

2. To reach 250 females:

	Cycles	Population	Years
a.	61	253.2	15.25
b.	69	250.9	17.25
c.	76	252.9	19.00
d.	42	257.2	10.5

3. a. 4 cycles, 56.65, −1.31%
      5 cycles, 59.35, 4.77%
      6 cycles, 61.76, 4.06%

   b. Population appears to decline, then increase again.

   c. $P_{25} = 108.488$, $P_{26} = 111.789$, $P_{27} = 115.191$; growth rates: 3.04%, 3.04%.

4. a. The long-term growth rate of the total population is 3.04% in each case.

   b. The initial population does not effect the long-term growth rate.

Chapter 3 Review

2. No. Answers will vary. Possible answer: The matrices must be conformable.

3. a. $\begin{array}{cccc} & M & C & S & D \\ L = [35 & 6 & 6 & 12] \end{array}$

   b. $L_2 =$ number of bags of chips ordered
      $L_4 =$ number of six-packs of drinks

   c. $216.60

4. a.

	Lodging	Food	Rec.
Crystal	13.00	20.00	5.00
Springs	12.50	19.50	7.50
Bear	20.00	18.00	0.00
Beaver	40.00	0.00	0.00

b. $C_{22} = 19.50$
$C_{43} = 0$

c. C_{13} = cost for recreation at Crystal Lodge
C_{31} = cost for lodging at Bear Lodge

5. a.

	System	Cart.	Case
Z-Mart	39.50	24.50	8.50
Base	49.90	29.95	12.50

b.

	System	Cart.	Case
Z-Mart	35.55	22.05	7.65
Base	39.92	23.96	10.00

c.

	System	Cart.	Case
Z-Mart	3.95	2.45	0.85
Base	9.98	5.99	2.50

d.

	System	Cart.	Case
Z-Mart	142.20	88.20	30.60
Base	159.68	95.84	40.00

6. a.

	Plate	Large	Small
No. [5	3	7]	

b.

	Ebony	Walnut	Rose	Maple
Plate	100	800	600	400
Large	200	1200	1000	800
Small	50	500	450	400

c.

Ebony	Walnut	Rose	Maple
[1450	11,000	9,150	7,200].

d.

	Plate	Large	Small
No. [5	3	7]	

	Weeks		Weeks
Plate	3	= No. [15 + 12 + 14]	
Large	4	= [41]	
Small	2		

7.
Tennis   Golf   Soccer   Return   Return

$$[\$50{,}000 \quad \$100{,}000 \quad \$75{,}000] \quad \begin{matrix} \text{Tennis} \\ \text{Golf} \\ \text{Soccer} \end{matrix} \begin{bmatrix} 0.082 \\ 0.065 \\ 0.075 \end{bmatrix} = [\$16{,}225]$$

8.
Jazz   Symp.   Orch.

$$\$[300.00 \quad 335.00 \quad 373.50]$$

9. $A^T = \begin{bmatrix} 4 & 5 \\ 2 & 1 \\ 6 & 3 \end{bmatrix}$

10. a. 3×2

    b. 4×3

    c. Not possible.

    d. 4×3

11. a. $M^2 = \begin{bmatrix} 2 & 2 \\ 2 & 2 \end{bmatrix}$ $M^3 = \begin{bmatrix} 4 & 4 \\ 4 & 4 \end{bmatrix}$ $M^4 = \begin{bmatrix} 8 & 8 \\ 8 & 8 \end{bmatrix}$

    b. $M^5 = \begin{bmatrix} 16 & 16 \\ 16 & 16 \end{bmatrix}$

    c. $M^n = \begin{bmatrix} 2^{n-1} & 2^{n-1} \\ 2^{n-1} & 2^{n-1} \end{bmatrix}$

    e. $M^2 = \begin{bmatrix} 1 & 0 \\ 8 & 9 \end{bmatrix}$ $M^3 = \begin{bmatrix} 1 & 0 \\ 26 & 27 \end{bmatrix}$ $M^4 = \begin{bmatrix} 1 & 0 \\ 80 & 81 \end{bmatrix}$

      $M^5 = \begin{bmatrix} 1 & 0 \\ 242 & 243 \end{bmatrix}$ $M^n = \begin{bmatrix} 1 & 0 \\ 3^n - 1 & 3^n \end{bmatrix}$

12. Identity, a square matrix with ones along the diagonal and zeros elsewhere.

13. a. Yes. $AB = BA = I$

    b. Yes. $AB = BA = I$

    c. No. Not square matrices.

14. a.

$$\begin{array}{cc} & \text{Male} \quad \text{Female} \\ \begin{array}{c} A \\ B \\ C \end{array} & \left[\begin{array}{cc} 189 & 196 \\ 175 & 180 \\ 251 & 254 \end{array}\right] \end{array} \qquad \text{b. } \begin{array}{c} A \\ B \\ C \end{array} \left[\begin{array}{c} 385 \\ 355 \\ 505 \end{array}\right]$$

15. a. 24 months.

b.

$$L = \begin{bmatrix} 0.0 & 0.6 & 0.0 & 0.0 & 0.0 & 0.0 \\ 0.5 & 0.0 & 0.8 & 0.0 & 0.0 & 0.0 \\ 1.1 & 0.0 & 0.0 & 0.9 & 0.0 & 0.0 \\ 0.9 & 0.0 & 0.0 & 0.0 & 0.8 & 0.0 \\ 0.4 & 0.0 & 0.0 & 0.0 & 0.0 & 0.6 \\ 0.0 & 0.0 & 0.0 & 0.0 & 0.0 & 0.0 \end{bmatrix}$$

c. 10%

d. 28 months.

Chapter 4

Lesson 4.1

1.

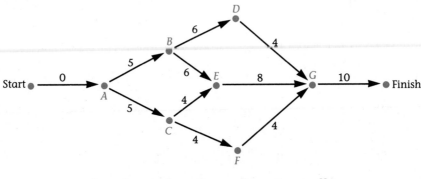

2.

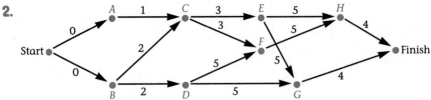

7. a. A: 2, none; B: 4, none; C: 3, A; D: 3, C and B; E: 2, B; F: 1, E; G: 4, D and F.

Lesson 4.2

1. EST for *C* through *G*: 7, 10, 11, 16, 23.
Min. project time: 26.
Critical path: Start—*ACEFG*—Finish.

3. EST for *D* through *I*: 6, 9, 8, 13, 11, 18.
Min. project time: 26.
Critical path: Start—*CFI*—Finish.

5. a.

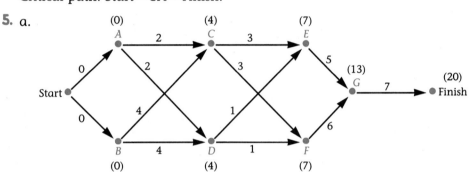

b. 22

c. Start—*ADG*—Finish.

d. The minimum time is reduced to 21 days, to 20 days.

e. No, below 8 days *A* is no longer on the critical path.

8. a. Day 16, day 17, day 18, both task *G* and the project will be delayed.

b. Day 11.

c. Day 5, day 6, day 5.

Lesson 4.3

1.

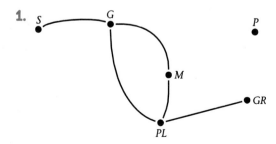

3. a. *A* and *F*, *B* and *F*, *B* and *C*, *A* and *C*, *A* and *D*, *A* and *E*, *B* and *E*, *B* and *D*, *F* and *C*, or *F* and *D*.

b. *FEDC*.

c. No, there is no path from *A* or *B* to the vertices *C*, *D*, *E*, or *F*.

d. No, not every pair of vertices is adjacent. For example, *B* and *C* are not adjacent.

5.

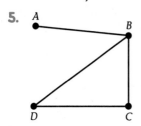

6.
$$\begin{bmatrix} 0 & 1 & 1 & 0 \\ 1 & 0 & 1 & 1 \\ 1 & 1 & 0 & 1 \\ 0 & 1 & 1 & 0 \end{bmatrix}$$

9. *V* : 3, *X* : 2, *Y* : 2, *Z* : 1

10. a. *B* : 2, *C* : 6, *D* : 3, *E* : 2

b.
$$\begin{bmatrix} 1 & 0 & 1 & 0 & 0 \\ 0 & 0 & 2 & 0 & 0 \\ 1 & 2 & 0 & 1 & 2 \\ 0 & 0 & 1 & 1 & 0 \\ 0 & 0 & 2 & 0 & 0 \end{bmatrix}$$

11. 4, 5, 6; 12, 20, 30;
$$T_4 = T_3 + 6 \qquad T_5 = T_4 + 8$$
$$T_6 = T_5 + 10 \qquad T_n = T_{n-1} + 2(n - 1)$$

Lesson 4.4

1. a. Both, the degrees of all vertices are even.

2. New circuit: *S*, *b*, *e*, *a*, *g*, *f*, *S*.
Final circuit: *S*, *e*, *f*, *a*, *b*, *c*, *S*, *b*, *e*, *a*, *g*, *f*, *S*.

3. Answers may vary. One possible circuit: *e, d, f, h, d, c, h, b, c, g, a, h, g, f, e.*

8. a.

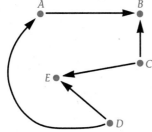

9. a. Yes. **b.** No. **c.** Yes.

11. a.

$$\begin{array}{c@{\quad}ccccc}
 & a & b & c & d & e \\
a & \begin{bmatrix} 0 & 0 & 1 & 0 & 1 \\ b & 1 & 0 & 0 & 1 & 0 \\ c & 0 & 1 & 0 & 0 & 0 \\ d & 0 & 0 & 1 & 0 & 0 \\ e & 0 & 1 & 0 & 1 & 0 \end{bmatrix}
\end{array}$$

Lesson 4.5

1. a. Yes.

b., c. The theorem does not apply.

8.

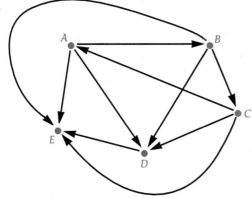

11. 6, 10, 15 $S_n = S_{n-1} + (n - 1)$

15. a.
$$M = \begin{bmatrix} 0 & 0 & 1 & 1 & 1 \\ 1 & 0 & 1 & 1 & 0 \\ 0 & 0 & 0 & 1 & 1 \\ 0 & 0 & 0 & 0 & 1 \\ 0 & 1 & 0 & 0 & 0 \end{bmatrix}$$

b.
$$M^2 = \begin{bmatrix} 0 & 1 & 0 & 1 & 2 \\ 0 & 0 & 1 & 2 & 3 \\ 0 & 1 & 0 & 0 & 1 \\ 0 & 1 & 0 & 0 & 0 \\ 1 & 0 & 1 & 1 & 0 \end{bmatrix}$$

c. The winner would be B.

Lesson 4.6

1. 4, 3, 2

3. a. List the vertices in order from the ones with the greatest degree to the ones with the least.

b. Those not adjacent to it or adjacent to one with that color.

5. a. 2, 3, 4, 5

9. 4

12. a.

Chapter 4 Review

2.

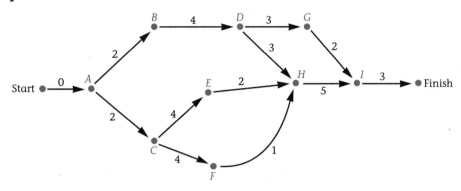

3.

Task	Time	Prerequisites
Start	0	—
A	2	None
B	3	None
C	4	A
D	4	A, B
E	2	B
F	3	C
G	5	D, E
H	7	F, G
Finish		

4. a. A, (0); B, (4); C, (4); D, (7); E, (6); F, (11); G, (10); H, (15); I, (16); J, (18).

b. Minimum project time: 23.

5. a. A, (0); B, (2); C, (2); D, (6); E, (6); F, (6); G, (9); H, (9); I, (14).

b. Critical path: Start—$ABDHI$—Finish.
Minimum project time: 17.

6. a. Yes, a path exists from each vertex to every other vertex.

b. No, not every pair of vertices is adjacent.

c. A, D, or C.

d. $BCDE$ or $BCAE$.

e. Deg(C) = 4.

f.
$$\begin{array}{c} \\ A \\ B \\ C \\ D \\ E \end{array}\begin{array}{ccccc} A & B & C & D & E \\ \begin{bmatrix} 0 & 0 & 1 & 0 & 1 \\ 0 & 0 & 1 & 0 & 0 \\ 1 & 1 & 0 & 1 & 1 \\ 0 & 0 & 1 & 0 & 1 \\ 1 & 0 & 1 & 1 & 0 \end{bmatrix} \end{array}$$

7. a. Euler path. Two vertices have odd degrees and the remaining vertices have even degrees.

b. Euler circuit. All vertices have even degrees.

8. a.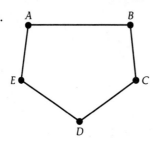

b.

9. a. No.

b. Yes.

c.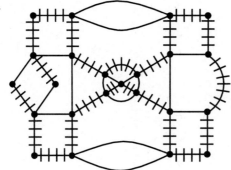

10. a. Yes.

b. No, the graph has exactly two odd vertices. You would have to begin at one of the vertices with an odd degree and end at the other.

11. a.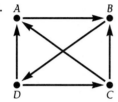

b. There is no Condorcet winner because none of the candidates can beat all of the other candidates in a one-on-one race.

c. *D, C, A, B; C, A, B, D; C, B, D, A; B, D, C, A; A, B, D, C.*

d. If the Hamiltonian path, *B, D, C, A* is chosen for a pairwise voting scheme, *B* will win. The path shows that for *B* to "survive," you need first to pair *A* and *C*. *C* wins. Next pair the winner, *C*, against *D*, and *D* wins. Finally, pair the winner, *D*, against *B* and *B* is the final winner of the election.

12. a.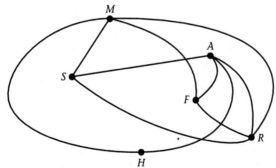

b. Three time slots. One possible schedule:

Time 1—Math and Art   Time 2—Reading and History
Time 3—Science and French

13. a.

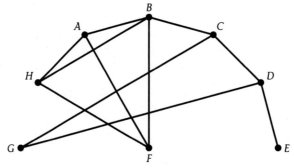

b. Four frequencies.

14. a. No, the outdegrees are not equal to the indegrees at each vertex.

b. Yes, the outdegree equals the indegree at all vertices but two. At one of those two vertices, the indegree is one greater than the outdegree and at the other vertex, the outdegree is one greater than the indegree.

Answers will vary, but the paths must begin at *B* and end at *D*.

15. a., b.

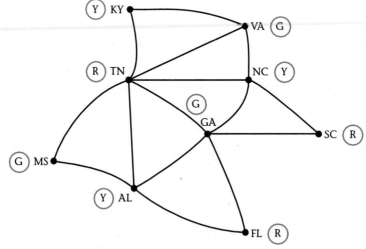

c. Three colors.

Chapter 5

Lesson 5.1

1. Planar.

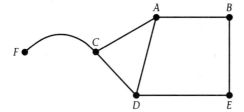

4. Hint—move A and B around.

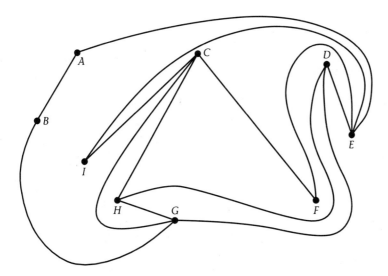

7.

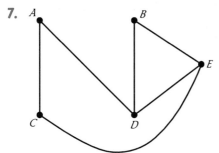

10. a.

$K_{2, 3}$

11. a. {A, B, C, D, E, F} and {G}

14. 6, 12, $m * n$

17. 30 handshakes, bipartite.

20. a and d, because all of the edges and vertices of the original graph are in a and d.

22. No. No.

Lesson 5.2

1. a., b.

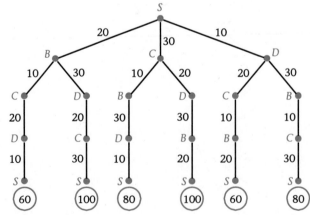

c. *SDCBS*   d. *SDCBS*   e. Yes.

5. a. 0.36 seconds, about 24 hours. b. 3.6×10^{-7} seconds, 0.09 seconds, about 676 hours.

Lesson 5.3

1. 6, 12, 11, *C*, *BC*; the shortest path from *A* to *E* is *AHGE*.

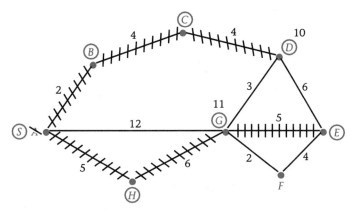

2. *ABECDF* (11)

5. a. Albany, *CEH*, Ladue

b. Albany, *BD*, Fenton, *GK*, Ladue. This problem yields a different solution than part a because you have to find two solutions and then add them: first you must find the shortest path from Albany to Fenton, then a path with Fenton as Start.

8.

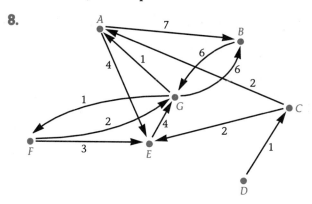

Shortest path: *DCAB*; least charge: 1 + 2 + 7 = 10.

Lesson 5.4

1. *BCEFB, CDEC, BCDEFB, BCFB, CEFC, CDEFC*

3. Five vertices.

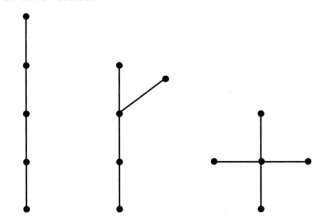

4.

Number of vertices	Number of edges
1	0
2	1
3	2
4	3
n	$n - 1$

a. 18 edges b. 16 vertices
c. The number of vertices = the number of edges -1.

7.

1	0	$S_1 = 0$
2	2	$S_2 = S_1 + 2$
3	4	$S_3 = S_2 + 2$
4	6	$S_4 = S_3 + 2$
5	8	$S_5 = S_4 + 2$
6	10	$S_6 = S_5 + 2$
		$S_n = S_{n-1} + 2$

9.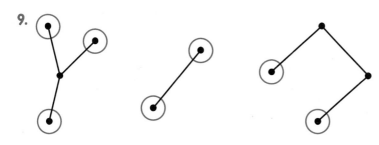

11. Rhombus.

Lesson 5.5

1. One possible spanning tree:

4. One possible spanning tree:

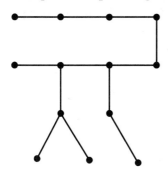

6. a. Yes.   b. *E* and *J*   c. *DE* and *DJ*   d. *F, K, I*

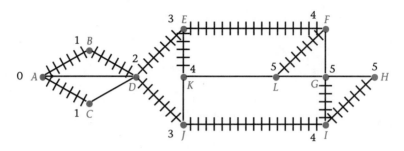

7. This is one of many possibilities.

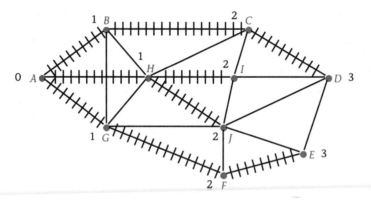

10. The minimum weight is 10.

13. $2,100

17. a. Weight 28.

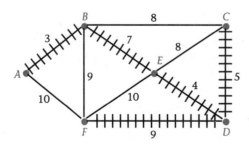

b. *A* to *F* is 10; *A* to *B* is 3; *A* to *C* is 15; *A* to *E* is 10; *A* to *D* is 14.

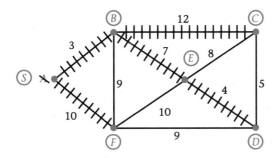

c. No. It is a spanning tree, but in this example, it is not minimal. Its total weight is 36, which is greater than 28.

Lesson 5.6

1.

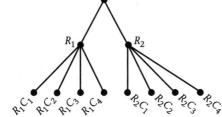

3. Binary tree.   a. *V* is level 2.   b. *C* is the parent.   c. *G* and *H* are children.

7. There are 17 questions in the book.

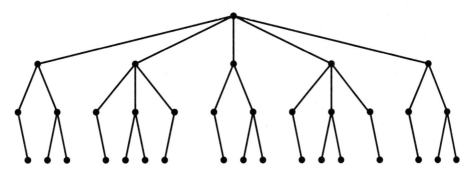

11.

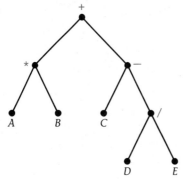

12. 3 2 * 8 2 3 * − +

15. a. 33   b. 7   c. 9   d. 14

16. a. 2 3 6 * + 4 1 + −

19. *ABDEGHCFI*

21. 17

22. a. 8

Chapter 5 Review

2.

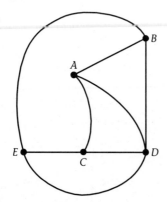

3. 2, 2, 2, 2

4. a. The vertices of the graph can be divided into two sets so that each edge of the graph has one endpoint in each set.

   b. Yes, all possible edges from one set of vertices to the other are drawn.

c. Yes.

d. 2.

5. a. *O–SCM–O.*

b. 314 ft.

c. Hamiltonian circuit.

6.

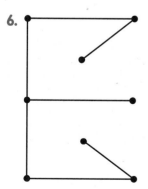

7. 3, 4, 5, n

8. a. Home–*T*–*P*–*G*–*H*–*BB*.

  b. 14 miles.

  c.

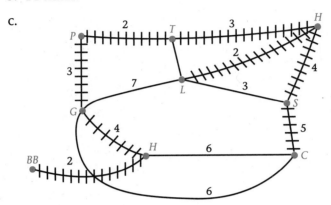

9. The total weight of a minimum spanning tree for the graph is 23 miles.

10. a. Yes, it is a connected graph with no cycles.

  b. Yes, it is a connected graph with no cycles.

  c. No, the graph contains a cycle.

11.

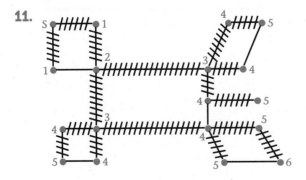

12. One possible solution:

13. a.

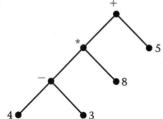

b. Total cost = $23,000.

14. Problems similar to those in Lesson 5.5.

15.

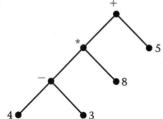

16. 18

17. Any expression is possible. The following is just one example:
The expression: 3 * (2 + 6) − 5
The expression tree:

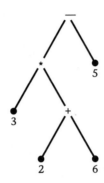

The postorder listing: 3 2 6 + * 5 −

Chapter 6

Lesson 6.1

1. li, lo, ln, ls, il, io, in, is, ol, oi, on, os, nl, ni, no, ns, sl, si, so, sn.

2. 5, 4, 5 × 4 = 20, 6 × 5 = 30

3. (1,2) (1,3) (1,4) (1,5) (1,6) (1,7) (1,8) (1,9) (2,3) (2,4) (2,5) (2,6) (2,7) (2,8) (2,9) (3,4) (3,5) (3,6) (3,7) (3,8) (3,9) (4,5) (4,6) (4,7) (4,8) (4,9) (5,6) (5,7) (5,8) (5,9) (6,7) (6,8) (6,9) (7,8) (7,9) (8,9)

4. 9, 8, $9 \times \frac{8}{2} = 36$

13.

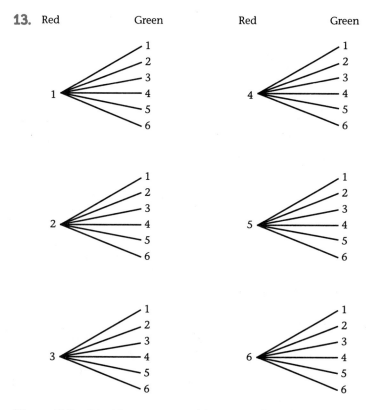

15. a. Win $1: 10 ways; win $2: 1 way; lose $1: 25 ways.

Lesson 6.2

1. $\dfrac{10!}{6!}$ 5040

3. a. 15

 b. A particular front sprocket and a particular rear sprocket.

5. a. 12,167,000

 b. $\dfrac{1}{12,617}$

7. a. 72

  b. $\frac{1}{2}$

  c. 0

  d. 1

10. a. 2704

  b. 2652

13. a. 30! or about 2.6525×10^{32}.

  b. About 1.0827×10^{28}; the number of seating arrangements is about 24,500 times as large.

14. a. 6

  b. A road from Claremont to Upland and a road from Upland to Pasadena.

  c. $3 \times 3 = 9$

18. a. 100,000 manufacturers.

  b. 100,000 products.

19. a. 1,048,576

  b. About 10 years; about 262 feet.

Lesson 6.3

3. a. 672

  b. 504

  c. 2,380. They are the same.

4. a. 210

  b. 210

  c. 1,024

5. a. 1,326

  b. 325

  c. $\dfrac{325}{1,326}$, or about 0.245.

8. a. 10

   b. 10

9. a. 7,059,052

   b. About 245 weeks, or a little less than 5 years.

   c. About 353 feet.

   d. 13,983,816

   e. $\dfrac{80,000}{13,983,816}$, or about .00572.

   f. The probability of winning in Virginia is nearly twice as good.

10. a. $C(6,5) \times C(38,1) = 228$

   b. 10,545

   c. 168,720

14. 255

19. a. 7

   b. 21

   c. 28

   e. $\frac{7}{28}$ or $\frac{1}{4}$

   f. 91

Lesson 6.4

1. a. $\dfrac{520}{1,000} = .52$

   b. $\dfrac{196}{360}$, or about .544.

   c. No, but they are fairly close.

   d. $\dfrac{360}{1,000} = .36$

   e. $\dfrac{196}{1,000} = .196$. The product is .1872.

f. .544 × .36 = .196. They are the same.

g. No.

3. a. $\dfrac{684}{1{,}000} = .684$

b. .36 + .52 = .88, which is larger than .684.

c. No.

5. a.

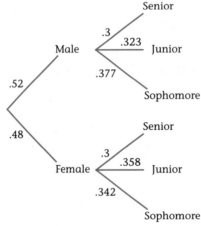

b. 1

9. a. .54

b. That the outcomes of the two games are independent.

10. a.

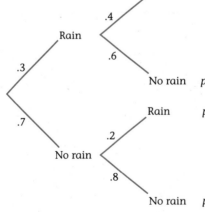

    b. .12

    c. 32

    d. .56

    e. No.

13. $\dfrac{1}{216}$

16. a. $.977^6$, or about .870

    b. The first is .870, and the second is $1 - \frac{1}{6} = \frac{5}{6}$, or about .83.

    c. .000529

    d. .999471, .99683

19. a. $\frac{1}{80}$

    b. That the two events are independent.

    c. The probability of selecting a man with red hair is .1; the probability of selecting a man who owns a blue car is .125; and the probability of selecting a man who has red hair and owns a blue car is about .0124. The product of the first two probabilities is .0125, so the events are quite close to being independent.

20. a. 36

    b. $\frac{1}{36}$

    c. $\frac{1}{1,296}$

    d. $\frac{35}{36}$

    e. $\frac{1,225}{1,296}$

    f. $\left(\frac{35}{36}\right)^{21}$, or about .5534.

Lesson 6.5

1. a. The probability of each outcome is $\frac{1}{2}$, and successive applications are independent.

    b. The probability of choosing the third answer is $\frac{1}{2}$, while the probability of choosing each of the others is $\frac{1}{4}$; successive applications are independent.

    c. Each possibility probably has a $\frac{1}{4}$ chance of occurring, but successive applications may not be independent if, for example, one finger is injured.

    d. Each possibility has the same chance of occurring, and successive applications are independent.

2. a. 10

    b. .3125

    c. .03125, .15625, .3125, .3125, .15625, .03125

    d. .00243, .02835, .1323, .3087, .36015, .16807

3. a. .2051

    b. .1172

    c. .0439

    d. .0098

    e. .00098

    f. .3770

5. a. .2765

    b.

Number of women	0	1	2	3	4	5	6
Probability	.0466	.1866	.3110	.2765	.1382	.0369	.0041

8. a. $\dfrac{1}{7,059,052}, \dfrac{7,059,051}{7,059,052}$

    b. $2.82, but this assumes that the jackpot is not shared with another party.

a. Amount won	27,000,000	−5,000,000
Probability	$\dfrac{5,000,000}{7,059,052}$	$\dfrac{2,059,052}{7,059,052}$

The expectation is $19,124,380, but this assumes the jackpot is not shared.

10. a.

Amount won	−1	1	20
Probability	$\frac{21}{36}$	$\frac{14}{36}$	$\frac{1}{36}$

b. $0.36

c. No, the council would lose about 36 cents per play. One way of correcting this would be to give no prize for matching a single number. In fact, the jackpot could then be increased but should be kept under $35.

13. a. $\frac{1}{4}$

b.

Amount won	−$0.50	$1.00
Probability	$\frac{3}{4}$	$\frac{1}{4}$

c. −$0.50

d. Yes, about 50 cents per play.

Chapter 6 Review

2. a. $\frac{7,900}{46,900}$, or about .168.

b. $\frac{2,300}{13,700}$, or about .168.

c. $\frac{13,700}{46,900}$, or about .292.

d. $\frac{2,300}{46,900}$, or about .049.

e. $\frac{19,300}{46,900}$, or about .412.

f. Yes.

g. No.

3. $C(5,1) = 5$, $C(5,2) = 10$, $C(5,3) = 10$, $C(5,4) = 5$, $C(5,5) = 1$

4. a. $\frac{1}{16}$

   b. $\frac{1}{16}$

   c. $\frac{1}{8}$

5. a. 720

   b. 48

   c. $\frac{1}{15}$

   d. The math books can be in positions 1 and 2, or in positions 2 and 3, or in positions 3 and 4, or in positions 4 and 5, or in positions 5 and 6. $5 \times 48 = 240$.

   e. $\frac{1}{3}$

6. a.

Amount won	$2	$1	−$1
Probability	$\frac{1}{4}$	$\frac{1}{4}$	$\frac{1}{2}$

   b. $\frac{1}{4}$

   c. Win about $25.

7. 720, about 28

8. There are $C(39,5) = 575{,}757$ different winning tickets possible in the first and $C(36,6) = 1{,}947{,}792$ ways of winning in the second, so the probability of winning in the first is between three and four times as great as in the second. About five in the first and one or two in the second.

9. a. $\frac{1}{4}$

   b. $\frac{4}{17}$

   c. $\frac{1}{17}$

   d.

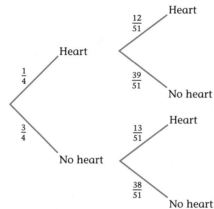

Heart $p(\text{Heart and Heart}) = \frac{1}{4} \times \frac{12}{51} = \frac{12}{204}$

No heart $p(\text{Heart and No heart}) = \frac{1}{4} \times \frac{39}{51} = \frac{39}{204}$

Heart $p(\text{No heart and Heart}) = \frac{3}{4} \times \frac{13}{51} = \frac{39}{204}$

No heart $p(\text{No heart and No heart}) = \frac{3}{4} \times \frac{38}{51} = \frac{114}{204}$

   e. No.

10. a. 165   **d.** 135

   b. 75

     e. $\frac{135}{165}$

   c. $\frac{75}{165}$

11. a.

Ticket $p(\text{Good driver and Ticket}) = .045$

No ticket $p(\text{Good driver and No ticket}) = .855$

Ticket $p(\text{Bad driver and Ticket}) = .07$

No ticket $p(\text{Bad driver and No ticket}) = .03$

    b. 5,750

    c. 3,500

    d. $\dfrac{3,500}{5,750}$

12. a. $\dfrac{1}{1,000}$

    b. $\dfrac{729}{1,000}$

    c. $\dfrac{504}{1,000}$

13. a. 100

    b. 45

14. a. Approximately .134.

    b. Approximately .804.

15. 15

16. a. .2

    b. .1

    c. .6

    d. .2

    e. No.

    f. Yes.

17. a. Approximately 5%.

    b. Approximately 93%.

    c. About 1.

18. Mutually exclusive: rolling a number divisible by 5 and rolling a number divisible by 3, rolling a number divisible by 5 and rolling a number divisible by 2. Independent: rolling a number divisible by 2 and rolling a number divisible by 3.

Chapter 7

Lesson 7.1

1. 25; 475; 45; 855; 2,000; 1,900; 5,000; 4,750; 2,500; 125; 7,500; 375; $0.05P$; $P - 0.05P$

2. a. 2%

   b.

   c. $D = P - 0.02P$

   d. $P = \$20,408$

3. a.

   b.
   $$C = \begin{matrix} & \text{Chips} & \text{Computers} \\ \text{Chips} & \begin{bmatrix} 0.02 \\ \text{Computers} & 0.01 \end{bmatrix} & \begin{matrix} 0.20 \\ 0.03 \end{matrix} \end{matrix}$$

   c. $20 chips, $10 computers.

   d. $150 computers, $1,000 chips.

   e. $25,510

   f. $51,546

4. a.

   b. 5 cents, 4 cents, 1 cent, 20 cents.

   c. $1 million, $0.8 million.

   d. $0.4 million, $8 million.

   e. $11 million, $38.8 million.

Lesson 7.2

1. a.

$$C = \begin{matrix} \text{Chips} \\ \text{Computers} \end{matrix} \begin{matrix} \text{Chips} & \text{Computers} \\ \begin{bmatrix} 0.02 & 0.20 \\ 0.01 & 0.03 \end{bmatrix} \end{matrix}$$

b.

$$P = \begin{matrix} \text{Chips} \\ \text{Computers} \end{matrix} \begin{bmatrix} \$40,000 \\ \$50,000 \end{bmatrix}$$

c.

$$CP = \begin{matrix} \text{Chips} \\ \text{Computers} \end{matrix} \begin{bmatrix} \$10,800 \\ \$1,900 \end{bmatrix}$$

d.

$$D = \begin{matrix} \text{Chips} \\ \text{Computers} \end{matrix} \begin{bmatrix} \$29,200 \\ \$48,100 \end{bmatrix}$$

e.

$$P = \begin{matrix} \text{Chips} \\ \text{Computers} \end{matrix} \begin{bmatrix} \$35,210 \\ \$72,528 \end{bmatrix}$$

4. a.

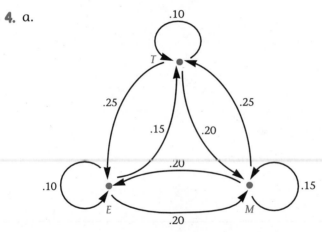

b.

$$C = \begin{matrix} T \\ E \\ M \end{matrix} \begin{matrix} T & E & M \\ \begin{bmatrix} 0.10 & 0.25 & 0.20 \\ 0.15 & 0.10 & 0.20 \\ 0.25 & 0.20 & 0.15 \end{bmatrix} \end{matrix}$$

Note: The entries in the matrices for parts c through f represent millions of dollars.

c.

$$P = \begin{matrix} T \\ E \\ M \end{matrix} \begin{bmatrix} 150 \\ 200 \\ 160 \end{bmatrix}$$

d.
$$CP = \begin{matrix} T \\ E \\ M \end{matrix} \begin{bmatrix} 97.0 \\ 74.5 \\ 101.5 \end{bmatrix}$$

e.
$$D = \begin{matrix} T \\ E \\ M \end{matrix} \begin{bmatrix} 53.0 \\ 125.5 \\ 58.5 \end{bmatrix}$$

f.
$$P = \begin{matrix} T \\ E \\ M \end{matrix} \begin{bmatrix} 218.60 \\ 195.24 \\ 239.65 \end{bmatrix}$$

5. a.

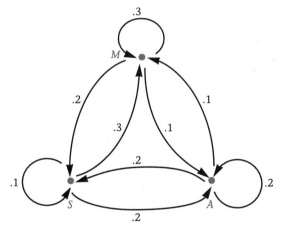

b. Manufacturing is most dependent on itself and service (0.3) and least dependent on agriculture (0.1).

c. $8 million from each.

Note: The entries in the matrices for parts d and e represent millions of dollars.

d.
$$CP = \begin{matrix} S \\ M \\ A \end{matrix} \begin{bmatrix} 12.5 \\ 13.0 \\ 9.5 \end{bmatrix}, \quad D = \begin{matrix} S \\ M \\ A \end{matrix} \begin{bmatrix} 7.5 \\ 12.0 \\ 5.5 \end{bmatrix}$$

e.
$$P = \begin{matrix} S \\ M \\ A \end{matrix} \begin{bmatrix} 11.03 \\ 11.61 \\ 9.21 \end{bmatrix}$$

Lesson 7.3

1. a. $D_0 = [.75 \quad .25]$ $\quad$ $D_1 = [.625 \quad .375]$ $\quad$ $D_2 = [.5875 \quad .4125]$
$D_3 = [.57625 \quad .42375]$ $\quad$ $D_4 = [.572875 \quad .427125]$

b. $D_{10} = [.571430 \quad .428570]$
$D_{15} = [.571429 \quad .428571]$

c. Based on these results the food director can expect 57% of the students to eat in the cafeteria on any given day in the long run.

e. No.

f. $T^{15} = \begin{bmatrix} .571429 & .428571 \\ .571429 & .428571 \end{bmatrix}$

The entries in each row in T^{15} are the same as those in D_{15}.

2. a. $[.571429 \quad .428571]$

3. a. $D_0 = [1 \quad 0]$ $\quad$ $D_1 = [.7 \quad .3]$ $\quad$ $D_2 = [.61 \quad .39]$ $\quad$ $D_3 = [.583 \quad .417]$
$D_4 = [.5749 \quad .4251]$

b. $D_{10} = [.571431 \quad .428569]$
$D_{15} = [.571429 \quad .428571]$

After several weeks, 57% of the students will be eating in the cafeteria.

4. a. No. The sum of the entries in row 2 is greater than 1.

b. No. The matrix is not a square matrix.

c. No. Entries must be probabilities (between 0 and 1 inclusive).

d. No. The matrix is not a square matrix and the sum of the entries in row 2 is less than 1.

e. Yes.

f. No. The sum of the entries in row 2 is less than 1.

5. a. .42.

b. .08.

c. [.6 .4].

d. $\begin{bmatrix} .7 & .3 \\ .2 & .8 \end{bmatrix}$.

e. $D_7 = [.40 \quad .60]$

f. 40%.

12. a. $D_0 = [1 \quad 0 \quad 0]$

b. $D_1 = [.8 \quad .2 \quad 0]$ $D_2 = [.66 \quad .28 \quad .06]$ $D_3 = [.556 \quad .300 \quad .144]$
$D_4 = [.4748 \quad .2912 \quad .2340]$

After four days there is a 47% probability that the rat will be well, 29% probability that it will be ill, and 23% probability that it will be dead.

Lesson 7.4

1. Best strategies	Player 1	Player 2	Strictly determined	Saddle point
a.	row 1	column 2	yes	8
b.	row 1	column 1	yes	0
c.	row 1 or 2	column 1	no	
d.	row 1	column 2	no	
e.	row 3	column 3	yes	4
f.	row 1	column 1	yes	0

2. a. Best strategies: Row 2, column 2, saddle point is 3.

b. Best strategies stay the same. Adds 4 to the saddle point.

c. Best strategies stay the same. Doubles the saddle point.

d. A conjecture: When a constant is added to each value in the payoff matrix of a strictly determined game, the best strategies stay the same and the saddle point is increased by the constant. If each value is multiplied by a constant, the best strategies stay the same and the saddle point is also multiplied by the constant.

4. a. Every other row dominates row B. Eliminate row B.
Column E dominates column G. Eliminate column G.
Row C dominates rows A and D. Eliminate rows A and D.
Column F dominates column E. Eliminate column E.
Best strategies: Row C and column F.

7.

$$
\begin{array}{c c}
 & \begin{array}{c c c} 1 & 2 & 3 \end{array} \\
\begin{array}{c} 1 \\ 2 \\ 3 \end{array} &
\left[\begin{array}{r r r}
10 & -10 & -10 \\
-20 & 20 & -20 \\
-30 & -30 & 30
\end{array}\right]
\end{array}
\quad \text{No saddle point.}
$$

Lesson 7.5

1. b. The probability that both Sol and Tina will show heads is .15.

c. The probability that Tina will show tails when Sol shows heads is .35.

d. The probability that Tina will show heads when Sol shows tails is .15.

e. The probability that both Sol and Tina will show tails is .35.

f.

Outcome	HH	HT	TH	TT
Probability	.15	.35	.15	.35
Amount won	4	−2	−3	1

g. Sol's payoff expectation for this game is −.2. This means that Sol will lose 2 pennies every 10 plays.

h. It is the same.

2. a. $A = [.75 \quad .25]$, $C = \begin{bmatrix} .3 \\ .7 \end{bmatrix}$,

$$
ABC = [.75 \quad .25] \begin{bmatrix} 4 & -2 \\ -3 & 1 \end{bmatrix} \begin{bmatrix} .3 \\ .7 \end{bmatrix} = -.2
$$

b. Choose different values for the probabilities for Sol in matrix A. Calculate the expected payoff for Sol in each instance.

c. Let $A = [4 \quad .6]$ and choose different values representing probabilities for Tina in matrix C. Calculate the expected payoff for Sol in each instance.

6.

$$\text{Payoff matrix:} \quad \text{Player 1} \quad \begin{array}{cc} & \text{Player 2} \\ & \begin{array}{cc} 1 & 2 \end{array} \\ \begin{array}{c} 1 \\ 2 \end{array} & \begin{bmatrix} 2 & -3 \\ -3 & 4 \end{bmatrix} \end{array}$$

Best strategies: Both row and column players play 1 finger seven-twelfths of the time and 2 fingers five-twelfths of the time. The row player can expect to lose 1 cent in 12 plays or about 8 cents in 100 plays. Not a fair game since the row player will lose. In a fair game the expected payoff for both players will be the same.

8. Row 2 dominates row 1. Eliminate row 1 (making phone calls). Column 2 dominates column 1. Eliminate column 1 (making phone calls). Group against should send out mailings three-fourths of the time and go door-to-door one-fourth of the time. Group in favor should send out mailings one-half of the time and go door-to-door one-half of the time. Opposing group only gets 250 signatures (not enough to get the issue on the ballot).

Chapter 7 Review

2. a.

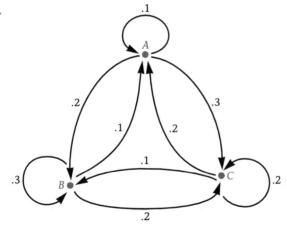

b.
$$CP = \begin{array}{c} A \\ B \\ C \end{array} \begin{bmatrix} 7.7 \\ 7.4 \\ 5.8 \end{bmatrix} \qquad D = \begin{array}{c} A \\ B \\ C \end{array} \begin{bmatrix} 0.3 \\ 4.6 \\ 9.2 \end{bmatrix}$$

c.
$$P = \begin{bmatrix} 18.6 \\ 20.4 \\ 22.2 \end{bmatrix}$$

3. Mike's best strategy is not to bluff. The saddle point for this matrix is -2.

4. a. Mike's best strategy is to play his black card seven-twentieths of the time and his red card thirteen-twentieths of the time.
Nancy's best strategy is to play her black card nine-twentieths of the time and her red card eleven-twentieths of the time.

c.

Outcome	BB	BR	RB	RR
Probability	$\frac{63}{400}$	$\frac{77}{400}$	$\frac{117}{400}$	$\frac{143}{400}$
Amount won	7	-6	-4	3

d. The expected value of the game for Mike is $\frac{-3}{20}$.

e. Mike can expect to lose three cents every twenty hands played.

5. a. Probability of another quiz on Friday is 29%.

b. Students should expect that the teacher will start class with a quiz one-third of the time.

c. She will start class with a quickie review 40% of the time.

6. a.
$$C = \begin{bmatrix} .20 & .25 & .55 \\ .45 & .35 & .20 \\ .20 & .25 & .55 \end{bmatrix}$$

b. 28.5%

c. O: 27%, I: 28%, B: 45%.

d. Answers will vary. A possible answer: The clerk will know how many of each type of sandwich to order each day.

7. a.

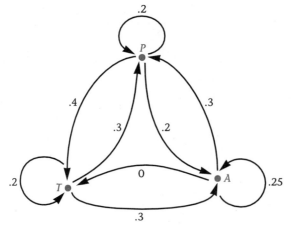

b.

$$
C = \begin{array}{c} \\ T \\ P \\ A \end{array} \begin{array}{ccc} T & P & A \\ \left[\begin{array}{ccc} 0.2 & 0.3 & 0.3 \\ 0.4 & 0.2 & 0.2 \\ 0.0 & 0.3 & 0.25 \end{array} \right] \end{array}
$$

c. Transportation is most dependent on petroleum (0.4) and least dependent on agriculture (0.0).

d. $1.08 million from petroleum. $1.35 million from agriculture.

e.

$$
CP = \begin{array}{c} T \\ P \\ A \end{array} \left[\begin{array}{c} 16.00 \\ 16.00 \\ 11.25 \end{array} \right] \quad D = \left[\begin{array}{c} 4.00 \\ 9.00 \\ 3.75 \end{array} \right] \quad \text{(in millions of dollars)}
$$

f.

$$
P = \left[\begin{array}{c} 16.4 \\ 17.5 \\ 11.0 \end{array} \right] \quad \text{(in millions of dollars)}
$$

8. The best strategy for both companies is to focus on school district A.

9. The Democrats' best strategy is to go with strategy A one-fourth of the time and strategy B three-fourths of the time. The Republicans' best strategy is to go with strategy C one-half of the time and strategy D one-half of the time.

Expectation for the Democrats: 45% of undecided voters joining them.

10. a.

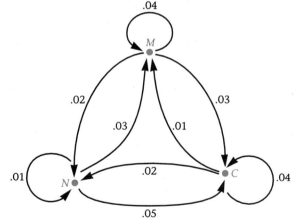

b.

$$
P = \left[\begin{array}{c} \$54,136 \\ \$34,112 \\ \$42,941 \end{array} \right]
$$

c.
$$CP = \begin{bmatrix} \$4,136 \\ \$4,112 \\ \$2,941 \end{bmatrix}$$

d. Internal consumption change: Production change:

$$\begin{bmatrix} \$1,005 \\ \$1,061 \\ \$804 \end{bmatrix} \qquad \begin{bmatrix} \$11,005 \\ \$9,061 \\ \$12,803 \end{bmatrix}$$

11. Store A: Lower prices. Store B: No change.

Chapter 8

Lesson 8.1

1. b. 1, 4, 9

c.

Number of couples	Number of handshakes	Recurrence relation
1	1	—
2	4	$H_2 = H_1 + 3$
3	9	$H_3 = H_2 + 5$
4	16	$H_4 = H_3 + 7$
5	25	$H_5 = H_4 + 9$

2. a.

Couples	Handshakes
1	0
2	2
3	6
4	12

b. $2n - 2$

c. $H_n = H_{n-1} + 2n - 2$

3. a. i. $H_n = H_{n-1} + 3$

ii. $H_n = (H_{n-1}) \cdot 2$

b. 16, 64, 36, 4320

5. a. 4

   c. 0

7. a.

Term number	Number of handshakes	First differences	Second differences
1	0	—	—
2	1	1	—
3	3	2	1
4	6	3	1
5	10	4	1
6	15	5	1
7	21	6	1
8	28	7	1

10. a.

Term number	Number of bees
0	5,000
1	5,600
2	6,272
3	7,024.64
4	7,867.60

   b. $B_n = 1.12B_{n-1}$

   c. After 27 years, in 2014.

Lesson 8.2

2. a. $H_n = 2n^2 - 5n$

   b. $H_n = 0.29 + 0.23(n - 1)$

   d. $H_n = 3^{n-1}$

3. a. $T_n = T_{n-1} + n - 1$

   b. 0

4. a. 1, -2, -11, -38, -119, -362

   b. -2.5

6. a.

Row number	Number of seats	Total seats
1	24	24
2	26	50
3	28	78
4	30	108
5	32	140
6	34	174

b. $S_n = S_{n-1} + 2$

c. $S_n = 24 + 2n - 2$

d. 37

f. $T_n = T_{n-1} + 24 + 2n - 2$

g. $T_n = n^2 + 23n$

Lesson 8.3

1. b. i. $H_n = H_{n-1} + 3$

    iii. $H_n = (H_{n-1}) \cdot 1.2$

    iv. $H_n = H_{n-1} + H_{n-2}$

c. i. $H_n = 2 + 3(n - 1)$

  iii. $H_n = 10(1.2^{n-1})$

5. c. $5,755.11

g. $5,772.76

6.

Year	4.8% Monthly	5% Yearly
0	$5,000.00	$5,000.00
1	$5,245.35	$5,025.00
2	$5,502.74	$5,050.13
3	$5,772.76	$5,075.38

7. a. 74; 585

b. 63.25; 817.5

10. $117,463.15

11. Approximately 6.5%.

15. About $32,500.

    b. A little over $10,100.

16. a. 1.14471

    b. 2,318

Lesson 8.4

1. a. $93,070.22

    b. $48,000

    c. $45,070.22

    d. $202,107.52, $93,070.22

5. a.

T (in months)	T_n
0	12,000
1	11,802
2	11,602.22
3	11,400.64

    b. $A_n = (A_{n-1}) \cdot 1.007 - 286$

    c. It takes 50 months.

7. a. $M_n = (M_{n-1}) \cdot 1.1 - 1000$

    b. At the end of the third year, $3,345.

    c. 10%

    d. The fixed point is $10,000.

8. $10,206.75

Lesson 8.5

1. a. -3; $T_n = 4(2^{n-1}) - 3$; 2.5353×10^{30}

   b. 3.5; $T_n = 1.5(3^{n-1}) + 3.5$; 2.5769×10^{47}

2. a. $B_n = (B_{n-1})\left(1 + \dfrac{0.08}{12}\right) + 150$

   b.

Month	Balance
0	150
1	301
2	453.01
3	606.03
4	760.07

   c. $-22,500$

   d. $22,801\left(1 + \dfrac{0.08}{12}\right)^{n-1} - 22,500$, or $22,650\left(1 + \dfrac{0.08}{12}\right)^{n} - 22,500$.

   e. $\$225,194.28$

   f. 472 months.

   g. Approximately $\$333$.

3. a. $B_n = (B_{n-1}) \cdot 1.01 - 220$

   c. 22000.

   e. $\$6763.18$.

   f. 61 months.

Lesson 8.6

1. a.

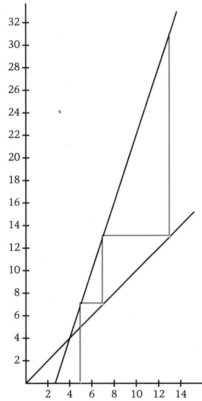

b.

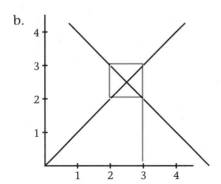

c.

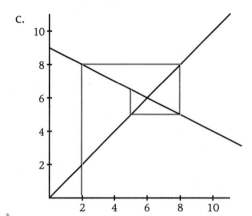

d.

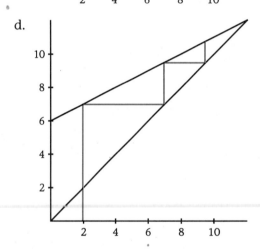

4. a. 1

b. 3

c.

n	t_n
1	1
2	5
3	13

d. $t_n = 2t_{n-1} + 3$

6. b.

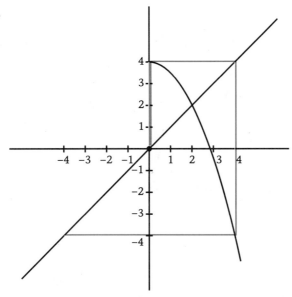

c. The behavior is unpredictable.

f. 2, −4

Chapter 8 Review

2. a. $H_n = H_{n-1} + 4$; $H_n = 2 + 4(n - 1)$; 398

b. $H_n = 3H_{n-1} - 1$; $H_n = \frac{5}{2} 3^{n-1} + \frac{1}{2}$; $4.29 \cdot 10^{47}$

c. $H_n = (H_{n-1}) \cdot 2$; $H_n = 3(2^{n-1})$; $1.9014759 \cdot 10^{30}$

3. a. Arithmetic.

b. Neither.

c. Geometric.

4. a. 0; 0, 0, 0, 0

b. −0.75; −0.75, −0.75, −0.75, −0.75

c. No fixed point.

5. a. $H_n = 2(5^{n-1})$; $3.1554436 \cdot 10^{69}$

b. $H_n = 2.75(5^{n-1}) - 0.75$; $4.338735 \cdot 10^{69}$

c. $H_n = 2 + (-3)(n - 1)$; -295

6. a. Geometric.

b. Neither.

c. Arithmetic.

7. a.

N	S_n	First differences	Second differences	Third differences
1	1	—	—	—
2	5	4	—	—
3	14	9	5	—
4	30	16	7	2
5	55	25	9	2

c. $\dfrac{n^3}{3} + \dfrac{n^2}{2} + \dfrac{n}{6}$; 204

8. Second degree; $H_n = n^2 - 6$

9. a.

Day	Gifts that day	Total gifts
1	1	1
2	3	4
3	6	10
4	10	20
5	15	35
6	21	56

b. $G_n = G_{n-1} + n$

$T_n = T_{n-1} + \dfrac{n^2}{2} + \dfrac{n}{2}$

c. $G_n = \dfrac{n^2}{2} + \dfrac{n}{2}$

$T_n = \dfrac{n^3}{3} + \dfrac{n^2}{2} + \dfrac{n}{6}$

10. a. $P_n = P_{n-1} + 0.23$

b. $P_n = 0.32 + 0.23(n-1)$

11. a.

Month	Balance
0	$1,000
1	$1,004
2	$1,008.02
3	$1,012.05

b. $B_n = 1.004(B_{n-1})$

c. $B_n = 1000(1.004)^n$

d. 14 years, 6 months.

12. a.

Month	Balance
0	$5,000
1	$5,126.67
2	$5,254.01
3	$5,382.03

b. $B_n = (B_{n-1})\left(1 + \dfrac{0.064}{12}\right) + 100$

c. $B_n = 23{,}750\left(1 + \dfrac{0.064}{12}\right)^n - 18{,}750$

d. $13,928.99

e. 200 months.

13. a. $B_n = 1.008(B_{n-1}) - 230$

b. $B_n = -17{,}750(1.008)^n + 28{,}750$

c. 61 months.

d. $3,030

e. $352.88

14. a. $V_n = 1.5V_{n-1} + 4{,}000$

b. $V_n = 16{,}000(1.5)^{n-1} - 8000$

15.

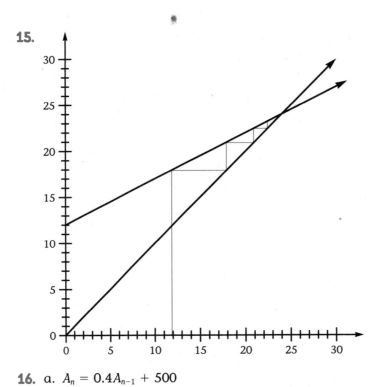

16. a. $A_n = 0.4A_{n-1} + 500$

b. The amount of medication in the body stabilizes at 833 mg.

c. The cobweb would be attracted to the point (833.33, 833.33), which is the intersection of $y = x$ and $y = 0.4x + 500$, as shown in this figure.

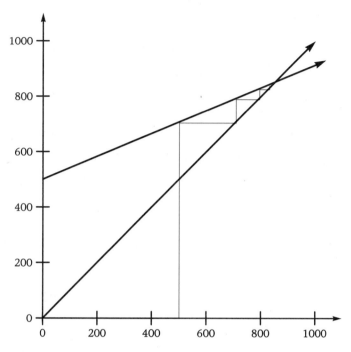

d. The amount in the body reaches the stable value (833 mg in this case) more quickly. The stable value is probably near the optimal dosage of the drug.

Illustration Credits

Cartoons on pages 62, 78, 105, 139, 165, 222, 283, 308, 365, 374, 409, and 432 are by Tom Durfee.

Page 1: Jerry Ohlinger's Movie Material Store.

Page 5: David Longstreath/AP Photo.

Page 8: Left: The Collections of the Library of Congress; center: Corbis; right: Chicago Historical Society.

Page 12: Stock Boston.

Page 13: Emile Wamsteker/AP Photo.

Page 18: Bettmann.

Page 45: NASA.

Page 50: Culver Pictures.

Page 53: Lynn Johnston Productions, Inc./Distributed by United Feature Syndicate, Inc.

Page 60: Corbis.

Page 61: Office of the Curator, The White House.

Page 69: Richard Bloom/SABA.

Page 75: Bottom: Dennis Brack/Blackstar.

Page 82: Reprinted by permission of United Feature Syndicate.

Page 86: Lambert/Archive Photos.

Page 96: Professor Aleksander Weron.

Page 97: Top: Steven J. Brams, New York University; bottom: Alan D. Taylor, Union College, Schenectady, New York.

Page 101: Lambert/Archive Photos.

Page 103: Corbis/Bettmann-UPI.

Page 110: Doug Mills/AP Photo.

Page 120: Stephen Frisch/Stock Boston.

Page 125: Annabelle Breakey.

Page 141: Chip Clark/Smithsonian Institution.

Page 151: Bob Daemmrich/Stock Boston.

Page 157: Archive Photos.

Page 159: Peter Vandermark/Stock Boston.

Page 173: Culver Pictures.

Page 193: Billy E. Barnes/Stock Boston.

Index